# Course Book in General Botany

COURSE BOOK IN
# General Botany

JOHN D. DODD

ILLUSTRATIONS BY *Kathleen Thomas O'Sullivan*
AND THE AUTHOR

THE IOWA STATE UNIVERSITY PRESS / *Ames*

TO THE MEMORY OF MY PARENTS
*Ernest S. Dodd* AND *Louise A. Dodd*

JOHN D. DODD is Professor of Botany and Plant Pathology at Iowa State University. He holds the Doctor of Philosophy degree from Columbia University and the Master's degree from the University of Vermont. His undergraduate work was done at New York State College of Forestry.

© 1962, 1977 Iowa State University Press
All rights reserved.
Composed and printed by the
Iowa State University Press,
Ames, Iowa, U.S.A.

First edition, 1962

Second printing, 1966
Third printing, 1968

Second edition, 1977
Second printing, 1978

First edition published under the title
*Form and Function in Plants.*

**Library of Congress Cataloging in Publication Data**

Dodd, John Durrance.
    Course book in general botany.
    Originally published in 1962 under title: Form and function in plants.
    Includes index.
    1. Botany.   I. Title.
QK47.D58    1976    581    76-50076
**ISBN 0-8138-0690-9**

# CONTENTS

*Preface,* **vi**

*Introduction:* THE EUCARYOTIC PLANT CELL: A REVIEW, **viii**

1 TRANSPIRATION: THE PRICE OF LIFE ON LAND, **2**

2 THE PLANT BODY OF VASCULAR PLANTS: GENERAL STRUCTURE, **12**

3 THE PLANT BODY OF VASCULAR PLANTS: VARIATIONS AND EVOLUTIONARY ORIGINS, **46**

4 REPRODUCTION IN THE LOWER GREEN LAND PLANTS, **74**

5 REPRODUCTION AND GROWTH IN THE HIGHER GREEN LAND PLANTS, **104**

6 THE GREEN ALGAE, **136**

7 EVOLUTIONARY DIRECTIONS: GREEN PLANTS, AN OVERVIEW, **156**

8 LIFE WITHOUT CHLOROPHYLL: THE FUNGI, **170**

9 LIFE WITHOUT CHLOROPHYLL: THE BACTERIA, **198**

10 THE NONGREEN ALGAE, **208**

11 THE CLASSIFICATION OF PLANTS, **236**

*Index,* **247**

# PREFACE

THE predecessor to this book, *Form and Function in Plants,* originated more than a decade ago as a possible solution to the task of developing a rational approach to the one term course in general botany without burying student interest in an overwhelming mass of detail.

Since then numerous profound changes in the teaching of basic biology courses have materialized, some of them rooted in long-standing student objections to repeated elementary treatment of several topics in various beginning biology courses. It is now common for such topics as cell structure, respiration, digestion, photosynthesis, membrane functions, mitosis and cytokinesis, meiosis, basic concepts of genetics, an introduction to molecular biology, and the like, to be treated as part of a general biology course offered to all students, with special sections of one sort or another included for future biology majors.

Since these topics are all treated again in greater depth in advanced courses, the role of a general botany course may be structured so as to assume an intelligent awareness of them. For this reason, it seems fitting that general botany revert to an original function of dealing with organisms and structures and be so organized that the student becomes aware of the numerous kinds of living plants existing in natural environments. Many teachers may welcome this approach to teaching general botany since it encourages students to consider plants as whole organisms existing in natural environments before they begin to explore more deeply the subject of molecular biology.

The Introduction is intended only as a review for those students who have had a considerable time lapse between taking general biology and the onset of the present course. It is in no way an adequate substitute for the biology course but merely attempts to set forth a desirable level of proficiency at which the general botany course outlined herein may be initiated.

In the original version, only those green algae were treated that provided useful introductions to concepts related to the higher green plants. However, with the diversion of many general botany topics into general biology it has become possible to expand the treatments of algae to include several other groups. It is recognized that not all of these can be covered in a one term course, and the additional material on algal groups is conveniently placed in a final chapter where portions of it may be used as desirable.

Another departure from the previous book is the inclusion of a brief discussion of bacteria which is intended primarily to stimulate an interest in the multiplicity of roles bacteria play in natural environments, roles that extend far beyond their vast significance in medicine and industry. However, the discussions of viruses, rickettsias, and mycoplasms presented in general biology courses need not be enlarged upon in a beginning botany course.

Many new illustrations have been added and I am pleased to acknowledge the assistance of Carol Peck and Michael Metzler in converting my pencil sketches to line drawings. To Dr. Lois Tiffany, Dr. Cecil Stewart, and Dr. Paul Hartman I owe a particular debt of gratitude for advice on certain critical points.

Figure 10.6 is from *The Nature-Printed British Seaweeds, Vol. II*, 1859, Johnstone, W. G. and Alexander Croall.

Certain illustrations in the text are taken from Asa Gray's *The Elements of Botany*, 1887. These are indicated in the legends by (Gray). Others are courtesy of the Chicago Natural History Museum and are indicated by (CNHM). Figure I.22 is courtesy of Dr. C. C. Bowen, Fig. 7.15 is from a slide by Dr. R. W. Pohl, and Fig. 6.26 is from the unpublished notes of the late Dr. Ada Hayden, all of the Iowa State Botany Department. Figure 8.39 was redrawn from Bragonier, W. H., Dodd, J. D., and Gilman, J. C., *Episodes in Plant Phylogeny*, 2d ed., 1955. Figures 10.7 and 10.22 were prepared by John Koppen and Carl Wiemers. Figure 5.56 is based on the work of Prof. E. B. Matzke. All other illustrations are original.

# INTRODUCTION

THE general nature of this course book presupposes a basic knowledge of plant cells but the inclusion of a brief review of cell structures and functions seems useful in establishing a frame of reference for many related topics to be discussed. In addition, the review should be helpful to those students who begin a course such as this when considerable time has elapsed since taking an introductory biology course. However, it should not be considered an adequate substitute for the introductory course.

## THE PLANT CELL

The plant cell may be thought of as a structural unit consisting of a small mass of protoplasm, the *protoplast,* in a containing layer, the *cell wall.* With significant exceptions among the lower plants, cellulose is a major component of the plant cell wall. In many-celled plants the primary cell wall is the joint product of adjacent cells and in essence is a three-layered structure. The middle layer *(middle lamella)* consists of an organic substance that acts as a cement holding the outer layers together (Fig. I.1).

*Eucaryotic cells* differ from *prokaryotic cells* in a number of ways, but the name refers particularly to differences in the organization of the nucleus. The term *karyon* is an old name referring to a nut such as hickory; reference to the nucleus is derived from comments of early microscopists who thought of the nucleus as a little nut lying in the cell.

The nucleus of the eucaryotic cell is membrane limited and the genetic material becomes organized into distinct chromosomes during the division process. Discussions of the fact that neither of these characteristics apply to prokaryotic cells are included in the chapters dealing with blue-green algae and bacteria. Higher plants, true fungi, and most groups of algae have eucaryotic cells.

*Protoplasm* is the general name for the substance of living matter; its major subdivisions are the *nucleus* and the *cytoplasm.* Much basic information about protoplasm is obtainable through observations with the light microscope but researchers using the electron microscope are constantly adding new information about its fine structure.

Optically clear cytoplasm, i.e., cytoplasm without granules, is sometimes called *hyaloplasm.* In physical and chemical terms it is a complex colloidal system containing various proteins, fats, carbohydrates, mineral salts, and a high percentage of water. Its general consistency has been likened to that of egg white. At times it exists in a fluid or *sol* form while at others it may exist as a semisolid or *gel.* Changes in the physical state of cytoplasm between sols and gels are reversible in the living condition.

One important feature of cytoplasm is the formation of surface membranes wherever it comes in contact with other cell parts. These *cytoplasmic membranes* play vital roles in cell functions.

The membrane adjacent to the cell wall is commonly termed the *plasma membrane,* while the *nuclear membrane* separates the cytoplasm from the nucleus. Plastids and mitochondria are also bounded by membranes. The organelle called the *Golgi apparatus* consists of a stack of flattened membranes but the stack itself is not surrounded by a membrane and this organelle generally is not visible in the light microscope.

# The Eucaryotic Plant Cell: A Review

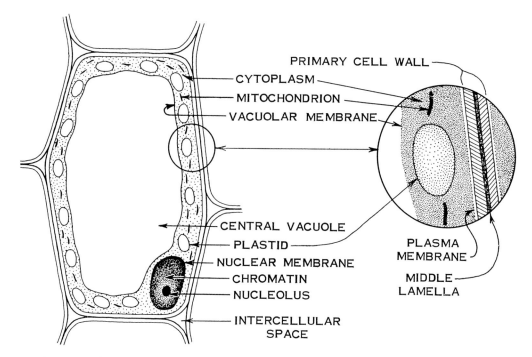

Fig. I.1. Diagrammatic representation of a mature green cell (of a higher plant) as seen in a thin section. The inset drawing is an enlargement of a portion of the cell wall and adjoining cytoplasm.

In mature green cells of the higher plants (Fig. I.1) the cytoplasm is spread into a very thin layer that completely lines the cell wall and surrounds a large central vacuole from which it is separated by a *vacuolar membrane* (see below). Occasionally, delicate strands of cytoplasm may cross the vacuole from one side to the other or from the walls to a centrally suspended mass in which the nucleus is located.

When living cells are observed with the compound microscope the top and bottom walls of the cell and the cytoplasm adjacent to them do not absorb enough light to be readily visible. The side walls and adjacent cytoplasm which are viewed on edge absorb more light and are much more visible. This may result in an optical illusion that cells are two-dimensional unless the fine adjustment of the microscope is manipulated to permit observation of the same cell in several planes.

When cytoplasm is in the fluid or sol condition it may move about the cell in a flowing manner. This process is called *cyclosis*. Often the chloroplasts are moved about by such movements and even the more massive nucleus may be displaced.

Under the microscope the rate of movement often seems rapid but if it is measured and then converted to a macroscopic frame of reference the rate seems very slow.

## THE CYTOPLASMIC ENCLOSURES

The basically clear substance of the cytoplasm usually contains a variety of objects that alter its appearance.

*Granules* may be stored food, minute crystals, or organic particles of uncertain nature. Plant cells tend to accumulate water in small droplets or *vacuoles* within the cytoplasm. These are separated from the cytoplasm by vacuolar membranes. Usually the cell sap in the vacuoles contains low concentrations of soluble inorganic salts and organic substances. Sometimes it may be colored due to the presence of water soluble pigments such as *anthocyanin*.

Much of the visible growth of plants is due to the increasing size of vacuoles as they absorb water. Commonly there is a gradual fusion of many small vacuoles into a few large ones and eventually these merge into a single large central vacuole (Fig. I.2).

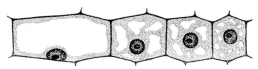

Fig. I.2. A series of drawings illustrating cell enlargement. Note the fusion of vacuoles and the relation between increasing vacuole size and cell enlargement.

By the time this happens the cytoplasm has been stretched so thin that it is barely visible.

Normally vacuoles tend to absorb more water than they can contain and this results in an outward pressure *(turgor)* on the cytoplasm and cell wall. This pressure is responsible in large part for increases in cell size and for the maintenance of the shapes of plant structures such as leaves and young stems. The crispness of celery and the firm textures of many other fresh vegetables and fruits result from the interaction of many cells each of which is exerting pressure against its neighbors.

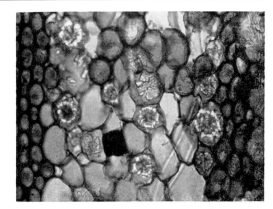

Fig. I.3. Crystals in cells (as seen in a transverse section of a basswood twig).

Vacuoles may contain *crystals* of many sizes and shapes (Fig. I.3); some of them are large enough to distort the shape of the cell. Crystals represent a method for disposal of excess materials that might have a deleterious effect on cell functions. Usually they are salts of organic acids and calcium or magnesium.

Minute protoplasmic connections *(plasmodesmata)* frequently exist between adjacent cells (Fig. I.4). They are barely visible with the light microscope and may allow transport of certain substances across cell boundaries.

*Mitochondria* are relatively small rod-shaped structures in the cytoplasm. They are not particularly distinct in living cytoplasm but can be rendered visible through the light microscope by appropriate techniques for killing and staining cells during the process of preparing them for examination. Their structure is more apparent in electron photomicrographs. Mitochondria play a very important role in cell functions. Most of the chemical reactions in living matter involve organic catalysts called enzymes and it is probable that mitochondria contain series of enzymes that control a major portion of the cell's respiratory mechanism. Each mitochondrion is bounded by a double membrane, with the inner membrane being highly convoluted. The narrow inward-pointing lobes are referred to as *cristae* (Fig. I.4).

The stacks of membranes referred to

# The Eucaryotic Plant Cell

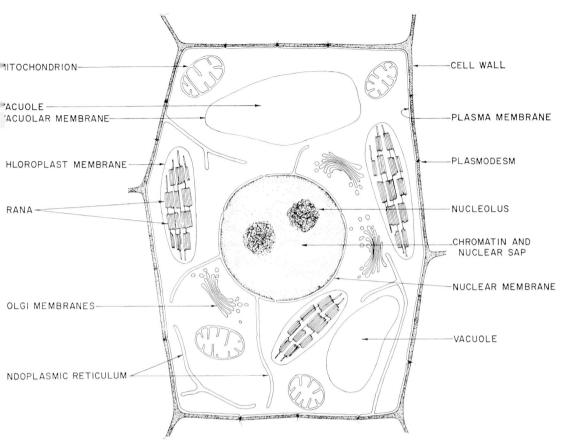

Fig. I.4. Schematic sectional view of a plant cell as seen in the electron microscope. (Minute bodies such as ribosomes and spherosomes have been excluded from the sketch.)

above as the *Golgi apparatus* are illustrated also in Fig. I.4. As is evident, the stacks are not enclosed by a bounding membrane. Seemingly, the Golgi membranes contribute to the process of cytoplasmic division as well as to certain types of secretory activities. This organelle is often called the *dictysome* by botanists.

The cytoplasm also contains an elaborate three-dimensional maze of delicate internal membranes that are contiguous with the more prominent surface membranes discussed above. This maze is called the *endoplasmic reticulum*. Its enormous total surface area seemingly provides working space for many of the chemical reactions that are carried on in living cytoplasm. Sectional fragments of these membranes are included in Fig. I.4.

*Ribosomes* are numerous minute particles which are visible only with the electron microscope. They contain RNA and their general significance is discussed subsequently in relation to the genetic control of cell functions. Individual ribosomes are not enclosed in membranes but a number of membrane-limited particles in the same general size range have become intensely interesting to research workers. Terms applied to some of them include *microbodies*, *spherosomes*, and *microtubules*.

*Plastids* are discrete membrane-limited units of cytoplasm having a denser appearance than undifferentiated cytoplasm. They are clearly visible with the light microscope and may be colorless or variously colored. Their functions are varied and significant. One of their functions is the enzymatic formation of starch from simple sugar. This is actually a separate function from the process of photosynthesis which can only be carried on by those plastids that contain chlorophyll. The ability to convert sugar to starch is more or less independent of the color of the plastid. It can occur in the green plastids of a leaf cell but it can occur also in the colorless plastids of roots or underground stems such as the potato tuber.

In some cases the plastids of higher plants contain minute masses of protein that seem to be centers of starch formation. They are analogous to the *pyrenoids* found in the chloroplasts of many algae but may not be exactly homologous with them.

The background substance of a plastid is the *stroma;* in it are suspended the *photosynthetic lamellae* or *thylakoids* (Fig. I.4). Each of these may be likened to an inflated balloon or a vesicle which has been so compressed that the flattened sides are parallel and almost in contact. The pigment molecules involved in photosynthesis are assumed to be attached to or partially imbedded in the thylakoids.

The number of thylakoids varies with growth conditions. Sometimes they proliferate so rapidly that they become infolded and arranged in stacks which may be visible with the light microscope. The stacks are called *grana*.

It is a point of some theoretical interest that photosynthetic cells in the vast majority of green land plants contain many small chloroplasts rather than one or a few large ones. As will be seen in Chapter 6, this is in considerable contrast to the green algae, which have a wide variety of chloroplast types.

It is generally true also that mature green cells of land plants have conspicuous central vacuoles and relatively thin cellulose walls. Some of these features are illustrated in drawings of cells from the green tissues of mosses, ferns, pines, corn, and apple trees (Fig. I.5).

Three major plastid types have colors that are determined by the pigments present.

1. *Chloroplasts*—chlorophylls *a* and *b*, car-

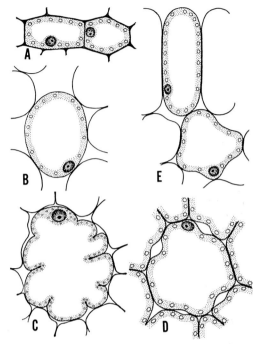

Fig. I.5. Green cells from the leaves of various higher plants drawn schematically to illustrate their basic similarities: (A) moss, (B) fern, (C) pine, (D) corn, and (E) apple.

otenes, and xanthophylls are present and the normal color is grass green.
2. *Chromoplasts*—only carotenes and xanthophylls are present and the normal color is some other color than green.
3. *Leucoplasts*—pigments are present in extremely low concentrations and these plastids appear colorless.

(Amyloplasts are plastids, usually leucoplasts, that store conspicuous amounts of starch.)

The term *chromatophore* should be mentioned here since it is often used with reference to the plastids of algae, particularly those that are not grass green even though they contain chlorophyll. Chromatophore means "color-bearer" and the term has a much wider range of meaning than chloroplast. Chloroplasts are chromatophores, but not all chromatophores are chloroplasts.

The three types of plastids listed above are more or less interconvertible. A chloroplast may become a chromoplast by losing its chlorophyll *a* and *b,* while retaining the more stable carotenes and xanthophylls. This happens commonly in the ripening of fruits and in autumnal leaf coloration. A chloroplast may become a leucoplast by losing most of its pigments. A chromoplast may become a chloroplast, as for example in carrots where "greening" occurs when portions of the root are exposed to light. Leucoplasts may become chloroplasts also, as in the case of the "greening" of potato tubers lying too close to the surface of the soil.

The chloroplasts in cells of higher plants are small and numerous with a somewhat flattened shape like a lens. Chromoplasts tend to be somewhat more irregular in shape.

## COLORATION IN PLANTS

Green plants are not necessarily green at all times. The leaves of certain widely cultivated horticultural varieties of plants are normally *variegated* (having a mixture of colors). Parts of such leaves are green while others may be white, red, or yellow. The leaves of a few plants appear bright red with no indication that chlorophyll is present. Plants grown in poor light or lacking certain chemical substances may have yellow leaves. The newly formed leaves of some plants may be yellow for some time before they become green. Algae growing at the surface of a pond where they are exposed to intense light may become yellowed while those underneath remain a bright green.

If the red leaf of an *Irisene* plant or a *Coleus* plant is placed in boiling water, it will change almost immediately to a bright green color. The slightly pink color of the water indicates that the red color is water soluble. The green, on the other hand, remains in the leaf indicating that it is not water soluble. The red pigment is an *anthocyanin.* It is dissolved in the cell sap and retained there because it does not diffuse through living cell membranes. Heating destroys these membranes and the red diffuses out.

Anthocyanins are of considerable interest even though their functions are poorly understood. Some plants contain them at all times while others manufacture them only under certain conditions. Many plants never form them at all. In the plants that do form them, they are an indication of considerable sugar in the cell sap since sugar is essential for their formation.

Their presence in some plants is an indication of an abnormal sugar supply. When tomato leaves turn red, for example, there may be a phosphorus deficiency in the soil. The lack of sufficient phosphorus interferes with normal starch formation and the accumulated sugar is used to form anthocyanin.

The anthocyanins are natural indicators of the acidity or alkalinity of the cell sap, being red under acid conditions and blue when basic. Many of the pink, red, purple, and blue flowers owe their colors to anthocyanins. White flowers lack pigments entirely.

The green color of leaves may be extracted by placing leaves in alcohol and heating. Other organic solvents such as acetone will also remove the green pigments.

Techniques such as those involving paper chromatography have made it possible to separate the various plastid pigments in relatively simple classroom demonstrations (Fig. I.6). These methods of pigment separation have shown us also that there are actually several kinds of carotenes and xanthophylls. However, the differences between them are small and their basic structures are similar. The most common carotene is a bright golden yellow while the xanthophylls have various shades of yellow, orange, red, or brown. The carotenoid pigment *lycopene* is responsible for the red color of ripe tomatoes, while a brown xanthophyll gives the characteristic color to a large group of sea weeds.

Many fruits are green while immature and a change in color is an indication of approaching ripeness. This color change is often due to a rapid disintegration of chlorophylls *a* and *b,* which normally mask the carotenes and xanthophylls. When the two green pigments disintegrate, the more

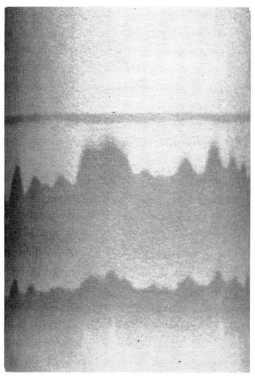

Fig. I.6. Separation of chloroplast pigments in a paper chromatogram. The pigments (from top to bottom) are carotene, chlorophyll *a*, and chlorophyll *b*. (The xanthophylls are not evident.)

stable carotenoid pigments become visible and the fruits take on shades of yellow and orange. In many cases the carotenes and xanthophylls later disintegrate, leaving the fruits with a brownish color.

In some mature fruits anthocyanins may also be present, giving tones of bright red, pink, purple, or blue to the fruits. The fruiting structures of algae frequently become brightly colored due to the accumulation of hematochrome pigments.

The colors of tree leaves in autumn are due in part to a similar series of changes. Disintegration of the green pigments permits the yellows and oranges to become visible for a brief period before they also disintegrate and the dead leaf turns brown. The bright reds and scarlets that turn whole hillsides to flame are due largely to the presence of anthocyanins. Degradation products of the various pigments add many subtle shades to the color scheme.

It is evident that the color spectacular of autumn is not due to freezing temperatures since it begins often before the first frost. Shorter days, a gradual lowering of the average temperature, and other related factors are responsible for the visible changes.

As though in preparation for the eventual leaf fall, a special *abcission layer* forms across the base of a petiole. The normal channels by which food is conducted away from the leaf are sealed off. In some plants, like the sugar maple, the accumulated products of photosynthesis are utilized in the formation of anthocyanin which turns the leaf red.

CHEMICAL NATURE OF PLANT PIGMENTS

The basic structure of the molecule of chlorophyll *a* has been known for a long time. It contains 55 carbon atoms, 72 hydrogen atoms, 5 oxygen atoms, 4 nitrogen atoms, and a single atom of magnesium. These atoms are arranged as 4 *pyrrol* rings centered about the single magnesium atom with a long chain alcohol, *phytol*, attached (Fig. I.7). The structure of chlorophyll *b* is basically the same, differing merely by the presence of 1 more oxygen atom and the absence of 2 of the hydrogens.

A carotene molecule may contain 40 carbon atoms and 56 hydrogen atoms. A chemist would describe it as an unsaturated hydrocarbon. As noted previously, the many different kinds of xanthophylls show a wide range of color. Xanthophylls are probably derived from carotene by the addition of one or more oxygen atoms and for this reason they are often called *oxy-*

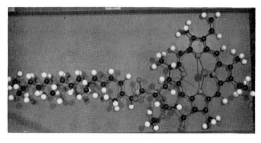

Fig. I.7. A molecular model of chlorophyll *a*.

*carotenes.* Their color characteristics seem to vary with the number of oxygen atoms present in the molecule.

The anthocyanins are glycosides. When they are broken down into component parts, one of these parts is always a sugar.

## THE MANUFACTURE OF FOOD

According to one definition, a *food* is any substance that can be broken down by living organisms to release energy. However, popular usage of the term *plant food* for various kinds of chemicals required for plant growth has led to a much broader interpretation of the word. Many authorities in the study of plant nutrition now use two terms, *organic nutrients* and *inorganic nutrients,* to distinguish between these basic concepts of food.

Organic nutrients fall into three general and familiar categories, *carbohydrates, fats,* and *proteins.* (A fourth category, that of the *nucleic acids,* should be added to this list. These important compounds are discussed briefly later in this introduction.)

Carbohydrates contain atoms of carbon, hydrogen, and oxygen in the ratio of 1:2:1. The basic unit of molecular structure of a carbohydrate can be written as $CH_2O$. In fact the term carbohydrate literally means *carbon water.* Units of $CH_2O$ are linked together in various ways to give a wide variety of different chemical compounds.

Fats are similar to carbohydrates in being composed of only carbon, hydrogen, and oxygen atoms but the proportion of oxygen atoms is much less.

All proteins contain carbon, hydrogen, oxygen, and nitrogen. Most of them, in addition, contain sulfur.

Carbon, hydrogen, oxygen, nitrogen, and phosphorus are significant in the structure of nucleic acids.

Organic nutrients are important to all living organisms since living matter requires a constant expenditure of energy in order to remain alive. Without an energy supply, protoplasm shortly becomes disorganized. We eat organic nutrients because we are hungry, but the hunger is part of a complex response to a reduction in the quantity of energy-containing substances available to our protoplasm.

The size of the blaze given off by a single mature corn plant when dry enough to burn may be imagined readily. Yet each corn plant grows from a single grain and the energy needed to support the imagined blaze was originally the energy of sunlight stored in the plant as a result of photosynthesis. The value of any crop plant depends to a large extent on the amount of energy-supplying foods that it contains. From a strictly mercenary point of view, therefore, it may be concluded that green plants have an enviable ability to convert sunlight into money.

When a corn plant or any other carbohydrate-containing substance is burned, several events occur simultaneously and rapidly. Gaseous oxygen from the atmosphere is used up. Another gas, carbon dioxide, is released to the atmosphere in equal volume to the oxygen consumed. Water in the form of water vapor is formed; energy in the form of heat is released. In other words, organic matter and oxygen are combined to form carbon dioxide and water with an accompanying release of energy. If the organic matter burned is a carbohydrate then the reaction may be summarized in the following way:

$$CH_2O + O_2 \longrightarrow CO_2 + H_2O + \text{energy}$$

Now suppose that this reaction were reversed:

$$CO_2 + H_2O + \text{energy} \longrightarrow CH_2O + O_2$$

This is basically the process that green plants are able to perform in sunlight. It is called *photosynthesis* and is the essential function of the green cell.

The simple sugar *glucose* is one of the products of photosynthesis that accumulates in plant cells and the above reaction can be rewritten in terms of production of glucose by the simple expedient of multiplying everything through by 6:

$$6\ CO_2 + 6\ H_2O \xrightarrow[\substack{\text{living plants} \\ \text{with chlorophyll}}]{\text{energy from light}} C_6H_{12}O_6 + 6\ O_2$$

When written in this way photosynthesis appears to be a very simple process. Actually, however, it occurs as a series of reactions, some of which are well understood while others are still being studied by research workers.

Photosynthesis not only produces the basic foods that sustain animal life but also releases molecular oxygen to the atmosphere, oxygen essential to the respiration process by which the stored energy in foods is made available for use in most living organisms. The planktonic (free-floating) algae of the ocean have a special significance in this connection. They are more abundant than the plants of the land and their total production of oxygen during photosynthesis is of enormous importance in maintaining the atmospheric balance. A few other natural phenomena result in a release of oxygen but the total amount involved is insignificant in comparison with that released during photosynthesis.

The normal concentration of oxygen in the atmosphere is slightly less than 20 percent. This must be replaced constantly because the supply is depleted by fires and other oxidative processes including the more common type of respiration carried on by plants and animals at all times.

## EXPERIMENTS WITH PHOTOSYNTHESIS

Many simple demonstrations rely on the presence of starch as proof that photosynthesis has occurred under the circumstances of the experiment even though starch is not a direct product of photosynthesis. Starch cannot be formed unless sugar is present and thus the presence of starch means that sugar was there previously. Starch is less mobile, easier to test for, and more revealing of the site of photosynthesis than is sugar.

The use of an iodine solution such as $I_2 \cdot KI$ will indicate the presence of starch by combining with it in such a way that a blue-black color results. Because the presence of chlorophyll may interfere with viewing this color it is necessary to bleach the leaf by soaking it in alcohol or acetone after first destroying cell membranes by immersing the leaf in boiling water.

The results of such a procedure are shown in Fig. I.8. A green and white leaf was used in this test to demonstrate the necessity of chlorophyll in photosynthesis. In a similar experiment designed to test the effect of growing plants with and without carbon dioxide the entire leaf from a plant grown without carbon dioxide would fail to show the starch reaction.

Fig. I.9 shows an albino pea seedling which has ceased growing after the food stored in the seed was used up. The albino condition results from an inability to form

Fig. I.8. Experiment demonstrating the necessity of chlorophyll in photosynthesis. (A) Living leaf of a variegated *Coleus* with the dark areas containing chlorophyll pigments. (B) The same leaf after pigments were removed and the leaf was treated with an iodine solution. The dark areas contain starch.

Fig. I.9. An albino (pigmentless) pea seedling among normal green seedlings.

chlorophyll and thus provides direct evidence of the significance of chlorophyll in food production.

The release of oxygen during photosynthesis may be used also as evidence of the occurrence of photosynthesis. Demonstrations of this type are easier to perform if water plants are used instead of land plants. Oxygen is soluble to a limited extent in water and there is a natural equilibrium between the oxygen in the atmosphere and that dissolved in the waters of the world. When the oxygen supply in water is increased beyond the equilibrium point the excess passes off into the air as a gas. The release of an oxygen bubble from a cut stem of an aquatic plant is shown in Fig. I.10. The accumulation of oxygen bubbles in tangled mats of aquatic algae often causes the mats to float to the surface of a pond (Fig. I.11).

Not all of the oxygen formed in photosynthesis escapes from the plant. Some of it is used in respiration which is carried on by the plant at all times. When light is absent or the light intensity is very low, not enough oxygen is formed to satisfy the respiration requirement and oxygen is absorbed from the environment. Under certain conditions the light may be of such an intensity that the respiration requirements

Fig. I.10. *Elodea* stem with a bubble of oxygen escaping from the cut end.

for oxygen are just balanced by the amounts produced in photosynthesis. During periods of maximum photosynthesis, however, the production of oxygen may be as much as 20 times that used in respiration.

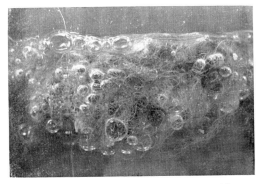

Fig. I.11. Mat of filamentous algae with numerous oxygen bubbles entrapped, causing it to rise to the surface.

## THE MECHANISM OF PHOTOSYNTHESIS

One of the most important research investigations in the study of photosynthesis involved the use of the isotope of oxygen known as heavy oxygen. Plants were grown in one set of experiments with the isotope as part of the carbon dioxide supply. In another set of experiments the isotope was part of the water added to the plants. Chemical analyses showed that heavy oxygen was released during photosynthesis only when it had been part of the water.

The results of these experiments have been interpreted to mean that water is split apart during photosynthesis. Therefore, in order to account for the known volume of oxygen produced it is necessary to assume that at least twice as many water molecules enter into the reaction as had been assumed previously. Furthermore, some water is reformed during the process. On the basis of this information the summary reaction for photosynthesis has been modified as follows:

$$6\ CO_2 + 12\ H_2O \xrightarrow[\text{living green plant}]{\text{light energy}} C_6H_{12}O_6 + 6\ H_2O + 6\ O_2$$

The splitting of water can be represented for discussion purposes by a simple formula:

$$H_2O \longrightarrow H + OH$$

The H units and accompanying high energy electrons (see below) are attached to compounds referred to as *hydrogen acceptors,* while the OH units ultimately function as the source of the oxygen molecules released during photosynthesis. The hydrogen acceptors are chemically reduced by the process and thus their energy level is increased.

The question arises as to how the water is split apart in such a way that energy is stored in the hydrogen acceptor. Very probably this is accomplished by the combined action of light and chlorophyll. It represents a conversion of the energy of visible light into chemical energy and, as such, is a key to the maintenance of life.

Apparently, a photon of any visible light can be absorbed by chlorophyll and take part in the splitting apart of a water molecule. This suggests that there may be energy left over from the impact of a high energy photon such as that of blue light and that this energy might emerge from chlorophyll as a photon with a lower energy value than the photon which entered the reaction. Such a happening would account for a property of chlorophyll called *fluorescence.* When chlorophyll is exposed to sunlight and observed from an angle that excludes transmitted light it will fluoresce with a dull red color. When ultraviolet light is used in place of sunlight, chlorophyll glows with a bright red color.

When white light is passed through a prism or a diffraction grating the various colors are separated into the familiar spectrum of the rainbow: red, orange, yellow, green, blue, indigo, and violet. If the white light passes through a chlorophyll extract before entering the prism it will be observed that most of the blue light and a large portion of the red light are missing. This indicates an especially large percentage of absorption of blue light and red light by chlorophyll.

Such evidence might be interpreted as meaning that red and blue light were the most effective in photosynthesis. However, the proportion of green light in natural sunlight is higher than the proportions of red and blue light. The percentage absorption of green light by chlorophyll in living plants is admittedly small but a small percentage of a large amount results in a significant total and it is possible that green light is effective in photosynthesis under natural conditions.

The proportionately higher absorption of red and blue light, however, serves to accentuate the greenness of light that is transmitted through a leaf and thus accounts for the normal color of vegetation.

Existing evidence suggests that chlorophyll molecules are excited to emit electrons upon absorption of light energy.

These *high-energy electrons* may be passed from compound to compound by an electron transport system, imparting chemical energy to several reactions along the way. A significant step in this procedure is the storage of energy by reducing NADP *(nicotinamide-adenine dinucleotide phosphate)* to NADPH.

Under natural conditions a very small percentage of water molecules dissociate into $H^+$ and $^-OH$ ions. The electron previously shared between them remains with the $^-OH$ and the $H^+$ lacks an electron. A molecule of NADP simultaneously accepts the high energy electron from the light-excited chlorophyll and takes up a unit of hydrogen to become NADPH which thereby has a greatly increased energy level.

Electrons must return ultimately to chlorophyll molecules in order for the emission process to be repeated. The returning electrons are stripped from the hydroxyl (OH) ions when the latter are freed from their attraction to those hydrogens that are taken up by the NADP. When freed of their extra electrons in this way the OH units may then unite in pairs or become attached to certain organic compounds to form peroxides:

$$OH + OH \longrightarrow H_2O_2$$

The breakdown of peroxides is the source of the molecular oxygen which is a characteristic byproduct of photosynthesis.

$$H_2O_2 \longrightarrow H_2O + \tfrac{1}{2} O_2$$

Numerous researchers have clearly demonstrated that the reduction of $CO_2$ to the carbohydrate level in plant cells may take place *in the dark* provided the cells contain sufficient quantities of such potent reducing agents as NADPH and *adenosine triphosphate* (ATP). The reduction is not direct, of course. Carbon dioxide is first incorporated into an existing organic molecule and the reduction takes place in a series of steps. These steps are well known but discussions of them are appropriately part of a course in plant physiology.

The significance of ATP in this discussion arises from evidence that ATP may also be formed in chloroplasts exposed to light. It has been suggested that some of the electrons emitted from chlorophyll enter into a series of reactions during which ATP is formed. These reactions occur in a cyclic fashion with the electron eventually returning to the chlorophyll at a much lower energy level. It is significant that water is not required as an electron donor in this cycle of events, which is called *photophosphorylation* rather than *photosynthesis*.

Most of our knowledge concerning the importance of ATP was obtained originally through investigations into the oxidative breakdown of sugar in respiration. In this process ADP (adenosine diphosphate) is converted to ATP by the addition of a third phosphate group. Much of the energy liberated in the breakdown of sugar is used to force this third phosphate group into position. The linkage that holds it in position has been likened to a spring under tension and is known as a high energy phosphate bond. When the third phosphate group is removed, thus releasing the spring, the stored energy becomes available for use in other chemical reactions.

## THE THEORY OF LIMITING FACTORS

This theory is an important biological concept illustrated by the problems of photosynthesis. Simply stated, the theory implies that when a given reaction is conditioned by several factors it will proceed at a rate determined by that one factor which is momentarily present in the least or limiting quantity. An appreciation of this theory is vital to understanding the vast interplay of forces that controls biological phenomena.

For example, in midmorning of a typical growing day in the Corn Belt, the weather will be bright and warm and the soil will contain sufficient water and nutrients to satisfy the immediate plant requirements. Yet the rate of food formation during the day will follow a certain set pattern, reaching a peak in the morning and often halting long before light intensity diminishes. What limits or holds back the productivity?

Probably in the morning it is the $CO_2$ supply. No matter how far from limiting the other factors might be, the process can go on no faster than the rate at which $CO_2$ enters the plant.

Light intensity is a limiting factor during large parts of each 24 hr period. Artificial increases in light intensity, especially as a greenhouse practice during winter months, have resulted in increased production rates in some cases. But the effects of light intensity and of light duration are not limited to photosynthesis alone. Several other functions of plants may be affected by such artificial increases and the total result becomes complex.

The reasons for an early afternoon slowdown in photosynthesis are more difficult to interpret since neither light intensity nor atmospheric supplies of $CO_2$ would be altered significantly from conditions prevailing during the late morning peak of production. Possibly the accumulation of the end products of photosynthesis may have an effect on the rate at which the process continues.

## UTILIZATION OF THE PRIMARY PRODUCTS OF PHOTOSYNTHESIS

The simple organic nutrients produced during photosynthesis are used in many ways but these may be grouped together loosely under three general headings:

1. building materials and storage products
2. sources of energy
3. effects on the water relations of cells

### BUILDING MATERIALS AND STORAGE PRODUCTS

Simple sugars such as *glucose* are the basic units from which other carbohydrates are constructed. A *sucrose* molecule, for instance, is formed by the linkage of a glucose molecule with a molecule of another simple sugar, *fructose*.

glucose + fructose $\longrightarrow$ sucrose (+ water)

Starch is a storage product of extreme importance to people. As a storage product it is of importance to the plants that produce it; another important benefit to plants is that starch formation removes the temporary excess of sugar (formed during photosynthesis) from solution. There is a definite relation between the amount of substances in solution in a cell and the tendency of water to enter the cell. Without the starch storage mechanism a green cell might accumulate so much soluble sugar that the inward movement of water would cause the cell to burst.

The formation of starch is essentially a condensation process. Starch is a *polysaccharide* since large numbers of glucose molecules are linked together to form each large molecule. The units are arranged in linear series; some kinds of starch molecules are branched. Each time a sugar unit is added to the chain a molecule of water is removed. A summary of this reaction may be written as follows:

$n$ sugar molecules $\longrightarrow$ 1 starch molecule + $(n-1)$ water molecules

($n$ is a large number but is not always the same.)

This chemical linkage is not a simple process. Each sugar molecule must enter into a temporary union with a phosphate group *(phosphorylation)* before it can be added to the chain. In addition the proper enzymes to catalyze the various reactions involved must be present.

Starch formation is not a continuously uniform process and starch is often deposited in layers of slightly different density. (Figure I.12 is a photomicrograph of starch grains from a potato tuber.)

Starch grains inside a plastid often become many times larger than the plastid itself. This may be observed clearly in sections of old green stems of a plant called *Pellionia*. In such sections the chloroplasts appear as small green caps attached to the sides of the much larger starch grains that have grown eccentrically within them (Fig. I.13).

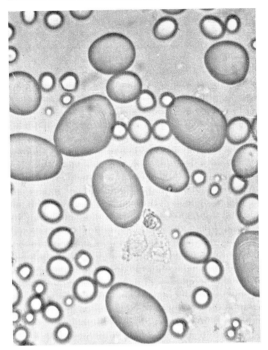

Fig. I.12. Starch grains of potato.

The reversal of starch formation is starch *digestion*. This process is also facilitated by a series of enzymes. Starch digestion is a hydrolytic reaction since a molecule of water must be added each time the linear chain of glucose units is broken.

*Cellulose* is a major component of the cell wall in the vast majority of green

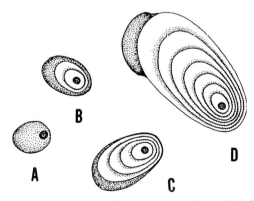

Fig. I.13. Plastids from stem sections of *Pellionia* containing starch grains in various stages of development. The volume of the starch grain increases from A to D while the volume of the plastid remains relatively unchanged.

plants and is thus a structural product rather than a storage product. It is of direct importance to humanity in such products as wood, paper pulp, and cellulose derivatives. People are not able to use cellulose as a food but herbivorous animals have organisms in their digestive tracts that digest cellulose and make it usable.

Cellulose is much like starch in that each molecule is made up of a series of glucose units linked together in a linear series. The type of chemical linkage that holds the units together is different, however, so that the physical and chemical characteristics of cellulose differ from those of starch.

Cellulose molecules are bound together in submicroscopic bundles called *micelles* or *microfibrils* distributed in a matrix of pectic substances. Apparently the bundling process is rather untidy in the primary wall and parts of some molecules extend from one microfibril to another. The resulting interconnected framework is both elastic and plastic to a certain extent.

Cellulose has an affinity for water and layers of bound water exist around the microfibrils. This adherent water has a further physical attraction for other molecules of water which are pulled into the spaces between the micelles, thus forcing them apart. Tremendous pressures can be generated when water is imbibed in this way. Rocks can be lifted or split apart by the swelling of dried wooden wedges soaked in water and warping wood can pull out deeply imbedded spikes. Normally the interior living cells of plants have walls that are saturated with water, but under certain conditions the replacement of water lost from cell walls can have an important effect on the total functioning of a plant.

The primary cell wall is stretched by cell enlargement to dimensions many times greater than its original area. Undoubtedly cellulose is added to the primary wall during this growth. Later, when stretching is complete, secondary wall layers are formed. These are not as plastic as the primary wall and the microfibrils are arranged in a much more orderly fashion. In addition other substances such as *lignin, suberin,* and *cutin* may be deposited in the wall. The nature of such depositions and their effects on cell characteristics are discussed in Chapters 1 and 2.

The individual cells of many-celled plants are cemented together by a chemical derivative of a carbohydrate substance called *pectin*. This is combined with calcium to form a very thin layer between the primary wall layers formed by adjacent cells. This layer is called the *middle lamella*. Its formation is associated with the development of a cell plate that accomplishes the division of the cytoplasm, particularly in higher plants.

*Proteins* are very complex molecules built up by the union of many simpler molecules called *amino acids*. The amino acids are constructed in plant cells from basic units of sugar which are highly modified and then combined with one or more amino groups ($NH_2^+$). One simple and important amino acid is *glycine* which has this molecular structure:

$$\begin{array}{c} H \quad\quad H \quad\quad O \\ \diagdown \quad | \quad\quad \diagup \\ N - C - C - OH \\ \diagup \quad | \\ H \quad\quad H \end{array}$$

Notice that at one end of this molecule there is an atom of nitrogen that is linked to an atom of carbon. The nitrogen atom and the two hydrogens attached to it constitute an amino group ($NH_2^+$). At the other end of the molecule is a carboxyl group (-COOH).

The various other amino acids have more carbon atoms than the glycine shown above and some of them have amino groups or carboxyl groups attached to other carbons of the chain. Amino acids can be united (by enzyme action) to form larger molecules. Two molecules of glycine, for instance, may be united to form a molecule of glycyl glycine:

$$\begin{array}{c} H \quad\quad H \quad\quad O \quad\quad H \quad\quad O \\ \diagdown \quad | \quad\quad \diagup \quad\quad | \quad\quad \diagup \\ N - C - C - N - C - C - OH \\ \diagup \quad | \quad\quad\quad | \quad\quad | \\ H \quad\quad H \quad\quad\quad H \quad\quad H \end{array}$$

Note that the larger molecule still has a free amino group at one end and a carboxyl group at the other, thus making possible the addition of more amino acids. The importance of these observations is that they reveal something of how protein molecules are constructed.

There are at least 22 naturally occurring amino acids. For the most part they are formed originally in plants. But once formed they may be used over and over again. The protoplasm of both plants and animals constantly builds proteins from amino acids and digests them back to amino acids.

It should be noted again that all proteins contain carbon, hydrogen, oxygen, and nitrogen and that many of them contain sulfur. Certain very important compounds, the nucleoproteins, consist of proteins united with nucleic acids in which phosphorus is a significant element.

*Fats* consist only of carbon, hydrogen, and oxygen; the proportion of oxygen is much lower than it is in carbohydrates. They are formed by a chemical union between *glycerine* and compounds called *fatty acids*. Glycerine is a three carbon compound. The fatty acids involved are usually long chain hydrocarbon compounds with a carboxyl group attached at one end. When fats are formed three molecules of fatty acids become attached to each molecule of glycerine. Digestion of fats breaks the chemical linkages between glycerine and the fatty acids.

Fatty substances are important as storage products, in the normal functioning of protoplasm, and as components of the outer protective layers of stems and leaves.

CARBOHYDRATES AS SOURCES OF ENERGY

A major value of a compound like glucose which occurs so commonly in living organisms is that it is a stable, soluble, movable, storage container of energy. Protoplasm is a complex system that requires a continual release of energy within itself in order to maintain its living qualities, and glucose is the most common immediate source of this energy. The process by which it is released is called *respiration*.

In many respects respiration is similar

to burning. Both processes form carbon dioxide and release energy and the more common type of respiration uses up oxygen. In fact the chemical reaction for the burning of a carbohydrate and a summary of the respiration of glucose are identical:

$$C_6H_{12}O_6 + 6\ O_2 \longrightarrow$$
$$6\ CO_2 + 6\ H_2O + 675 \text{ calories of energy}$$

However, respiration is a stepwise process during which a great many reactions take place in series, each one controlled by a specific enzyme that acts on the products of the preceding step. Also, at least part of the energy released is in the form of usable chemical energy rather than heat. The remarkable chemical compound, *adenosine triphosphate* (ATP), is, however, the recipient of much of the energy released during respiration and by entering into numerous other reactions in living organisms it releases the energy necessary to make these reactions go to completion.

Plants are not perfect in transferring all of the energy of glucose to other compounds. Some of the energy does escape as heat and sometimes the escaping heat of respiration of improperly stored grain or silage may accumulate to cause explosions or fires.

In some organisms respiration may occur in the absence of molecular oxygen. This *anaerobic respiration* is a process that releases energy in much smaller amounts than does aerobic respiration. Also, the end products of this reaction are different. *Alcoholic fermentation*, for example, is an anaerobic process that produces alcohol and carbon dioxide instead of water and carbon dioxide:

$$C_6H_{12}O_6 \longrightarrow$$
$$2\ C_2H_5OH + 2\ CO_2 + 25 \text{ calories of energy}$$

In both aerobic and anaerobic respiration, glucose, a 6-carbon compound, is broken down into two 3-carbon compounds which are further modified to form *pyruvic acid*. This process is called *glycolysis*. If molecular oxygen is available, the 3-carbon compounds then pass through a cycle of reactions involving a series of organic acids that break them down completely to $CO_2$ and water with a release of large amounts of energy.

If molecular oxygen is not available, however, the 3-carbon compounds pass through a different cycle of reactions in which they are not completely oxidized to carbon dioxide and water. The total amount of energy released is comparatively small and the end products are compounds such as alcohols and organic acids.

## THE EFFECTS OF SUGAR FORMATION ON WATER RELATIONS OF PLANT CELLS

In a preceding discussion it was noted that a significant benefit of starch formation to plants is the removal of sugar from solution since the presence of sugar in solution in cells has a direct effect on the tendency of water to enter cells. It is possible to demonstrate that cells with a high concentration of sugar and ready access to water can increase in size to the point of bursting.

In order to understand how this might be it is necessary to consider some of the factors affecting the movements of substances in solution. The first such factor is *diffusion*.

### DIFFUSION

Molecules of substances are constantly in motion. This motion tends to be in a straight line and given molecules travel indefinitely in straight lines providing nothing interferes with the motion. Within the limits of the physical phenomena of biology it is unlikely that molecules ever get such an opportunity. Instead, they are deflected repeatedly and move in irregular zig-zag patterns.

Since molecules can move furthest in directions where there is the least interference with their motion, diffusion involves a movement of molecules away from regions of their own greater concentration to regions of their own lesser concentration. Although concentration differences exert the major influence on the direction of diffusion, other influences such as temperature and pressure differences have a bearing on the process.

We will be concerned largely with the diffusion of gases and of substances in solution in liquids. The series of drawings in Fig. I.14 illustrates the diffusion of gas molecules in a closed container. If the gas were $CO_2$ and the bowl contained a solution of potassium hydroxide (KOH), gas molecules would be removed constantly by going into solution and reacting with the solute. Diffusion would tend to bring about a uniform concentration but no equilibrium would be possible as long as the solution continued to absorb the gas.

When a cube of sugar is placed in water, the concentration of sugar molecules in solution is greatest immediately around the cube (Fig. I.15). Their interference with each other is greater nearer the cube than farther away and thus those on the periphery of the molecular swarm are able to move farther before being deflected. Eventually there will be an even distribution of sugar molecules throughout the solution.

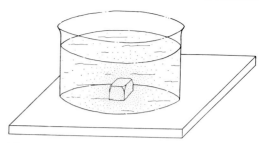

Fig. I.15. Schematic representation of the diffusion of dissolved sugar molecules away from the immediate vicinity of a sugar cube immersed in water.

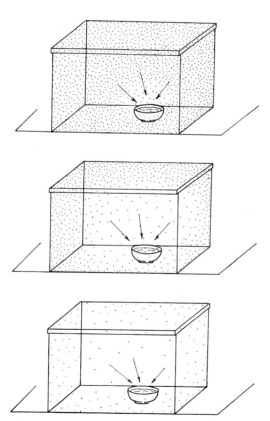

Fig. I.14. Schematic representation of a diffusion experiment.

OSMOSIS

When a membrane is placed in the path of diffusing substances, diffusion rates of the solvent and the individual substances dissolved in it may be altered. Consequently volume changes may occur in the solutions on opposite sides of the membrane. This is particularly true when the membrane is freely permeable to the solvent but only partially permeable or impermeable to the substances which are dissolved in it (solutes). The diffusion of the solvent through such a differentially permeable membrane is called *osmosis*.

Figure I.16 illustrates a widely used and familiar device for demonstrating osmosis. The sausage casing membrane contains a sugar solution and is immersed in distilled water. The open tube allows the solution to rise as its volume increases. The membrane allows free movement of water molecules in both directions but greatly inhibits the diffusion of sugar molecules.

According to the definition of diffusion, both the sugar and the water tend to move away from regions of their own greater concentration. The water succeeds in doing this easily but the sugar does not. The inward diffusion of water causes the volume of solution inside the membrane to increase and thus the solution rises in the tube. There is an actual hydrostatic pressure developed in such a situation and

# The Eucaryotic Plant Cell

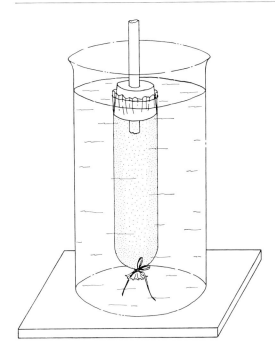

Fig. I.16. Laboratory demonstration of osmosis.

it has a considerable magnitude. Pressures in excess of 20 kg per cm² are well within the range of possibility for laboratory demonstrations of osmosis.

It is important to note that cell membranes such as the plasma membrane and the vacuolar membrane are differentially permeable and affect the rates of diffusion of different substances in different ways. In some cases the effect is so slight as to be unmeasurable. In others it amounts to virtual exclusion of particular solutes. In between the extreme cases the existence of a graded series of effects on diffusion rates is probable.

The term *osmotic pressure* is an expression of the potential maximum pressure which might be generated under ideal conditions when a solution is separated from the pure solvent by a membrane that is permeable only to the solvent. Such ideal conditions do not prevail in living cells and the actual pressures generated as a result of osmosis are always less than the potential maximum. The actual pressure is exerted by the cell sap against the cytoplasm, and through it, against the cell wall. It is called *turgor pressure* and when a positive turgor pressure exists in a cell, the cell is said to be *turgid*. Turgid cells tend to swell and, in growing tissues, this is a factor in cell enlargement. In mature cells, however, the outward pressure is eventually balanced by a resistance of the cell wall to further swelling and by the turgor pressures of adjacent cells.

PLASMOLYSIS

The concentration of soluble materials in the cell sap varies from cell to cell and from plant to plant. Ordinarily it is quite low and may be much less than 5 percent, being a mixture of sugars, salts, amino acids, other organic acids, and pigments. Some of these pass through the cytoplasmic membranes readily, some pass through slowly, and others are restricted to the cell sap. The water concentration is, conversely, very high, perhaps 95 percent or more. Thus when a cell is placed in a solution with a higher solute concentration and therefore a lower concentration of water, water diffuses out of the cell. When this happens the vacuole decreases in volume and the whole protoplast shrinks away from the cell wall. This phenomenon is called *plasmolysis*. It may be defined as the shrinkage in size of the protoplast due to a loss of water by osmosis.

Plasmolysis may be observed readily with water plants such as *Elodea* (Fig. I.17) or the cells of an alga like *Spirogyra* (Fig.

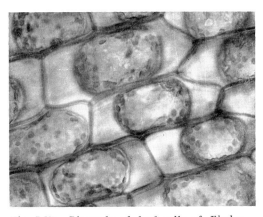

Fig. I.17. Plasmolyzed leaf cells of *Elodea*.

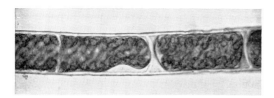

Fig. I.18. Plasmolyzed cells of the green alga, *Spirogyra*.

I.18) which have large central vacuoles. If plasmolysis is observed in cells containing the red pigment anthocyanin it will be noted that the color intensity increases as the relative concentration of these pigments is increased by the withdrawal of water.

When a cell has lost so much water by outward diffusion that the cell contents no longer exert a positive pressure on the cell wall the cell is said to be *flaccid*. This term has a meaning directly opposite to the term turgid.

A simple demonstration of how plant form is affected by relative turgidity or flaccidity of cells can be performed with two strips of tissue cut from the same potato (Fig. I.19). Both strips should be the same length and one should be placed in distilled water while the other is placed in a salt solution. In the distilled water, the cells of the potato tend to become larger due to the inward diffusion of water. The piece of potato becomes slightly longer and crisper. If bent between the fingers it will snap in two rather than bend. The other piece of potato in the salt solution becomes slightly shorter. Each of the cells has decreased in volume slightly due to the loss of water by outward diffusion. The cells no longer push against each other and the strip becomes so limber that it will bend without breaking.

THE NATURE OF CYTOPLASMIC MEMBRANES IN RELATION TO DIFFUSION

Cytoplasm has been described as having a basic protein framework and it is probable that somewhat flattened meshworks of protein molecules form the basic structure of the cytoplasmic membranes. Certain parts of protein molecules are hydrophilic; i.e., they hold a layer of water molecules about them by tenacious adhesive forces. Water molecules held in this way tend to be oriented similarly with similarly charged poles outward. Other water molecules are attracted to these poles and become bound as a second layer. Several layers of oriented water molecules may exist around a protein molecule but there is some unbound water in the microcapillary spaces between the protein molecules. The adhesive forces that attract the water molecules are very powerful but when water is freely available they are soon satisfied.

When water diffuses through a membrane it is unlikely that a single molecule goes all the way through at once. As one molecule escapes from one side of a membrane another molecule is absorbed from the water on the other side to replace it. The direction of diffusion is determined by escape from the membrane into a solution with a lower concentration of water and replacement from a solution with a higher concentration.

The problem of diffusion of solutes through a membrane is more complicated. Certainly the membrane should not be pictured as a sieve that keeps out the larger molecules. Many molecules move through by simple diffusion in the unbound water existing between the protein molecules. If these molecules are nonpolarized, their rate of diffusion may depend entirely on concentration differences. If they are polarized or have polarized groups attached to them their progress may be greatly affected by electromagnetic forces in the membrane. The same thing applies even more strongly to ionic particles.

Fig. I.19. Two strips of potato, originally of the same length, after immersion in distilled water (above), and in a 3 percent salt solution (below).

One of the complexities of the membrane problem is that certain ions tend to pass through membranes against the normal diffusion gradient and to accumulate in the cell sap in enormous concentrations compared to those in the external solution. It is thought that such accumulations require the expenditure of energy and they have been correlated with high respiration rates.

Another consideration is the diffusion of substances that are fat soluble rather than water soluble. Fatty substances tend to accumulate in membranes where they are dispersed as colloidal droplets in water. Under certain circumstances there may be a change in the membrane that causes the fatty substances to become the continuous phase while the unbound water becomes dispersed as separate droplets of colloidal size. Fat-soluble compounds are then able to diffuse in the continuous phase while water-soluble substances normally entering by simple diffusion are excluded. The movement of water is not affected particularly since the protein framework continues the process of giving up water molecules to one side and replacing them from the other.

IMBIBITION

Throughout the preceding discussions several references have been made to the taking up of water by substances that bind water to their molecules by adhesive forces. The bound water molecules then attract other molecules into polarized layers in which their motion is inhibited. The net effect is to force the attracting molecules apart and the substance is observed to swell. This phenomenon may be referred to as *imbibition*. During the swelling enormous forces are generated but once the swelling is completed the forces diminish rapidly and do not become effective again unless the substances become dehydrated.

## THE NUCLEUS

The nucleus tends to assume a spherical shape, especially in cells that lack conspicuous central vacuoles. In highly vacuolated cells where the cytoplasm becomes stretched into a thin peripheral layer the nucleus becomes flattened somewhat into a hemispherical shape. This is due to turgor pressure exerted by the cell sap against the cytoplasm in which the nucleus is always enclosed.

The structure of the nucleus seems easily described at first. At the nuclear-cytoplasm boundary is the *nuclear membrane*. Within this there is a large quantity of a fluid substance, the *nuclear sap*, and one or more large spherical bodies called *nucleoli*. Distributed throughout the nucleus is a three-dimensional network of finely divided material called *chromatin*.

The apparent simplicity vanishes when further study of the nucleus is undertaken. The nature and function of the nucleoli, for example, are still not clear. The physical distribution of the chromatin in an undisturbed living nucleus is difficult to determine. Certainly the brightly stained network seen in a prepared slide is not truly characteristic of the living nucleus. When, for instance, the nuclear division process is studied in stained slides it appears as though the chromatin network becomes organized into rod-shaped bodies called *chromosomes*. It is probable, however, that the chromosomes actually exist as discrete bodies in the nondividing nucleus as well, even though the boundaries between them are rarely detectable.

Nuclear and cytoplasmic divisions occur repeatedly in regions called meristems and sections through the apical meristem of an onion root (Fig. I.20) are used extensively in introductory studies of these processes. The roots are killed quickly and with a minimum amount of distortion before the sections are made. This means that the division process in any particular cell is stopped at the stage it had reached at the moment of fixation. By examining many cells stopped in various stages it is possible to obtain a concept of the entire process of nuclear and cytoplasmic division. However, this method of study may create an impression that the cell jumps from stage to stage while actually there is a gradual progression between the more obvious stages that increases in speed from start to finish.

xxviii INTRODUCTION

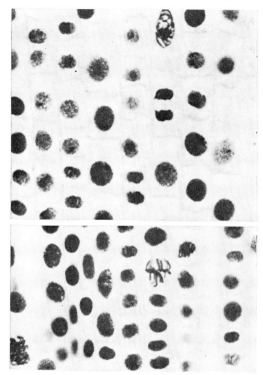

Fig. I.20. Section of an onion root tip showing cells in various stages of division. (Compare Figs. I.21 and I.22.)

## NUCLEAR DIVISION (MITOSIS)

The division of one nucleus to form two is called *mitosis* and in stained sections the first evidence that mitosis has begun is an increased brilliance of the stain. In a nondividing nucleus, on the other hand, the chromatin is very finely dispersed and does not stain brightly. The increased stainability is correlated with an aggregation of the chromatin into coarser threads as mitosis begins. The main events in the process of mitosis have been named as phases (Figs. I.21, I.22). They will be discussed in order beginning with the early stages called the *prophase*.

### PROPHASE

During the early portions of this phase many changes occur:

1. The chromatin becomes more concentrated and takes the shape of long thin strands called *chromosomes*. These are at first tangled together like a ball of yarn. It is known that each chromosome is a double structure and each of the lengthwise parts is called a *chromatid*. The threadlike core *(chromonema)* of each chromatid is loosely coiled at this time. As prophase continues the chromosomes continue to become shorter and thicker and to stain more deeply. These changes are associated with a more pronounced coiling of the chromonemata.
2. The nucleolus gradually disappears.
3. When the chromosomes have shortened considerably the nuclear membrane rather abruptly disappears, allowing the chromosomes to lie free in the cytoplasm.
4. At about the time that the nuclear membrane disappears some very fine threadlike structures called *spindle fibers* appear in the cytoplasm. These occur in two groups at opposite poles of the cell. They converge to indefinite points in the cytoplasm at opposite poles of the cell but they are not attached to any definite structure; i.e., there is no centrosome as in animal cells.

Some of the spindle fibers become attached to the chromosomes at definite points called *kinetochores*. A fiber from one pole attaches to one chromatid of a chromosome while a fiber from the opposite pole attaches to the other chromatid. Other fibers are unattached and pass through the chromosome mass, extending from pole to pole.

### METAPHASE

The chromosomes soon become rearranged so that they lie in a flat plane across the cell. This *equatorial plate* lies midway between the two poles of the cell. The movement of the chromosomes is associated with an adjustment of the lengths of the attached spindle fibers; whether this is cause or effect is difficult to prove. In lengthwise sections of root tips the equatorial plate is normally seen in edge view only. Research workers interested in the numbers of chromosomes per cell cut cross sections so

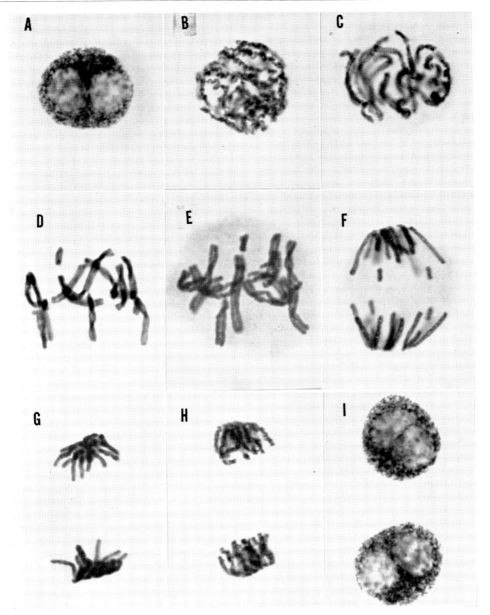

Fig. I.21. A series of photomicrographs of nuclei undergoing mitosis. The series of drawings in Fig. I.22 was conceived from these photomicrographs. In preparing the slides from which these pictures were taken a staining technique was used which is specific only for chromosomal substances.

that they can view the equatorial plates from the poles. When seen in this way the chromosomes are spread out evenly in approximately the same plane and can be counted readily.

ANAPHASE

The anaphase is a phase of movement. The two chromatids of each chromosome separate and move toward opposite poles. Some authorities maintain that this is due

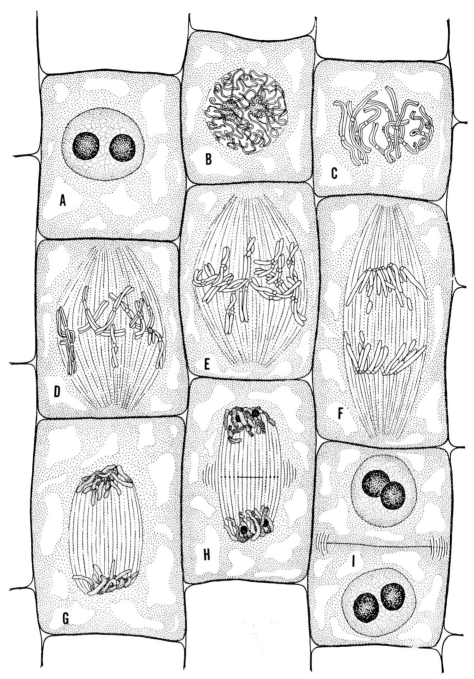

Fig. I.22. Schematic representations of stages in mitosis and cell division based on the photomicrographs in Fig. I.21. A. Interphase. B. Early prophase. C. Later prophase. D. Rearrangement of chromosomes just prior to metaphase. E. End of metaphase (chromatids beginning to separate). F. Anaphase. G. Late anaphase. H. Telophase of mitosis and beginning of cytokinesis. I. Daughter nuclei formed, cytokinesis almost complete.

to a contraction of the attached spindle fibers that causes the chromatids to be pulled apart. The separated chromatids are now called *daughter chromosomes*. It is important that the two separated groups of daughter chromosomes are alike with respect to numbers and kinds and that the number and kind in each group are identical with those of the nucleus that began the division. In this way, the process of mitosis maintains a constant chromosome number in the body cells of a growing organism.

TELOPHASE

Anaphase is completed when all of the chromosomes have reached the poles of the cell. Telophase is the end phase during which daughter nuclei are reconstructed from the clumps of chromosomes at each pole. In each clump the individual chromosomes begin to lose their clearly separate identity and a nuclear membrane is formed around the mass. Then this daughter nucleus begins to swell through the absorption of water and other substances. The chromatin becomes more and more diffuse and loses its brilliant stainability. During this reconstruction process the nucleoli reappear.

## CYTOPLASMIC DIVISION (CYTOKINESIS)

While the division of the cytoplasm is not truly a part of mitosis it is a related process. In the cells of higher plants, it involves a residual part of the mechanism of mitosis and usually begins during telophase. In higher plants cytoplasmic division is initiated by a thin, membranous *cell plate* that forms in the approximate position and plane previously occupied by the chromosomes at metaphase. The cell plate seems to be the result of the fusion of minute vesicles derived from the Golgi membranes which accumulate in an equatorial plane across the spindle-shaped mass of fibers extending from pole to pole. When first formed the cell plate does not completely cross the cell. Additional peripheral fibers appear in the cytoplasm around the plate. Vesicles accumulate on these fibers also and fuse with the existing cell plate. In this way the plate grows outward and eventually separates the cytoplasm into two units, each one of which contains a daughter nucleus. Each of the new protoplasts soon begins to deposit cellulose on its own side of this layer. Thus the primary cell wall is the joint product of two adjacent cells. Secondary wall thickenings will be discussed later since they do not normally begin to form in meristematic tissues.

INTERPHASE

Following nuclear and cytoplasmic divisions the daughter cells grow for some time before they divide again. This growth involves both increase in size due to the absorption of food and water and assimilation of some of this material into new protoplasm. It is known that the amounts of those chemicals that are significant in the structure of chromosomes are doubled during interphase and from this it has been inferred that chromatid duplication occurs during this phase.

## NATURE OF CHROMOSOMES

Scientists are led to the conclusion that chromosomes are made up of large numbers of smaller units called *genes*. The interaction of large numbers of genes insures that a bean seed will give rise to a bean plant but it is probable also that a few genes or even a single gene may influence the final expression of a single adult characteristic such as flower color.

As a generality with significant exceptions, it can be stated that all living cells of the body of a plant have the same numbers and kinds of chromosomes. However, when one realizes how many different kinds of cells there are in a plant the question arises as to how cells with the same inheritance can become so vastly different. Some indication of the answer is given by the fact that no two cells in a plant exist in exactly the same environment. Epidermal cells of a leaf and the green cells inside are exposed to different environmental conditions. Cells of a root have different conditions of life than do stem cells. The interplay of large numbers of genes in the control of cell physiology permits an infinite variety of cellular response. Thus, neighboring cells with identical sets of chromosomes and genes may mature into entirely different kinds of cells because their immediate environments are different.

Students of the mechanisms of heredity have learned much about genes in the past several decades. In some organisms investigators can point to an exact spot in a particular chromosome where exists a gene that affects the expression of a single adult character. Yet most of these investigators hesitate to say exactly what a gene is beyond acknowledging that *deoxyribose nucleic acid (DNA)* is intimately involved in its structure.

It is assumed that the student is basically familiar with the nature of DNA as well as with the related compound *ribonucleic acid (RNA)*. To review, the DNA molecule is composed of numerous units called *nucleotides*, each of which consists of one molecule of the sugar *deoxyribose*, a phosphate group, and one of four *nitrogen bases: adenine, thymine, cytosine,* or *guanine*. The nucleotides are joined in linear series by chemical bonds between sugar and phosphate groups. Also, they are united laterally in *base pairs*, with thymine always paired with adenine and cytosine always paired with guanine. The resultant molecule is double stranded with a helical twist similar to that of a twisted rope ladder.

While the side to side unions of nitrogen bases are limited to the two possibilities noted above, the linear orientations are not fixed and in fact the various linear combinations of the bases make possible the *genetic code*.

During interphase of mitosis it is assumed that DNA molecules are duplicated and the mechanism of mitosis serves to separate the duplicates with each of the daughter nuclei receiving an identical set.

Additional *single strand* duplicates called *messenger RNA* are produced in the nucleus with the DNA molecules serving as templates for their construction. Thus the order of nitrogen bases in the messenger RNA is the same as in the DNA that was copied. However, two curious differences exist between DNA and RNA: the nitrogen base *uracil* replaces thymine and the sugar *ribose* replaces deoxyribose. Once formed the molecules of messenger RNA migrate into the cytoplasm where they become attached to submicroscopic particles called *ribosomes*.

Other smaller units of RNA are formed also in the nucleus and migrate into the cytoplasm. These are called *transfer RNA* and each unit is able to attach a specific amino acid to itself, as determined in some way by the three bases that are exposed at one end of the molecule.

Ultimately a matching place is exposed on the messenger RNA that has the same catenation of 3 nitrogen bases. The amino acid is then unlocked from the transfer RNA by enzymatic action and is locked into position on a growing chain of amino acids that eventually will become a protein molecule.

Enzymes are essentially protein in nature and the described procedure accounts for the fact that many different kinds of enzymes can be created in a given cell, each one assembled from amino acids in a pattern defined by the order of nitrogen bases in a given RNA molecule. This pattern, determined by the base order in a DNA molecule, is replicated during each nuclear division cycle.

Each enzyme apparently serves to control a specific chemical reaction and thus the enzymes present in a cell determine most of the chemical reactions that cell can achieve. In this way they influence how the cell will react in a given microchemical environment.

In a sense therefore the genetic control operates through the DNA-RNA relationship which provides the coded information for the construction of specific enzymes. If, for instance, coded information is present in a plant to direct the formation of the enzyme that puts pieces of anthocyanin together it will be possible for the plant to have red flowers. Conversely, if the coded information is absent the enzyme will not be created, anthocyanin cannot be formed, and the flowers cannot be red.

## THE BASIC PLANT LIFE CYCLE

It has long been recognized that living organisms have sets of chromosomes in their nuclei that in all members of a given species are essentially alike. The number of chromosomes in a set is indicated by the symbol $n$ which permits designation of a fixed number for each species. (Among the known species of living things $n$ varies from one to several hundred.)

The body cells of most higher plants and animals have 2 sets of chromosomes per nucleus; this condition may be referred to as $2n$ or *diploid*. Sex cells *(gametes)* characteristically have but 1 set of chromosomes in their nuclei and this condition is termed *n, haploid,* or *monoploid*. In most plant life cycles there exists an additional type of reproductive cell which also has the haploid chromosome number. This cell is called a *spore* (more specifically a *meiospore*). It is not a sex cell since it does not unite with other cells.

In plant species that produce both gametes and spores the spore usually germinates to form a plant body in which all of the cells are haploid. The function of this plant is to produce gametes and it is called the *gametophyte*.

When 2 gametes unite, as in fertilization, the resultant cell is a *zygote*. It is the first diploid cell in the life cycle and in the green land plants at least it divides repeatedly by mitosis, giving rise to a plant body in which all cells are diploid and genetically similar. At maturity certain special cells of the plant called *spore mother cells* or *sporocytes* undergo meiosis to produce tetrads of haploid *spores*. This spore-producing generation is appropriately called the *sporophyte*. The spores are the first haploid cells in the life cycle and, as noted above, each one of them may germinate to give rise to a gametophyte.

The basic plant life cycle is thus seen to involve an endless *alternation of generations*, with one generation being haploid and the other diploid. Changes in chromosome numbers occur at only two points in the life cycle. The number is doubled when 2 gametes unite during fertilization and is cut in half again when the spore mother cells undergo meiosis to form the spores.

Meiosis can occur only in cells having 2 sets of like or homologous chromosomes, the 2 sets having been brought together previously at the time of fertilization. As a result of the 2 divisions of meiosis, the haploid chromosome number is restored and each of the 4 nuclei has 1 set of chromosomes.

### SYNOPSIS OF NUCLEAR EVENTS IN PLANT LIFE CYCLES

Because an understanding of meiosis is often attained without a corresponding awareness of where and when it occurs, the following summary of nuclear events in the life cycles of plants has been included to place both meiosis and mitosis in proper perspective.

Mitosis in haploid cells occurs during growth of gametophytes from spore germination to maturity, including the divisions that lead to gamete formation.

Mitosis in diploid cells occurs during growth of sporophytes from zygote germination to maturity, including the divisions that lead to the formation of the spore mother cells.

Mitosis in triploid ($3n$) cells occurs during the formation of the endosperm from the primary endosperm cell in the ovules of flowering plants. (See Chapter 5.)

Meiosis in the zygote occurs in the life cycles of many algae, but not in all. (See Chapter 6.)

Meiosis in spore mother cells occurs in all groups having an alternation of diploid spore-producing plants with haploid gamete-producing plants.

Meiosis prior to gamete formation occurs in certain algae and in most animals. In such cases the cells formed as a result of meiosis become modified into gametes instead of spores.

The discussion of the details of mitosis was illustrated with diploid cells from the meristematic region of an onion root tip.

However, the basic pattern of mitosis is the same in haploid cells and triploid cells. Similarly, the basic features of meiosis are the same whether the process is studied in zygotes or in spore mother cells (Figs. I.23, I.24). Also, despite the fact that the two divisions of meiosis have features that distinguish this process from mitosis, it is apparent that the basic mechanism is the same in both processes.

## MEIOSIS

The two divisions of meiosis may be referred to as meiosis I and meiosis II and the phases of their divisions bear the same names as in mitosis.

### Prophase of Meiosis I

In very early prophase of this division the chromosomes appear as long, slender, twisted threads much as in a prophase of mitosis. It is evident that the normal duplication of chromatids has not been completed since each chromosome appears to have only 1 chromatid instead of 2 (Fig. I.25).

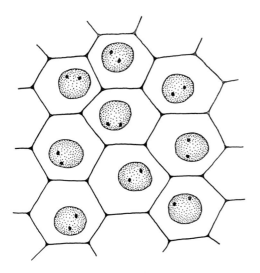

Fig. I.23. Mass of spore mother cells prior to meiosis. (Compare Fig. 5.28.)

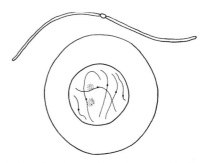

Fig. I.25. Prophase of meiosis I. Early stage when each chromosome consists of but a single chromatid. The chromosome number here and subsequently is $2n = 6$.

As prophase continues the chromosomes no longer behave independently as they do in mitosis. Instead the homologous chromosomes pair with each other intimately and as a result the apparent number of chromosomes is reduced to one-half the normal diploid number (Fig. I.26).

Shortly thereafter, the appearance of a lengthwise split in each of the paired chromosomes indicates the belated completion of the process of chromatid duplication. Since each of the chromosomes now has 2 evident chromatids there are 4 chromatids in each of the joined chromosome pairs (Fig I.27). It takes both the first and the second division of meiosis to separate these

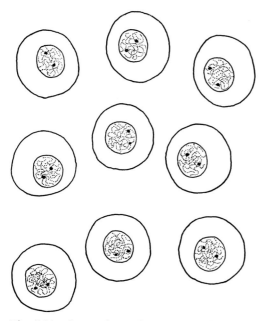

Fig. I.24. Separation of spore mother cells which frequently occurs as meiosis begins. Each one tends to assume a spherical shape.

# The Eucaryotic Plant Cell

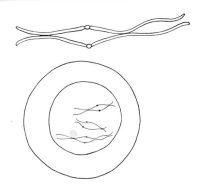

Fig. I.26. Prophase of meiosis I. The pairing of homologous chromosomes has occurred.

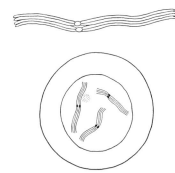

Fig. I.27. Prophase of meiosis I. Each of the paired, homologous chromosomes has completed a chromatid duplication. Each chromosome now has 2 chromatids and, thus, there are 4 chromatids in each chromosome pair.

4 chromatids and each one of them becomes part of a different nucleus.

As prophase I continues the homologous chromosomes begin to twist around each other as they shorten and there may be frequent breaks in the chromatids. The fragments are quickly rejoined but as the result of slight shearing movements it often happens that the rejoining occurs between fragments of nonsister chromatids (Fig. I.28). These *chiasmata* have a significant influence on the final distribution of genes. Also when the paired chromosomes begin to repel each other in late prophase the chiasmata serve to keep them from becoming completely separated.

## Metaphase of Meiosis I

During late prophase the nuclear membrane disappears and spindle fibers are

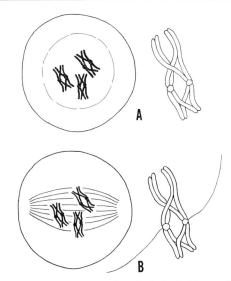

Fig. I.28. Prophase of meiosis I. A. The paired chromosomes have begun to repel each other. B. The nuclear membrane and the nucleoli have disappeared, spindle fibers have become attached at the kinetochores, and chiasmata are evident.

formed. Fibers from opposite poles become attached to the chromosomes at the kinetochores. Throughout the rest of the first division the 2 chromatids of each chromosome remain united at the kinetochore. The kinetochores of each member of a chromosome pair, however, are not united. Then the rearrangement occurs which distributes the paired chromosomes across the equatorial plate (Fig. I.29). This is metaphase I, and because of the denseness of the chromosomes, the apparent reduction in numbers, and their even distribution across the equatorial plate, it is an advan-

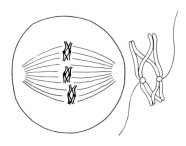

Fig. I.29. Metaphase of meiosis I. The chromosome pairs are distributed across the equatorial plate.

tageous stage for the determination of chromosome numbers.

### ANAPHASE OF MEIOSIS I

During this phase the whole chromosomes, which had been united in pairs and held together by the chiasmata, separate and move to opposite poles (Fig. I.30). The

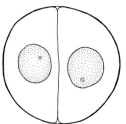

Fig. I.32. Interphase between meiosis I and meiosis II.

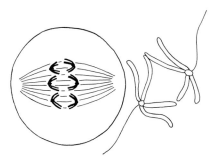

Fig. I.30. Anaphase of meiosis I. A separation of whole chromosomes, rather than chromatids, has occurred.

mechanics of this process are not much different from the same phase of mitosis. Each chromosome consists of 2 chromatids which are still united at the kinetochore and the chromosome number in each of the two separating groups is haploid.

### TELOPHASE OF MEIOSIS I

Except for the fact that each chromosome obviously has 2 chromatids there is no difference between the telophase of meiosis and the normal mitotic telophase. The chromosomes become clumped densely at the poles (Fig. I.31) and after the nuclear membrane is again formed they begin to swell and become diffuse. Depending on the plant being studied cell division may or may not occur after the first division of meiosis is complete (Fig. I.32).

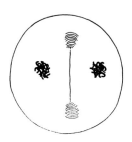

Fig. I.31. Telophase of meiosis I. The chromosomes are arranged in dense clumps, one at each pole of the cell. Cytokinesis is in progress.

### PROPHASE OF MEIOSIS II

This prophase follows rapidly after the telophase of the first division. The chromosomes each have 2 chromatids but aside from any crossovers which may have occurred these are the same 2 that were present in the first division; no chromatid reduplication has occurred during the brief interphase (Fig. I.33). The number of chromosomes is, of course, haploid. As the nuclear membrane disappears spindle fibers are formed and become attached to the chromatids at the kinetochores (Fig. I.34).

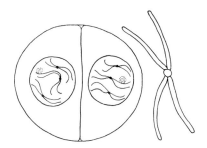

Fig. I.33. Prophase of meiosis II. Each nucleus has the $n$ number of chromosomes (in this case $n = 3$), and each chromosome has the same 2 chromatids it had at the end of meiosis I.

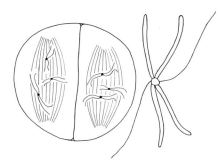

Fig. I.34. Prophase of meiosis II. The nuclear membranes and the nucleoli have disappeared and spindle fibers have become attached at the kinetochores.

## METAPHASE OF MEIOSIS II

This phase is no different from that of normal mitosis. The 2 nuclei formed in the preceding division usually divide in unison but frequently the plane of division of 1 member of the pair is oriented at right angles to the plane of division of its twin (Fig. I.35).

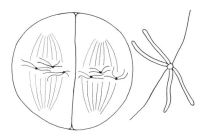

Fig. I.35. Metaphase of meiosis II. The chromosomes have become distributed across the equatorial plate.

## ANAPHASE OF MEIOSIS II

In this phase, the kinetochores divide and thus the 2 chromatids of each chromosome which had remained together through the whole of meiosis I finally are able to separate and become daughter chromosomes. The groups of daughter chromosomes approaching opposite poles have the haploid number. In the 2 simultaneous divisions there are 4 such groups and in each of them there is 1 chromatid from each of the pairs of homologous chromosomes in prophase I (Fig. I.36).

## TELOPHASE OF MEIOSIS II

The chromosomes in each of the four groups clump together (Fig. I.37) and then

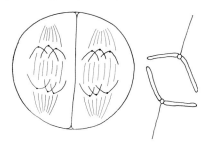

Fig. I.36. Anaphase of meiosis II. The chromatids of each chromosome have separated and become daughter chromosomes. Each of the 4 groups of chromosomes has the $n$ number.

begin to form nuclei. The 4 daughter nuclei are usually contained within the original cell wall. Then cell walls form between them so that a tetrad of spores is formed (Fig. I.38). Each spore has a haploid nucleus and can give rise to a gametophyte.

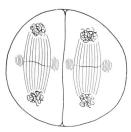

Fig. I.37. Telophase of meiosis II. The daughter chromosomes are now densely clumped at the poles and cytokinesis is in progress.

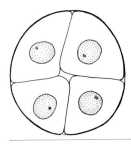

Fig. I.38. Tetrad of spores. With the completion of the second division of meiosis, and the associated cytokinesis, 4 spores are formed, each having a haploid nucleus.

In summary, meiosis is a process involving 2 successive and somewhat specialized nuclear divisions during which the chromosomes divide but once. As a result 4 nuclei are formed, each of them having one-half the chromosome number of the parent cell nucleus.

## BASIC MECHANISM OF INHERITANCE

An initial understanding of the relation between sexual life cycles and inheritance is best achieved by first considering plants which are genetically *homozygous*. Homozygosity is the condition in which both members of a given homologous pair of chromosomes of the diploid set contain identical genes for the characteristics being studied. In such a plant when pairing occurs during meiosis there is a perfect matching of these genes. When tetrads of spores are produced after meiosis, each one contains one of the identical genes for each characteristic. Consequently the resulting male and female gametophytes and the male and female gametes contain identical genes in their nuclei. When the gametes fuse at fertilization the diploid zygote receives two sets of chromosomes and thuse has two identical genes for each of the characters concerned. In this way the genetic constitution of a homozygous line is maintained.

If, however, the members of a gene pair are dissimilar the genetic constitution of the zygote would be *heterozygous* for the characteristics concerned and so would all the cells of the diploid plant which developed from the zygote.

A relatively simple example of how such a situation affects the gametophyte generation is offered by the liverwort *Marchantia* whose life cycle is discussed in Chapter 4. In this plant sperms are produced on one haploid plant and eggs on another. The maleness or femaleness of a given plant is controlled by genes. The zygote receives one set of chromosomes which contain a gene for maleness of the gametophyte and another set containing a gene for femaleness. The zygote and the resultant sporophyte are heterozygous with respect to sexuality of the gametophyte. Of course this has no significance in the development of the sporophyte but when meiosis occurs and spores are formed one-half of the spores receive the gene for maleness and the other half the gene for femaleness. Consequently, half of the succeeding gametophytes are male and the other half female.

Since the nuclei of the cells of the gametophytes contain only one set of chromosomes apiece the question of homozygosity or heterozygosity does not arise. Nor does the matter of dominance versus recessiveness (discussed later) of the members of a gene pair have any significance in the gametophyte generation. These topics, however, are of great concern when a study of inheritance in flowering plants is undertaken.

In nature and in practice it often happens that pollen from one plant is transferred to the stigma of another plant whose genetic history is different. When such pollinations occur varied circumstances may determine the eventual success or failure of the hybrid. If it is successful, however, the genetic nature of the new plant will be mixed or heterozygous for many characteristics and it will show varying degrees of mixture of the characteristics of the parents.

SIMPLE MONOHYBRID CROSS

In a standard problem on which this discussion is based, a white-flowered plant is crossed with a red-flowered plant. The progeny from this cross all develop into red-flowered plants since red flower color is dominant over white.

Certain symbols are used to indicate characteristics being studied in such problems. The homozygous red-flowered plant is indicated as *RR,* which means that its diploid cells contain two homologous chromosomes each one having a gene which determines red flower color. The homozygous white-flowered plant is indicated as *rr.* (Character symbols are derived from the first letter of the word describing the dominant character, in this case red. Capital letters are used to indicate dominance. The same letter is used for the recessive character but recessiveness is indicated by use of the small letter. A scheme such as this is necessary in order to keep track of gene pairs in complex problems involving several characteristics.)

The simple cross being discussed is summarized in the following diagram. The use of italicized symbols indicates that the chromosomes have two chromatids apiece while the roman symbols indicate the single chromatid condition that prevails briefly during anaphase II and telophase II. This is a temporary condition because normal chromatid duplication occurs before the nuclei concerned proceed with the next phase of the life cycle. Thus the italicized form will be used in discussions. (A familiarity with the flowering plant life cycle is assumed here. These terms are reviewed and explained in Chapter 5.)

THE EUCARYOTIC PLANT CELL                                                                                                xxxix

Red-flowered Parent   X   White-flowered Parent

Microspore Mother Cells      Megaspore Mother Cells

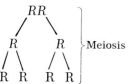

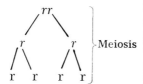

All four microspores can become pollen grains and eventually form pollen tubes with male gametes.

Three of the megaspores degenerate. One survives to produce the embryo sac with one egg.

such a case both the microspore mother cells and the megaspore mother cells are heterozygous and contain $Rr$. The eventual separation of this gene pair is indicated by the accompanying schematic diagram of meiosis in a spore mother cell which has the chromosome number $2n = 6$ (Figs. I.39, I.40, I.41, I.42). Selected stages in the meiotic process are used to indicate the distribution of genes for flower color during meiosis in spore mother cells. The dominant gene $R$ indicates red flower color while the recessive gene $r$ indicates white flower color. The labeled dots in the chromosomes are schematic indications of the position of these genes.

In this example it does not matter which of the pollen grains produces the successful pollen tube that enters an embryo sac. The male gamete can contain only $R$ and cannot contain $r$. Nor does it matter which of the four megaspores is the surviving one. The egg can contain only $r$ and cannot contain $R$.

As a result of fertilization the diploid nucleus of the zygote must contain both $R$ and $r$. The new sporophyte generation is designated as the $F_1$ and all of its cells have this $Rr$ constitution since they are derived from the zygote by mitosis.

It is now interesting to see what happens when the $F_1$ is self-pollinated. In

The significant feature here is that half of the nuclei formed by meiosis contain $R$ and the other half $r$. This means that one-half of the pollen grains will carry $R$ and the other half $r$. It also means that the surviving megaspore in any particular ovule will have an equal chance of containing $R$ or $r$.

Since there is no change in the genetic makeup of nuclei between spore formation and gamete formation there is an equal chance that any particular gamete, egg or sperm, will contain $R$ or $r$.

This situation may be presented in another way.

Fig. I.39. Early prophase of meiosis I. Before pairing of the homologous chromosomes and before the completion of chromatid duplication.

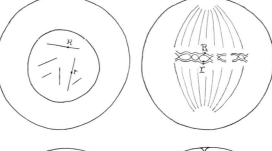

Fig. I.41. Metaphase of meiosis I.

Fig. I.40. Late prophase of meiosis I. After pairing of the homologous chromosomes. The double dots indicate gene duplication which was accomplished as a part of chromatid duplication.

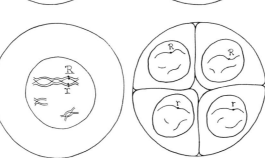

Fig. I.42. Tetrad of spores. The daughter chromosomes in the respective nuclei show the final distribution of the genes R and r after the completion of meiosis II.

### F₁ Plant (Self-pollinated)

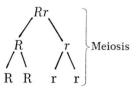

Microspore Mother Cells

It is a matter of chance which of these forms the pollen tube that enters the embryo sac.

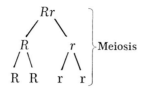

Megaspore Mother Cells

It is a matter of chance which of these forms the embryo sac.

From the above it is evident that:
1. The female gamete can contain $R$ or $r$.
2. The male gamete can contain $R$ or $r$.

A simple checkerboard diagram shows the possible recombinations which might occur in the resulting zygotes that give rise to the F₂ generation. In plants that produce large numbers of seeds it is usual to find all of these possible combinations in the predicted ratios.

|  |  | Female Gametes | |
|---|---|---|---|
|  |  | $R$ | $r$ |
| Male | $R$ | $RR$ | $Rr$ |
| Gametes | $r$ | $Rr$ | $rr$ |

1. The zygote $RR$ would result in a mature plant that was homozygous for red flower color.
2. The zygote $rr$ would result in a plant that was homozygous for white flower color.
3. The 2 zygotes $Rr$ would result in heterozygous plants like the F₁ generation.

There are three *genotypes* (genetic types) above and they occur in a ratio of 1:2:1.

Nothing has been said so far concerning the color of the flowers in this cross. The F₁ plants were all red-flowered because, as stipulated at the start, the red gene is dominant over the white, although situations do exist where partial dominance might result in pink flowers. In the F₂ generation the $RR$ plant would be red-flowered and so would the 2 $Rr$ plants and only the $rr$ plant would be white-flowered.

## INTRODUCTION

If a large number of plants were grown the observable results would be 3:1 on the average in favor of red-flowered plants.

The observable ratios are spoken of as *phenotype* ratios. Often the true nature of the actual gene constitutions (genotypes) can be derived mathematically from the observed phenotype ratio.

In order to determine which of the red-flowered plants were homozygous or heterozygous for flower color it would be necessary to self-pollinate each and grow the seeds to mature plants of the F₃ generation. The homozygous ($RR$) plants would give rise only to red-flowered progeny, while the heterozygous ones $Rr$) would give rise to red-flowered plants and white-flowered plants in the ratio of about 3:1.

### SIMPLE DIHYBRID CROSS

In the following example plants are crossed that have 2 gene differences. In this example the following points are stipulated:

1. Flower color is influenced by a single gene. Red flower color ($R$) is dominant over white ($r$).
2. Height of the plant is influenced by a single gene. Tall ($T$) is dominant over dwarf ($t$).
3. The genes for height and color are located on different chromosomes.

Under such circumstances the number of possible recombinations in the F₂ generation is increased as will be seen below.

Tall, Red-flowered X Dwarf, White-flowered
Plant                          Plant

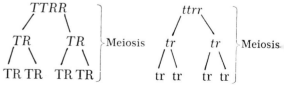

Megaspores, embryo sacs, and eggs can contain only $tr$ in their nuclei. Microspores, pollen grains, pollen tubes, and male gametes can contain only $TR$ in their nuclei. The zygote therefore will contain $TtRr$ and the resulting $F_1$ generation will be all tall and red-flowered. However, when the microspore mother cells and megaspore mother cells in the flowers of the $F_1$ generation undergo meiosis it normally happens that some of the resulting haploid nuclei contain gene combinations that had not existed together before in the haploid nuclei of gametes. The explanation for such recombinations of genes can best be undertaken with diagrams illustrating the possible paired chromosomes at metaphase I since it is there that the distribution of hereditary factors is governed by chance (Fig. I.43). The orientations of $T$, $t$, $R$, and $r$ shown at B and BB (of Fig. I.43) are the only ones possible and the

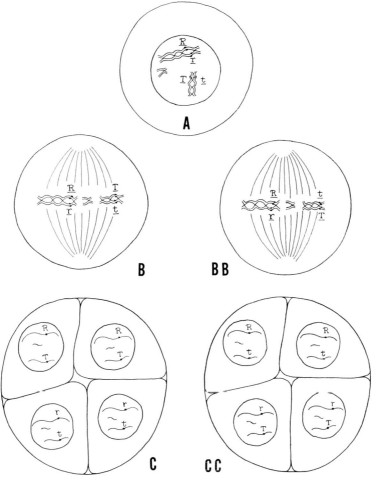

Fig. I.43. Selected stages of meiosis in a spore mother cell from the $F_1$ generation of a plant heterozygous for flower color and height. The dominant gene R indicates red flower color while the recessive gene r indicates white flower color. The dominant gene T indicates tallness and the recessive gene t indicates dwarfness. Genes for flower color and height are on separate chromosomes. The labeled dots on the chromosomes are schematic indications of the positions of the genes in question.
A. Prophase of meiosis I. At a stage when pairing of homologous chromosomes and chromatid duplication have been completed.
B and BB. Metaphase of meiosis I. Showing the two possible orientations (with respect to each other) of the paired chromosomes containing the gene pairs, $Rr$ and $Tt$.
C and CC. Tetrads of spores. With daughter chromosomes shown in the respective nuclei. The distribution of the possible combinations, RT, Rt, rT, and rt, is indicated.

chance that one or the other arrangement will occur in a given situation is equal.

It will be noted the series A-B-C produces 2 nuclei containing $TR$ and 2 nuclei containing $tr$. In the other (A-BB-CC) 2 nuclei with $tR$ and 2 nuclei with $Tr$ are formed. The latter are the new gene combinations which, as indicated above, had not existed together previously in haploid nuclei.

The orientation of the chromosomes at metaphase I is a matter of chance only and thus when large numbers of microspores are produced they will contain approximately equal numbers of $TR$, $Tr$, $tR$, or $tr$. The surviving megaspore in any particular ovule will contain 1 of these 4 possibilities also, and when large numbers of ovules are considered there will be an approximately equal distribution of the 4 possible types among them.

Any particular spore, be it microspore or megaspore, has an equal chance of containing $TR$, $Tr$, $tR$, or $tr$. Thus the gametes that are formed later, be they eggs or sperms, also have equal chances of containing 1 of the 4. When the gametes fuse the zygote could contain any of 9 different gene combinations as indicated in the checkerboard below.

*Female Gametes*

|  |  | TR | Tr | tR | tr |
|---|---|---|---|---|---|
| Male Gametes | TR | TTRR | TTRr | TtRR | TtRr |
| | Tr | TTRr | TTrr | TtRr | Ttrr |
| | tR | TtRR | TtRr | ttRR | ttRr |
| | tr | TtRr | Ttrr | ttRr | ttrr |

The information from the checker board can be summarized as follows:

| Genotype | Number of Occurrences | Phenotype |
|---|---|---|
| *TTRR | 1 ⎫ | Tall Red-flowered |
| TTRr | 2 ⎬ 9 | Tall Red-flowered |
| TtRR | 2 ⎬ | Tall Red-flowered |
| TtRr | 4 ⎭ | Tall Red-flowered |
| *TTrr | 1 ⎫ 3 | Tall White-flowered |
| Ttrr | 2 ⎭ | Tall White-flowered |
| *ttRR | 1 ⎫ 3 | Dwarf Red-flowered |
| ttRr | 2 ⎭ | Dwarf Red-flowered |
| *ttrr | 1 ⎬ 1 | Dwarf White-flowered |

In the $F_2$ generation, on an average basis, 9 out of every 16 plants would be tall and red-flowered, 3 would be tall and white-flowered, 3 would be dwarf and red-flowered, and 1 would be dwarf and white-flowered. This phenotype ratio of 9:3:3:1 is typical of simple dihybrid crosses.

The 4 combinations that have been asterisked (*) are homozygous. Two of them, $TTRR$ and $ttrr$, are identical with the original parents. The other 2, $TTrr$ and $ttRR$, are new homozygous individuals that did not exist previously. Each of them would continue to breed true if self-pollinated. This is a most important point because such recombinations of characters into new homozygous lines play a part in the natural processes by which new organisms evolve. It is also a basic part of the technique used by plant breeders to fix desirable combinations of genetic characteristics in new varieties.

SIMPLE MENDELIAN GENETICS

The examples used in the preceding discussion are based on those discussed in a classical paper published in 1866 by Gregor Mendel, who worked with varieties of the common garden pea

At the time that his work was published no scientists had any concept of chromosomes or genes. Mendel inferred that unit factors were involved but he could not have had any specific knowledge of what they were.

Modern genetic studies show that many problems of inheritance are much more complicated than indicated by simple Mendelian ratios. A few of the complicating factors are listed here.

1. Two or more of the characters to be studied might be controlled by genes located on the same chromosome.
2. Genes might exert a partial dominance only.
3. Many genes might influence the final expression of a single characteristic.
4. Chiasmata occuring in meiosis I might influence distribution of the genes being studied.
5. Recessive lethal genes could result in the death of homozygous individuals.
6. Gene mutations (changes) might occur during the course of an experiment.
7. The cytoplasm has been shown to have an influence on heredity.
8. The environment may influence the expression of a given gene.

## MECHANISMS OF EVOLUTION

The word evolution implies change and change in any system can occur in many ways. Sometimes the original system undergoes so many changes that the modern version scarcely resembles it at all. For example, many people have witnessed the dramatic evolution of the modern automobile and know of the relationship between a Model T Ford and a modern Thunderbird. Yet it would be difficult for a child who had not been informed of the changes to pick these two automobiles from an assemblage of all models and makes of cars which exist, and identify them as belonging to the same family of cars. Changes such as these resemble hereditary changes in living organisms. They differ from environmentally induced changes, like dents in fenders, which affect only the present individual and are not inherited.

Living organisms tend to produce many more offspring than can survive to maturity. This leads to intense competition for space and food both between members of the same species and among different species. If all the tree seeds that fall in a back yard each spring should grow to maturity, the homeowner would soon be faced with an impassable jungle. The few which do survive have overcome enormous odds. They have not been cut by a lawnmower. They have escaped the blade of a hoe. They have not been chewed off by a rabbit. They have not been killed by a plant disease. Their roots have managed to penetrate into soil already crowded with roots of other plants. Their leaves have managed to receive enough light for photosynthesis despite the shade of larger plants. And they have managed to crowd out their relatives that landed in the same spots.

Fitness for survival is certainly important but the matter of chance plays its part, too, in the survival of an individual. Many of the seedlings cut by a lawnmower may have been just as fit as the one that survived in the lilac hedge, but they were not as fortunate. So it is not true that only the fittest survive. Instead it can be stated that the more fit survive in greater numbers than the less fit and over periods of time the progeny of the more fit gradually supplant the progeny of the less fit. This, in brief, is the basic theory of evolution synthesized by Charles Darwin (and independently by Alfred Wallace) and published in the year 1859.

Changes in the inherited nature of living organisms may have one or more of the following effects on the progeny which inherit the changes:

1. Most frequently the ability of the individual to survive under conditions of natural competition in the normal environment weakens, in which case the individual or its progeny eventually disappears.
2. The ability of the individual to survive under conditions of natural competition in the normal environment might increase. In this case the progeny of the individual would eventually replace their less well-endowed relatives.

3. The ability of the individual to compete in an environment where the species had not been successful before might increase. Providing a mechanism exists by which it can be transported, the species might successfully invade this new environment.
4. There might be no noticeable effect on the ability of the individual to survive in the normal environment.

Such changes would tend to become randomly distributed in a population since they would not be affected by natural selection.

The above discussion emphasizes that the inherited changes must occur before the environment can act in the process of natural selection. When an individual organism is faced with the struggle for survival, it has certain inherited limits to its success in the struggle. The environment cannot change these limits but if certain individuals among the many have inherited changes that increase their ability to compete it is natural that they will be more successful and produce more progeny than the rest of the population.

All such heritable changes are frequently called *mutations,* although there is considerable latitude in the exact usage of this term. Sometimes, but not always, a simple doubling of the chromosome number by natural or artificial means will have the effect of noticeably changing the nature of the organism. Many cases are known wherein sterile hybrids have been made capable of producing fertile seeds in this way, making the organism capable of at least entering the struggle for survival.

More commonly the word mutation is used in connection with changes in single genes. These changes vary from complete obliteration of a gene, to rearrangement of a few of its many atoms, or even to a change in the position of a gene in a chromosome. Such mutations result frequently from natural radiation. Most of them probably occur in vegetative cells and are neither known nor significant. A few of them occur in cells that are, or become, part of reproductive tissues. Most of these result in death of the cell or its incapacity to take part in normal meiosis. Thus the mutation cannot proceed past this point. A smaller number pass through meiosis but the spores that contain them are unable to produce normal gametophytes. Of the few which are able to do so, most give rise to gametophytes that are unable to form fertile gametes. Occasionally a fertile zygote is formed and gives rise to an embryo but then the seed is unable to mature or to germinate. In the remote instances when the seed does germinate the seedling is likely to be weak and unable to survive normal competition. Moreover, if it does grow to maturity it will very likely be sterile.

Finally, if all of the obstacles are overcome and a fertile individual with a mutated gene is produced, then begins the long process of natural selection which eventually determines whether or not the mutation has any positive survival value. Most mutated genes turn out to be recessive and it is only when the chance recombinations of genes bring together two of the recessive genes that the ultimate test can be undertaken.

It is small wonder that evolution is a process which requires long periods of time. But, then, time is of no importance except to human beings.

COURSE BOOK IN GENERAL BOTANY

# ONE

THE organization of this chapter is influenced largely by the evolutionary changes that made it possible for green cells to exist out of the aquatic environments in which they had their origins. The leaf is a major achievement in this aspect of plant evolution since it provides an extensive internal environment suitable to the existence of green cells and protects them from many of the adverse effects of the external environment. However, before discussing the possible evolutionary changes leading to the leaf we will review certain aspects of the relationship of green cells to the environment. In this connection the exchanges of carbon dioxide and oxygen between green cells and the environment are especially important.

## GAS EXCHANGES BETWEEN THE CELL AND ITS ENVIRONMENT

The gaseous carbon dioxide reservoir in the atmosphere amounts to approximately 0.03–0.04 percent of the total volume of the atmosphere. A somewhat larger reservoir of this gas occurs in the oceans and other waters of the world. In solution it often exists as a mixture of carbonates, bicarbonates, and dissolved molecular carbon dioxide. Certain physical and chemical laws govern the balanced relationship between the carbon dioxide in solution and that in the gaseous form.

If the carbon dioxide in any given volume of water is used up by the photosynthetic activity of water plants or by chemical reactions which form insoluble carbonates, it is replaced from the atmosphere. Conversely, when the percentage of carbon dioxide in the atmosphere is lowered, some of the dissolved gas goes back into the gaseous state.

The burning of organic matter, the respiration of living organisms, and the chemical weathering of insoluble carbonates in the earth's surface tend to increase the available supplies of carbon dioxide. On the other hand, photosynthesis and formation of insoluble carbonates tend to decrease the available supplies.

Another important factor in the carbon dioxide balance is the average temperature of the oceans. Cooler waters hold more carbon dioxide in solution than do warmer waters and a change in the average temperature of the oceans would theoretically result in a corresponding change in the total volume in solution. This would affect the amount of carbon dioxide in the atmosphere and in the light of a demonstrable relationship between carbon dioxide concentration and the absorption of infrared radiation by the atmosphere a change in the average temperature of the atmosphere might be induced.

It is unlikely that our present-day atmosphere contains the same concentration of carbon dioxide and other gases as did the atmosphere of the earth when living matter first appeared. Furthermore, many of the significant changes in the evolutionary development of plants and animals have been accompanied by major changes in the atmosphere itself.

It is probable, for instance, that the original atmosphere of the earth contained significant concentrations of hydrocarbon gases such as methane ($CH_4$). Electrical discharges and other forces accomplished chemical unions between these gases and others, particularly ammonia, to form a wide variety of organic chemicals which accumulated in the oceans. The postulated series of events by which these compounds became organized into aggregates and

# Transpiration: The Price of Life on Land

eventually evolved into protoplasm will not be dwelt on here since we are concerned mainly with photosynthetic organisms. Most modern concepts concerning the origins of life include the idea that photosynthesis is a sophisticated process and should not be considered as an attribute of original life forms.

However, the evolutionary advent of photosynthetic organisms must have had a tremendous impact on the evolution of all then existing life forms. The ability to tap the limitless supplies of energy available in sunlight meant an enormous increase in the rate of formation of basic food. Also, the increased availability of oxygen fostered the evolution of biochemical mechanisms that could more efficiently release the energy stored in these basic foods.

The oxygen formed during photosynthesis is in dissolved form. It diffuses outward through the cell membranes into the external water. When the concentration of oxygen in the water reaches the saturation point it becomes gaseous oxygen and passes off into the air.

The utilization of carbon dioxide in photosynthesis also depends on its solubility in water since it is only in the dissolved condition that carbon dioxide molecules or bicarbonate ions can diffuse through the cell membranes of plants and thus reach the actual site of their reduction to carbohydrate levels.

From this discussion it is evident that gaseous exchanges between any living organism and its environment require that the gases be in solution. How then do the green cells of the higher plants function when the plants exist in an atmosphere of relatively dry air? One answer to this question will become clear as the structure of higher plants is analyzed.

## THE RELATION OF LEAF STRUCTURE AND FUNCTION

The outermost layer of cells in leaves and younger parts of stems and roots is the *epidermis*. In leaves and stems the epidermal cells secrete a waxy substance, *cutin*, which forms a protective surface layer over these parts of the plant. To a greater or lesser degree this *cuticle* is impervious to the diffusion of water vapor and thus retards the drying out of the interior green cells. Another way of expressing this is that the cuticle is not easily wettable. Thus it does not absorb water from the living cells it covers and lose it to the atmosphere by evaporation.

However, a complete vapor seal would inhibit the very necessary exchanges of carbon dioxide and oxygen between the external atmosphere and the internal photosynthetic tissues. Thus we find that an evolutionary compromise exists in which the cutinized epidermis is perforated with minute openings called *stomates*. These openings are numerous and more or less evenly distributed (Fig. 1.1A). Depending on the plant, they may be equally numerous in both the upper and lower epidermis of the leaves or there may be decided differences in numbers between the two. Wheat, for instance, may have 2,000 stomates per $cm^2$ in the upper epidermis and 4,500 per $cm^2$ in the lower epidermis. Apple leaves have no stomates in the upper epidermis but often have 25,000 per $cm^2$ in the lower epidermis.

Each stomate is a lens-shaped opening between a pair of guard cells (Figs. 1.1B, 1.2A). Usually the guard cells are shaped

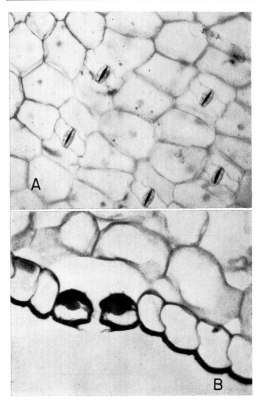

Fig. 1.1. Stomates. A. Epidermis of a *Zebrina* leaf showing distribution of stomates. B. Epidermis of a lily leaf in transverse section showing relation of guard cells to the stomate.

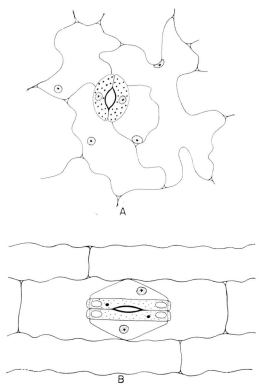

Fig. 1.2. Stomates. A. Stomate in the epidermis of a snapdragon leaf. B. Stomate in the epidermis of a corn leaf. The cells adjacent to the guard cells in this leaf are called accessory cells.

like beans, a major exception being the stomates of grasses which appear somewhat like dog bones (Fig. 1.2B).

The photosynthetic cells of a leaf are packed together in such a way that spaces occur between them, permitting air to circulate in an internal atmosphere (Fig. 1.3). The compactness of the packing varies between plants and between leaves on the same plant but, with few exceptions, some part of each green cell is in contact with the internal atmosphere. The total volume of internal atmosphere is important in determining the rate of gas exchange with the external atmosphere.

In many leaves the green cells next to the upper epidermis are elongated and cylindrical with only small spaces between them (Fig. 1.3). This tissue is called *palisade parenchyma*. The cells next to the lower epidermis are more loosely packed.

They have irregular shapes and larger spaces between them. This is the *spongy parenchyma*. Some leaves have a palisade layer next to the lower epidermis as well as the upper, and the spongy parenchyma is between these two layers in the middle of the leaf. Others may lack palisade tissue entirely (see Fig. 1.6).

The term *mesophyll* is used often for the soft inner tissues of a leaf, whether or not they are separable into spongy and palisade layers. Also, the term *chlorenchyma* (derived from a shortening of chloroplast-containing parenchyma) is useful in referring to photosynthetic tissues wherever they occur. Because chlorenchyma cells are in contact with internal air spaces and because the air spaces are connected to the external atmosphere through the open stomates the leaf mesophyll may

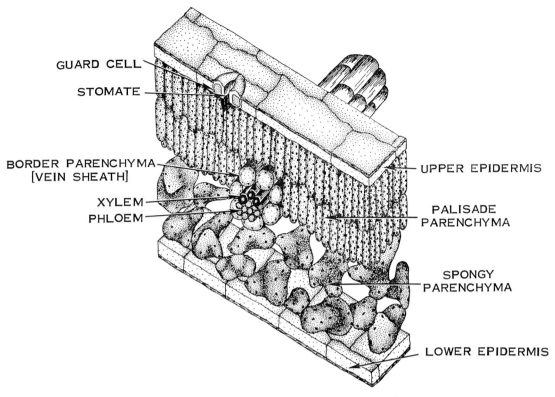

Fig. 1.3. Schematic three-dimensional representation of leaf structure.

be described as a *ventilated chlorenchyma*. The significance of the origin of this type of tissue is discussed in Chapter 3 as one of the major events in plant evolution.

Observation of a *leaf skeleton,* in which the mesophyll and epidermis have rotted away leaving only the vein system, shows that the veins make up an elaborately branched network penetrating all parts of the leaf (Fig. 1.4). They exhibit considerable variation in size and the smallest are generally invisible without the aid of a microscope. In corn and many similar plants the veins are mostly parallel and appear not to branch. However, close observation shows the presence of numerous, minute, cross-connecting veinlets (Fig. 1.5).

The small veins and vein endings are vitally important in the total functioning of a leaf because they accomplish most of the exchanges of substances between the green cells and the conducting system.

The veins contain two separate conducting tissues lying one above the other (Fig. 1.3). The upper tissue, *xylem,* conducts water and inorganic chemicals toward the ends of the veins. The water-conducting elements have thickened walls and are empty of living contents. They are discussed in greater detail in Chapter 2. The lower tissue, *phloem,* conducts organic foodstuffs in solution away from the green cells that manufacture them. The types of cells in phloem tissue also will be discussed later. The xylem and phloem of the leaf veins connect with similar tissues in the stem. These in turn connect with the xylem and phloem of the root system, giving rise to a continuous two-way conducting system from the top to the bottom of the plant.

The larger veins become surrounded by mechanical supporting tissue and thus form a framework for the leaf (Fig. 1.4).

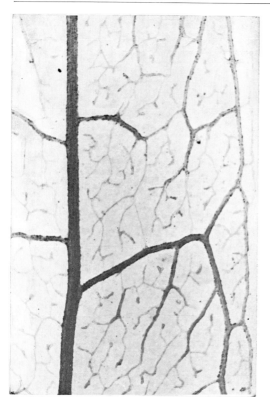

Fig. 1.4. Skeletonized leaf with only the vein system remaining.

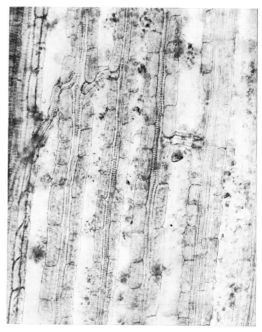

Fig. 1.5. Corn leaf tissue showing parallel veins and cross-connecting veinlets.

However, it is the mutual pressures of mesophyll cells that make leaves appear fresh and crisp.

The larger veins probably do not function at all in the actual exchanges between the vascular tissues and the green cells. This is accomplished only by the smaller veins which have thus a significance in leaf function similar to that of capillaries in the blood system.

The smaller veins are usually surrounded by a layer of thin-walled living cells called *border parenchyma* or sometimes vein sheath cells (Fig. 1.3). These may be seen to good advantage in many leaf sections but are particularly evident in cross sections of young corn leaves (Fig. 1.6).

The major significance of the open stomates is that they permit exchanges of carbon dioxide and oxygen between the external and the internal atmospheres. Once inside the leaf, however, carbon dioxide must go into solution in order to diffuse inward in the cells to the site of the photosynthetic reactions. The walls of the green cells that are exposed to the internal atmosphere contain a film of moisture. When carbon dioxide molecules in gaseous form come into contact with this water film some of them go into solution and then diffuse into the protoplasm.

The same sequence applies in reverse to the disposal of the oxygen formed during photosynthesis. The oxygen molecules in solution diffuse outward toward the water film in the cell wall. From there they evolve into the gaseous state in the internal atmosphere and proceed to diffuse into the external atmosphere through the open stomates.

It should be noted here that once the rate of photosynthesis falls below the rate of respiration the direction of diffusion for both carbon dioxide and oxygen is reversed. Respiration is a continous process in green cells as in all living cells, but it is not until photosynthesis ceases in a given

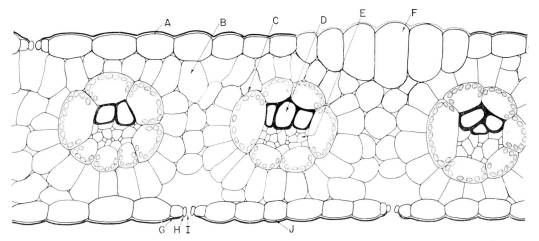

Fig. 1.6. Cross section of a portion of a corn leaf showing only smaller veins. A. Cutinized epidermis (upper). B. Mesophyll cell. C. Vein sheath cell (border parenchyma). D. Xylem. E. Phloem. F. Bulliform cell. G. Accessory cell. H. Guard cell. I. Stomate. J. Cutinized epidermis (lower).

day that the respiratory utilization of oxygen and the release of carbon dioxide can be demonstrated readily.

For both carbon dioxide and oxygen the significant point of transfer is the water film in the cell wall, since it is here that the change from the gaseous form to the soluble (or the reverse) must take place. When a film of water is in contact with atmosphere it is unavoidable that water molecules escape from the film into the air. This evaporation process is an inescapable result of leaf structure. The rate of evaporation depends on such factors as temperature, relative humidity of the internal atmosphere, osmotic concentration of the cell sap, and the colloidal organization of the cell itself.

If the stomates are open, water vapor molecules will diffuse into the external atmosphere. This loss of water in the vapor form from the aerial parts of the plant is known as *transpiration*. As noted above it is an unavoidable result of leaf structure.

## TRANSPIRATION

Transpiration involves two processes: (1) evaporation of water from moist cell walls into intercellular spaces, and (2) diffusion of water vapor molecules through the open stomates. The rate of outward diffusion depends largely on the difference between the concentrations of water vapor molecules inside and outside the stomates, i.e., the diffusion gradient of water vapor. When the stomates are closed the internal atmosphere soon becomes saturated with water vapor. This means that just as many water molecules return to the water films in the cell wall as escape from them, and there is no net loss of water from the cells.

WATER MOVEMENTS DUE TO TRANSPIRATION

When the stomates are open and outward diffusion occurs there is a net loss of water from the walls of the internal cells. This is replaced by water from the cell sap resulting in a lower relative concentration of water in the cell sap (and concomitantly an increase in the relative solute concentration). The water deficit in any given cell is made up by diffusion of water from adjacent cells and finally from the border parenchyma cells. The deficit in the border parenchyma is made up by diffusion from the xylem. The xylem solution normally has a negative pressure due to the downward pull of gravity and to some extent the frictional resistance to water movement in the xylem. However, the energy of water molecules in diffusion is greater than these opposing forces and the water diffuses out of the xylem into the border parenchyma cells. The xylem solution normally has a high concentration of water and a very low concentration of solutes compared to the adjacent border parenchyma cells which provide the osmotically active membranes in this process.

Water molecules are mutually attracted by intermolecular (cohesive) forces and when a molecule escapes from the xylem there tends to be a shrinkage of the water column. Due to the adherence of water to the cell wall such a shrinkage would tend to cause a collapse of the xylem but the cells of the xylem are internally reinforced to prevent such a collapse. Instead, the cohesive forces are satisfied by pulling up another water molecule from below. This transfers the deficiency downward through a vast number of almost simultaneous steps until it is finally satisfied by entrance of another molecule of water from the soil into the root.

The water in the xylem is in a sealed system and contains no gas bubbles. Hence it is extremely difficult to stretch or break the water column, which in a very crude sense can be compared to a wire being pulled upward The energy creating this *transpiration pull* is the kinetic energy of water molecules in the exposed cell walls of the leaf that enables them to escape into the internal atmosphere.

The magnitude of the pressures involved in transpiration pull can be roughly estimated in the following example. A redwood tree may extend 100 m into the air and have roots extending downward 10 m into the soil. The total height of the continuous water column would thus be 110 m. By way of comparison, a vacuum pump under ideal conditions can lift water only 10 m; i.e., the pressure of the atmosphere can support a column of water 10 m high. The upward movement of water in a 110 m column would require at least 10 times this pressure to overcome the force of gravity alone, to say nothing of the frictional resistance to water movement in the small pores of the conducting tissue.

RELATION OF STOMATES TO TRANSPIRATION

In the above discussion of stomates it was noted that each stomate is surrounded by a pair of guard cells (Fig. 1.2). The degree of opening of the stomates is determined directly by the turgidity of the guard cells. When they are fully turgid the stomates are wide open but as the turgidity of the guard cells decreases the walls of the guard cells come together and the stomates close.

Experiments with diffusion of gases through small pores have shown that the diffusion rate depends more on the circumference of the opening than on its area. Due to the shape of the stomates the circumference of the opening is not much altered by partial closing and, apparently, no significant reduction of transpiration occurs until the stomates are nearly closed.

Another significant point is that the rate of diffusion through a large number of minute openings such as stomates is much greater than the diffusion rate through one large opening having the same total area as all of the minute openings together.

The control of transpiration by plants is difficult to analyze. Although the cutinized epidermis reduces direct water loss to a minimum, direct cuticular transpiration might account for as much as 10 percent of the total transpiration rate. As has been established, the stomates permit diffusion of carbon dioxide and oxygen as a primary function and the escape of water vapor is unavoidable when the stomates are open. If the stomates should close whenever the transpiration rate became high the vital exchanges of carbon dioxide and oxygen with the atmosphere would be interrupted and the plant would suffer. In many plants stomates do not close until definite wilting has occurred, so the closing is something of an emergency measure. The availability of soil water, the extent of the root system. and the relative efficiency of the water transport system all have a bearing on stomate closing in such cases.

Some plants never close their stomates but such plants are usually restricted to wet environments. Other plants have definite cycles of stomatal behavior: opening in the morning, remaining open during most of the day, and closing in late afternoon. The stomates of some plants may close in the afternoon and then open again at night.

The effect of increasing light intensity on the opening of stomates in the morning has been much discussed. An increased turgidity of the guard cells results in the opening of the stomates, but the cause for the increase is not clear. The increased sugar content of the guard cells due to photosynthesis alone would not, apparently, be enough to account for the change. One ingenious theory suggests that the removal of carbon dioxide from solution by photosynthesis in the guard cells changes the acidity of the cell sap. This change affects the rate of enzymatic digestion of stored starch to sugar in the guard cells and thus a much greater amount of sugar is put into solution than could be produced by photosynthesis alone in a short time. This might well account for the rapid osmotic uptake of water which makes the guard cells turgid enough to open the stomates.

Recent evidence acquired with an electron microprobe points to rapid increases in potassium concentrations in guard cells during the period of stomatal opening. It is difficult to determine whether this indicates a hitherto unsuspected mechanism or more simply reflects a response to ionic imbalance brought about by the rapid withdrawal of bicarbonate ions in the photosynthetic process.

FACTORS AFFECTING TRANSPIRATION

The rate of vaporization inside the leaf is affected by the radiant energy to which the leaf is exposed and some cooling may result from this evaporation. Most of the heat that is absorbed is promptly re-radiated; in thin leaves the whole leaf tends to maintain a temperature close to that of the surrounding atmosphere. In thick leaves, on the other hand, the cooling effect may be significant if more energy is absorbed than can be effectively re-radiated.

Air temperature is important in another way since air can hold more water at higher temperatures than at lower ones. The term *relative humidity* which expresses this ability may be defined as the percentage of water vapor in the air at a given temperature compared to the maximum amount that it could hold at the same temperature. If the temperature rises and the actual amount of water vapor in the air remains unchanged then the relative humidity is lowered. This increases the steepness of the water vapor gradient between external and internal atmospheres of the leaf and the transpiration rate increases accordingly. Conversely, when the temperature falls and the actual water vapor content remains unchanged the relative humidity increases and the transpiration rate is lowered.

The internal structure of leaves also has an effect on transpiration rates. Leaves with loosely packed cells and large internal air spaces tend to lose water through open stomates at more rapid rates than leaves with more compactly arranged tissues and smaller air spaces.

Wind movements serve to carry water vapor away from the layers of atmosphere near the ground, thus affecting the relative humidity. Wind action also hastens the drying of the soil and in this way affects the rate of water absorption by roots. Another action of wind is the dispersal of minute clouds of water vapor which form over the open stomates. When present these clouds decrease the steepness of the diffusion gradient. Their removal by wind action removes this deterrent to transpiration and the rate increases.

The possible beneficial or harmful effects of transpiration on the life of plants have been topics of dissension for many years. Since transpiration is an unavoidable result of leaf structure and life on land, it follows that any beneficial effects would have had a secondary origin. Also, under certain conditions excessive transpiration does occur and plants suffer thereby.

One possible advantage is the cooling effect of evaporation in thick leaves. Another is the much disputed role of the transpiration stream in the upward movement of inorganic chemicals absorbed from the soil. There is scarcely any question that these chemicals do move in the xylem solution; the main point at issue seems to be whether or not the aerial parts of plants would receive enough of such chemicals if the transpiration stream were greatly reduced. This question has not been settled to the complete satisfaction of all interested authorities.

ENVIRONMENTAL SIGNIFICANCES
OF TRANSPIRATION

In some localities a single corn plant may absorb more than 200 l of water during the growing season and lose at least 98 percent of this amount to the air by transpiration. From an acre with 10,000 corn plants the total transpiration losses would be more than 2 million l, an amount roughly equivalent to 16 in of rainfall, which is more than half the annual total rainfall in many parts of the Corn Belt.

By another method of calculation one can determine that corn transpires several hundred l of water for each kg of dry organic material produced by photosynthesis.

Nor is the example of corn an extreme case. It is not at all uncommon for an acre of forest trees in temperate regions to transpire the equivalent of 30 in of rainfall in one growing season while trees in some tropical areas may lose the equivalent of 100 acre-in during a year's time.

Through natural selection in evolution each kind of land plant has arrived at a balance between photosynthetic efficiency, which is favored by high rates of gas exchanges with the environment, and the need to keep transpiration rates within bounds. This balance determines in large part where a given plant may grow. Humankind's early success in learning to cultivate desirable plants as crops was due in large part to a practical awareness of this balance.

# TWO

The growth of many-celled plant parts is the composite result of the growth of their individual cells. These cells are formed as a result of nuclear and cytoplasmic divisions. They grow larger because of water absorption and an increase in the volume of protoplasm. Finally they become modified in various ways during the complex process of differentiation.

In the simplest plants cells separate shortly after division and the whole plant consists of a single cell that carries on all of the essential functions. As many-celled plants evolved and became more complex, different organs took on specific functions. The leaf of the higher plants, for instance, has become the major photosynthetic organ; the root has become an organ for absorption and anchorage; and the flower has become specialized to facilitate the reproductive cycle.

One of the more significant changes was the segregation of cells responsible for new cell formation into special areas called *meristems*. Some of the more complex algae and all of the higher plants have a meristem at the tip of each growing axis.

In some cases, especially among the lower forms of land plants, the meristem consists of a single apical cell (Fig. 2.1). The cells derived from the apical cell are capable of further division but the numbers of such divisions are limited. The apical cell, however, is capable of dividing indefinitely as long as the plant remains in a strictly vegetative condition.

On the other hand, the apical meristems of the higher land plants consist of groups of apical initials rather than single apical cells. Also, in many of the higher plants the girth of the plant is increased as the result of cell divisions in lateral

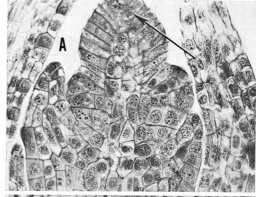
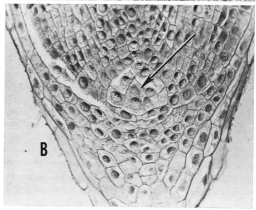

Fig. 2.1. Longitudinal sections of meristems with the conspicuous apical cells indicated by arrows. A. Stem tip of *Equisetum*. B. Fern root tip.

meristems (see the discussion of cambium later in this chapter).

## ORIGINS OF PRIMARY TISSUES IN STEMS AND LEAVES

The joints of stems where leaves are attached are called *nodes* and the stem segments between nodes are the *internodes*.

In order to understand the growth of a

# The Plant Body of Vascular Plants: General Structure

stem and the formation of the attached leaves it is necessary to investigate the tip region of a stem where new leaves are developed. Young corn plants are very useful in such an exercise. With a little care and patience the leaves can be removed one by one from the base upward. Each leaf is smaller and more delicate than the one below. The successive internodes are seen to be shorter and smaller in diameter as each leaf is unwrapped (Figs. 2.2, 2.3). Also the stems' growth in length by elongation of the internodes becomes readily apparent. Soon the leaves are so small that the aid of a hand lens or a dissecting microscope is needed to make the final manipulations. When the smallest leaf has been removed a minute dome-shaped structure is exposed at the tip of the stem. If the corn plant happened to be approaching the "knee-high" stage it would be usual to find a miniature tassel already formed at the tip of the stem. Corn stops producing new

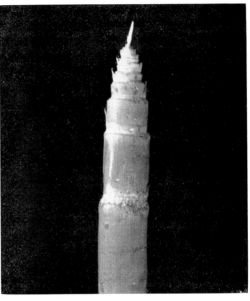

Fig. 2.3. Young corn stem with leaves removed to show nodes and internodes. The minute tassel at the tip is enlarged in Fig. 2.4.

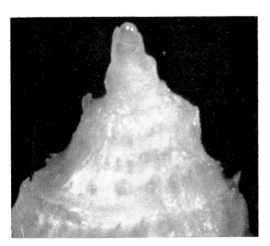

Fig. 2.2. Dissected stem tip of a young corn plant showing a dome-shaped apical meristem.

leaves prior to this stage and begins to form the flowers of the tassel (Fig. 2.4).

The dome-shaped structure at the tip of the stem is the *apical meristem* (Fig. 2.5). A small mass of cells in this tissue has the essential nature of remaining undifferentiated and capable of indefinite cell division at least during the phase of vegetative growth. These cells are sometimes called the apical initials and constitute the *promeristem* (Figs. 2.6, 2.7). They divide in several planes but their rate of division is not especially rapid. The number of cells in the promeristem itself remains relatively constant by a process of diverting peripheral members into surrounding tissue systems.

The first indication of a new leaf is a

Fig. 2.4. Photomicrograph of the newly formed tassel shown in position in Fig. 2.3.

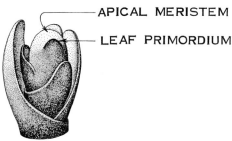

Fig. 2.5. Three-dimensional drawing of a stem tip showing the positional relationships of the leaf primordia and the apical meristem.

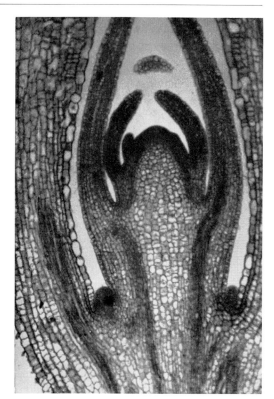

Fig. 2.6. Longitudinal section of the stem tip of flax. Note that leaves and buds of this stem would have opposite arrangement. (Compare with Fig. 2.7 for appropriate labels of parts.)

localized swelling on the sloping surface of the apical meristem (Figs. 2.5, 2.6, 2.7). This seems to be due to an increase in the division rate of cells under the surface at that point. The swelling is called a *leaf primordium*. The rate of formation of new leaf primordia and their positioning on the meristem follow definite patterns that are characteristic of the particular plant. Sometimes they are placed oppositely; in other plants they are arranged in spiral patterns exhibiting definite mathematical regularity.

In sectional view the very young leaves appear to be long and narrow but actually they soon take on in miniature scale the characteristic outlines of mature leaves. Commonly they curve up and over the meristem, with the older and larger ones protecting the smaller ones inside. Nodes and internodes do not become distinguishable until the internodal cells begin to elongate. Internodes many cm below the tip often continue to grow longer as may be observed by making periodic measurements of the internodes of immature stems.

When the young leaves reach a certain size *branch primordia* appear in the leaf axils ("armpits") (Figs. 2.6, 2.7). At first these are merely dome-shaped masses of cells like the apical meristem itself. Later they become active apical meristems and form branches of the stem system. In many plants this development is continuous but the woody plants of the temperate region form *dormant buds* (Fig. 2.8) in which no detectable growth occurs during the winter season. Each dormant bud contains unelongated stem segments with attached

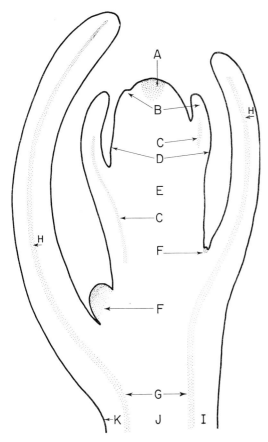

Fig. 2.7. Schematic diagram of a longitudinal section of a stem tip which would give rise to a stem with alternate arrangement of leaves and buds. A. Promeristem. B. Leaf primordia. C. Procambium. D. Protoderm. E. Ground meristem. F. Branch primordium. G. Vascular bundles of stem. H. Vein in young leaf. I. Cortex. J. Pith. K. Epidermis.

turn derived from cells that originate in the promeristem.) The *epidermis* is a single layer of cells that completely covers the primary plant body; it is derived from a primary meristem called the *protoderm*. The conducting tissues that make up the *primary vascular system* are derived from the *procambium*. Origins of procambium strands may be seen in the bases of leaf primordia. As both leaves and stems develop, procambium strands extend into the leaves to form the vein system and into the stem to form the vascular bundles. Thus the vascular system of leaves and stems is continuous from its inception. The remaining tissues that actually make up the bulk of the primary body are the fundamental or ground tissues; they are derived from the *ground meristem*. Some of the tissues included in this category are the pith and cortex of stems and the leaf mesophyll (Figs. 2.6, 2.7, 2.10).

The planes in which new cell walls are formed have an important effect on the resultant tissue systems (Fig. 2.10). For instance, cell divisions in the protoderm always occur in planes perpendicular to the surface of the plant part. This means that the epidermis is able to spread over the surface of the plant without increasing in thickness. Divisions in the procambium tend to be longitudinal, i.e., in planes parallel to the long axis of the plant part. Since the whole part is growing in length at the same time, cells in the procambium tend to be stretched longitudinally. Procambium cells do not actually penetrate newly developed plant parts. Rather cells already in position at the upper and lower ends of a procambium strand become modified into procambium cells, thus increasing the strand's linear dimensions.

As noted above, the primary plant body is mostly composed of ground tissues (pith, cortex, and mesophyll) and divisions in cells of the ground meristem are almost always transverse, i.e., at right angles to the long axis of growth. Therefore each new cell wall tends to be parallel to the walls formed in preceding divisions and as a result these tissues are composed of longitudinal rows of cells.

leaves, or immature flowers, or both. The whole structure is enclosed with small, tough, modified leaves called *bud scales*. It is possible in many cases to remove these bud scales and observe the immature leaves (Fig. 2.9) or flowers with the aid of a hand lens or a dissecting microscope. Careful observation of opening buds in the spring is a valuable exercise (Fig. 2.8).

There are three basic tissue systems in the primary body of plants and each one of them is derived from a specific *primary meristem*. (The primary meristems are in

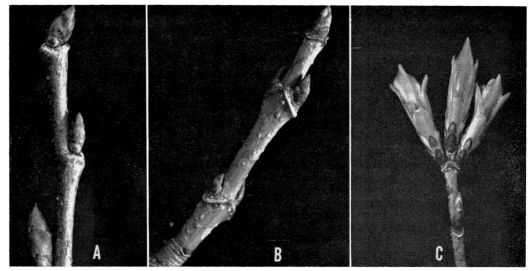

Fig. 2.8. Woody twigs with dormant buds. A. Elm. B. Maple. C. Maple buds as they begin to open in spring.

Fig. 2.9. Dormant bud of maple with bud scales removed to show next year's leaves.

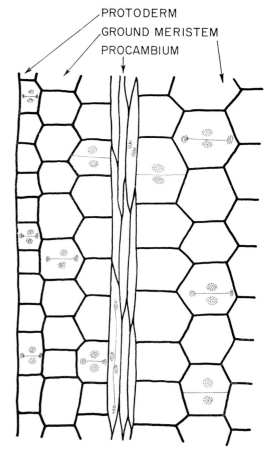

Fig. 2.10. Schematic illustration of the characteristic planes of cell division in the 3 primary meristems.

## CELL AND TISSUE TYPES

The various tissues occurring in plants contain a variety of cell types. Many of the cell types discussed below are found in several tissues while others may be limited to a particular one.

*Parenchyma* consists of cells that mature without major changes other than increase in volume (Fig. 2.11). They remain

Fig. 2.12. Sclerenchyma cells (bundle sheath fibers from a corn stem).

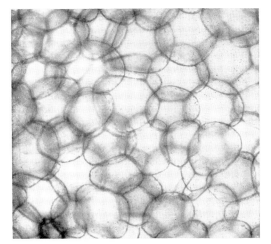

Fig. 2.11. Parenchyma cells (pith tissue of elderberry).

relatively thin-walled and have large central vacuoles. Commonly they store reserve foods and may provide a type of water reservoir during periods of water shortage. They may contain chloroplasts; the term *chlorenchyma* refers to green parenchyma whose major function is photosynthesis. The green cells of leaves and stems would be classified as chlorenchyma on this basis.

In young plant parts the parenchyma cells are highly turgid. Their mutual pressures give considerable mechanical support to such parts and in this way they aid in the maintenance of shape. The absorption of water into parenchyma cells is one of the main factors in the enlargement of the primary plant body. These cells may also become meristematic under certain conditions. There is some cell-to-cell conduction of substances in parenchyma tissue but this occurs at a relatively slow rate.

*Sclerenchyma* consists of cells whose walls are uniformly thickened (Fig. 2.12) and impregnated with the complex carbohydrate lignin. At maturity the walls may be so thick that the hollow center or lumen appears as small as a pinpoint in cross section. This type of cell is not specialized for conduction. Instead, it affords a semirigid mechanical support to the plant parts wherein it occurs. The places where sclerenchyma develops are usually precisely located. In some plants the angles of the stem are filled with sclerenchyma. In a plant with a smoothly cylindrical stem the sclerenchyma may occur as a cylinder inside the epidermis or as a strand *(bundle cap)* outside each of the vascular bundles. In leaves the major veins that form the leaf framework are heavily invested with sclerenchyma.

This type of cell is frequently long and pointed and is the source of many commercial fibers. In flax, for instance, each of the linen fibers is a bundle of sclerenchyma fibers that remain united.

The *stone cells* of pears that give a gritty texture to the fruit are nonelongated sclerenchyma cells. They are often objects of intricate design since the wall thickenings occur in layers that are crossed by many minute canals radiating from the lumen (Fig. 2.13).

*Collenchyma* consists of cells that also function in mechanical support. However, they differ from sclerenchyma in several ways: they provide an elastic rather than a semirigid support; the thickenings of the

Fig. 2.13. Stone cells of pear fruits (a type of sclerenchyma).

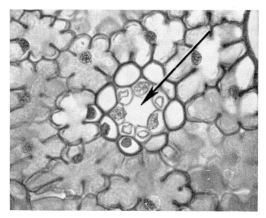

Fig. 2.15. Partial section of a pine needle with arrow pointing to a resin canal.

wall are of cellulose and pectin rather than lignin; and they are thickened unevenly instead of uniformly as are sclerenchyma cells. Cross sections of beet petioles offer a remarkable example of collenchyma cells (Fig. 2.14).

It should be noted that a clear distinction between parenchyma, sclerenchyma, and collenchyma cannot always be made.

In many plant parts groups of cells take on the specialized function of secretion. Many types of substances are secreted and some of them have important economic significance.

*Resin* is secreted in the resin ducts of many gymnosperms. These ducts are actually elongated spaces between the secretory cells. They may be observed to good advantage in cross sections of pine stems or needles (Fig. 2.15). The function of

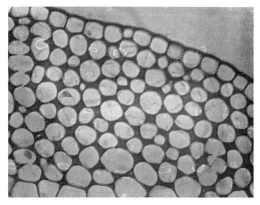

Fig. 2.14. Collenchyma cells from a beet petiole.

resin in the plant is not clear but it is an item of commerce, the source of turpentine and a number of other products.

Various kinds of *oil* are secreted in oil ducts which occur in various plants. Sections of sunflower stems show them to good advantage. Oils collected from members of the mint family are a valuable economic product.

*Nectar* is a sugary solution secreted or excreted by nectar glands in flowers. It attracts insects and thus aids in pollination. Bees are able to convert nectar into honey.

*Latex* is a milky-appearing fluid secreted in the latex tubes of rubber plants and many others. It is a source of natural rubber.

The aerial parts of plants are often partially covered by hairlike outgrowths from the epidermis. These *epidermal hairs* may be single celled or many celled and they may be branched or unbranched. Some of them, as in the common geranium (Fig. 2.16), have bulbous glands at the tips that contain a volatile oil. Interpreting the functions of such hairs presents problems. One might assume that they cut down transpiration, but this has been proved incorrect in several instances. Another possibility is that they interfere with the movements of small insects that feed on plant substances.

## VASCULAR PLANTS: GENERAL STRUCTURE

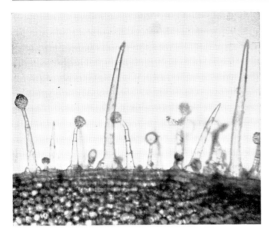

Fig. 2.16. Transverse section of a geranium stem showing epidermal hairs.

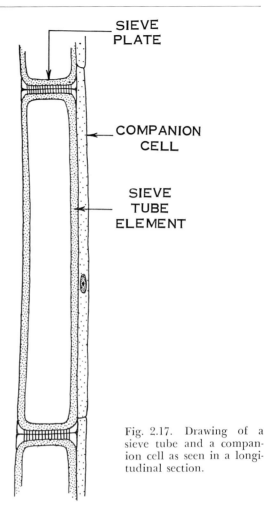

Fig. 2.17. Drawing of a sieve tube and a companion cell as seen in a longitudinal section.

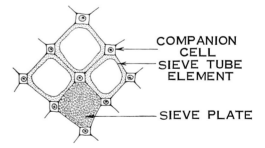

Fig. 2.18. Drawing of a portion of the phloem tissue from a large vein of a corn leaf, showing sieve tubes and companion cells. One transverse wall of a sieve tube element is included to show the sieve plate.

### PHLOEM

The tissue that is specialized for conduction of organic substances in solution is called *phloem*. In this tissue there may be four different kinds of cells which occur in various proportions in different kinds of plants. They are:

1. sieve tube elements
2. companion cells
3. parenchyma cells
4. fibers

The parenchyma cells of phloem are essentially similar to the parenchyma of nonvascular tissue, while the fibers are elongated sclerenchyma cells. The *sieve tube elements* are the particular cells that are specialized for conduction of organic foods in solution. In most angiosperms the individual sieve tube elements are connected end to end to form long *sieve tubes* (Fig. 2.17). The end walls that separate each cell of the tube from the next one appear to be perforated with many fine pores and it is possible that minute protoplasmic strands pass through them. When seen in end view, as in cross sections of stems (Fig. 2.18), the perforated walls have the appearance of sieves and this accounts for their name. Parts of the lateral walls between adjacent elements may be perforated also. When the phloem is functionally mature, i.e., when it is actually transporting foods, the cytoplasm of each sieve tube element is functional but the nucleus has disappeared.

*Companion cells* are intimately associated with the sieve tube elements (Figs. 2.17, 2.18) since they are sister cells derived by a longitudinal division of a preceding cell. The companion cells remain small in diameter and retain their nuclei, which appear to be very metabolically active. It is a somewhat obvious hypothesis that the companion cell has a great influence on the functioning of the sieve tube element. Such simple theories are often difficult to substantiate, however, especially when one realizes that companion cells do not occur at all in the phloem of gymnosperms.

The fact that most organic food transfer occurs in the phloem can be demonstrated by cutting a girdle around a stem in such a way that the phloem is completely severed. Plants girdled in this way eventually die because of starvation of the roots. However, there is usually an abundant supply of stored food in the roots so that a girdled tree might not die for several years. Also, the roots might obtain food through natural grafts with the roots of nearby trees.

In the spring of the year, before the leaves mature, there may be an upward movement of a solution containing soluble foods derived from digestion of stored foods in stems and roots. This solution seems to move in the xylem and is associated with an active absorption of water by the roots. If sugar maples are punctured with small holes into the wood at this time of year the sap will flow outward. Dehydration changes this sap into the the maple syrup and sugar of commerce.

## XYLEM

The major function of *xylem* is the conduction of water and dissolved inorganic chemicals throughout the plant. The general direction is upward from roots to stems to leaves. In xylem tissue there may be four different kinds of cells which occur in various proportions in different kinds of plants. These cells are:

1. tracheids
2. vessels
3. fibers
4. parenchyma cells

*Tracheids* and *vessels* are the important conducting elements of the xylem and will be discussed in some detail here. Parenchyma cells have been discussed earlier in this chapter. Fibers in the xylem are long and slender and have very thick walls.

*Tracheids* are long slender cells with tapering, chisel-shaped end walls (Fig. 2.19). In some plants they may be hundreds of times longer than wide. The walls are lignified and thickened but not uniformly. Thin areas or pits occur in the secondary wall layers; they permit rapid diffusion of water from one tracheid to another. The pits are not actual holes in the walls between cells since the primary wall remains intact.

Tracheids do not function in conduction until they are dead and empty of protoplasmic contents. It follows that they play only a passive role in the movement of water, serving primarily as passageways. Tracheids may be observed to good advantage in cross and longitudinal sections of pine wood (Fig. 2.19; see also Fig. 2.39).

*Vessels* are often larger in diameter than tracheids although it is difficult to tell them apart in cross sections. In lengthwise sections it may be observed that vessels are made up of longitudinal series of cells attached end to end. The cross walls between cells are eliminated by digestive processes (Fig. 2.20). In some plants a single vessel is composed of hundreds of segments forming a microscopic water pipe; the vessel may extend from root to stem to leaf. The vessel is a much more efficient water conductor than the tracheid and makes it possible for plants with high transpiration rates to replace the lost water quickly. Since high transpiration rates are the price that plants pay for efficient absorption of carbon dioxide, it follows that the vessel has played an important part in the success of many modern plants.

# Vascular Plants: General Structure

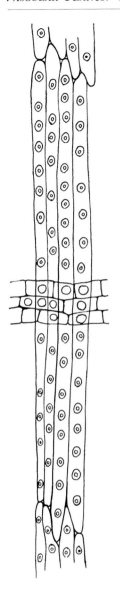

Fig. 2.19. Tracheids as seen in a longitudinal radial section. The transverse band of cells is a portion of a xylem ray.

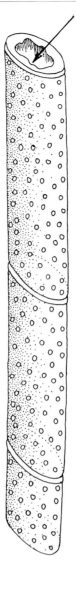

Fig. 2.20. Drawing of a xylem vessel in which the vessel segments are evident. The disintegration of a cross wall is indicated by the arrow.

The vessels of corn, as seen in sections, are excellent examples since each vascular bundle of the stem contains two vessels that are exceptionally large and well developed (see Fig. 2.26).

Many of the cell types which have been discussed can be seen in entirety in macerated tissues. Preparations of this sort are made by treating small segments of plant parts with strong acids that dissolve out the middle lamellae holding the cells together. The thickenings of wall layers and the open ends of vessel segments show up very clearly in such material.

The walls of vessels and tracheids become thickened due to the increase of cellulose in the secondary walls and the impregnation of the entire wall with lignin. These thickenings absorb stains readily and cause the xylem cells to stand out prominently in stained sections. The thickenings are important factors in the mechanical strength of stems, particularly woody stems. A more basic function may be the resistance to collapse which is afforded by these thickenings under conditions of negative pressure in the water column due to high transpiration rates.

At least four patterns of secondary wall thickenings are recognizable although they frequently intergrade. *Annular thickenings* consist of separate ringlike bands (Fig. 2.21). Spiral thickenings consist of helically wound bands of secondary wall substance which resemble coiled springs (Fig. 2.21).

*Scalariform thickenings* consist of transversely disposed rings of secondary wall substance vertically connected by numerous short segments of the same material. The uniform distribution of both components creates a ladderlike effect (Fig. 2.22).

An intermediate type of thickening that creates the appearance of a network is called *reticulate*.

*Pitted thickenings* are probably the most advanced type; the whole wall is thickened except for uniformly distributed thin spots which are called *pits* (Fig. 2.22).

Annular and spiral thickenings occur

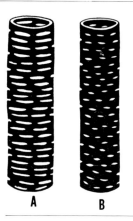

Fig. 2.22. Wall thickenings in metaxylem elements. A. Scalariform thickenings. B. Pitted thickenings.

in elements of the earliest-formed xylem (protoxylem). Their design permits the vessels and tracheids to elongate as the young plant parts grow in length. Scalariform and pitted thickenings occur in xylem cells *(metaxylem)* that mature after the plant part has finished its lengthwise growth.

*Simple pits* are merely areas of the cell wall that do not develop secondary thickenings (Fig. 2.23). They are not holes in the wall, since the primary wall is not perforated. *Bordered pits* are more complicated, since the secondary wall partially overgrows the simple pit to form a dome-

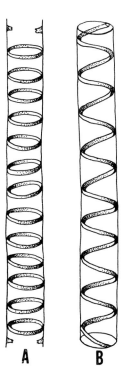

Fig. 2.21. Wall thickenings in protoxylem elements. A. Annular thickenings. B. Spiral thickenings.

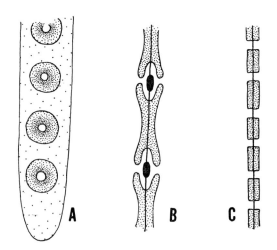

Fig. 2.23. Simple pits and bordered pits. A. Bordered pits as seen in the surface of a tracheid. B. Bordered pits as seen in a section of a tracheid wall. C. Pairs of simple pits in a cell wall section.

shaped structure with a tiny perforation in the middle (Fig. 2.23). In pine the central portion of the primary wall between a pair of bordered pits develops a thickening called a *torus* that acts as a valve membrane. When the pressures in two adjacent xylem cells are not much different the torus stays in the middle; water diffuses readily through the pit openings and the unthickened primary wall around the perimeter of the torus. But if there is a sudden change in the water pressure in one of the cells the torus between each pit pair is pushed over in such a way that it plugs the opening to the pit, thus slowing down water movement between the two cells. This neat and simple safety device may be observed clearly in sections of the tracheids of pine wood.

## ARRANGEMENT OF TISSUES IN PRIMITIVE STEMS

As will be discussed in greater detail in Chapter 3, the plant bodies of many primitive vascular plants consisted of branching green stems without leaves. From careful studies of fossil remains of these plants it has been determined that their vascular tissues occurred as a central core in each stem. The epidermis was cutinized and had stomates. Tissues underlying the epidermis were green and photosynthetic. Some details of the differentiation of tissues in such a stem are included here as being significant in laying a groundwork for a discussion of the tissues of higher plants.

The central core of tissue is frequently called a *stele;* it bears a functional relationship to veins of the leaf and vascular bundles in the stems of higher plants.

The bulk of tissue between the stele and the epidermis is called the *cortex* and is derived from the ground meristem. In the cortex, two types of tissue are recognizable in sections of some fossilized stems. The outer layer, next to the epidermis, was green and photosynthetic. Numerous air spaces existed between the green cells. Interior to the green cells were several layers of parenchyma that became thick walled and lignified into sclerenchyma tissue.

A highly specialized layer, one cell in thickness, that separates the cortex from the stele is the *endodermis.* The functional significance of this layer is discussed under the general heading of root structure.

All of the tissues inside the endodermis matured from the procambium which in these plants existed as a central strand of thin-walled elongated cells. The inner cells of this core matured as xylem while the cells immediately surrounding the xylem matured as phloem. The outermost cells of the procambium strand matured as parenchyma cells and comprise the *pericycle.* The functions of the pericycle in primitive stems remain obscure but the pericycle in the roots of higher plants has at least three important functions.

The direction of lateral differentiation of the xylem elements in the primitive stem is significant in the evolution of vascular plants. In these plants the outermost cells of the potential xylem tissue differentiated first to form a tissue known as the *protoxylem* (Fig. 2.24A).

Differentiation of the remaining central cells then proceeded inward (centripetally) with this later maturing tissue constituting the *metaxylem* (Fig. 2.24B). Commonly the metaxylem cells are larger in diameter than those of the protoxylem. (It must be noted here that in some species not all of the central cells matured as conducting elements.)

The stem that has been described is considered to have the most primitive type of arrangement of vascular tissues and this type is known as a *protostele.* In some living plants, including *Psilotum* and certain species of ferns and club mosses, the vascular tissue is arranged in this manner. In the flowering plants, however, the stem structure is of a more advanced type.

By way of comparison, the roots of all plants, including those of flowering plants, have vascular arrangements that are protostelic. In passing from a discussion of the vascular tissues of primitive vascular plants to those of the flowering plants it

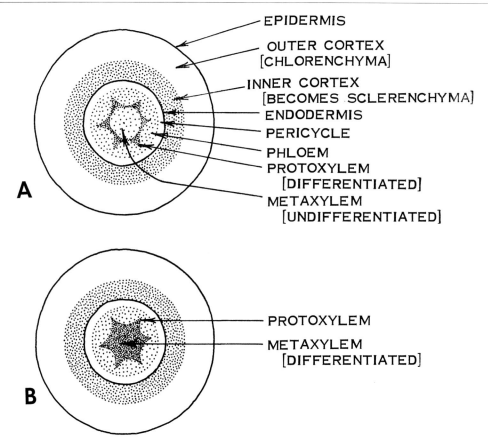

Fig. 2.24. Schematic representations of stem sections with exarch xylem; i.e., the protoxylem is outside the metaxylem indicating a centripetal direction of differentiation of the primary xylem. A. Early stage with only protoxylem differentiated. B. Later stage with both protoxylem and metaxylem differentiated.

seems desirable to begin the discussion with the leaf. Presumably the vein system of the leaf has been modified from the stem system of primitive plants while the stem of the higher plants is a complex structure which lacks a counterpart among the stems of primitive plants.

## DIFFERENTIATION OF VASCULAR TISSUES IN LEAVES

Sections of moderately large veins in corn leaves show xylem and phloem tissue which is almost diagrammatically clear; thus they make ideal subjects for microscopic study of these tissues. Furthermore, the veins are parallel and in cross sections they are oriented well for microscopic examination. In sections of young leaves of a corn seedling a whole series of developmental stages in the differentiation and maturation of vascular tissue may be observed.

In the very young leaf the future vascular tissues exist as strands of cells that are all much alike. These are the procambium strands. The component cells are elongated, thin-walled, densely protoplasmic, and small in cross-sectional area. The strands lie lengthwise in the leaf. Cells toward the lower epidermis eventually become phloem cells while those toward the upper epidermis become xylem.

The first cells in a procambium strand to differentiate are the ones along the lower and the upper surfaces of the strand

(Fig. 2.25). The first-formed phloem is called *protophloem* and consists of a few sieve tube elements. The first-formed xylem is the *protoxylem*. It consists of a few very small vessels with annular and spiral thickenings. In the corn leaf these first-formed xylem vessels appear to be in a radial row as seen in cross section. The cells of the strand which lie between the protoxylem and the protophloem remain undifferentiated for some time, although some of them increase in size and there may be a certain amount of cell division which increases the size of the strand.

Protoxylem and protophloem function during the period when the leaf is enlarging even though they are stretched by this growth. Eventually they are destroyed or become functionless.

Once elongation of a part of the leaf is complete the *metaphloem* and *metaxylem* begin to differentiate. In the metaphloem there is one final division in each of the phloem mother cells. This is an unequal division which results in one member of a cell pair being larger than the other. The large one differentiates into a sieve tube element while the small one becomes a companion cell. Thus the metaphloem has the appearance of a mosaic in transverse sections (Figs. 2.18, 2.26).

In the major veins of corn two of the metaxylem vessels become very large before they differentiate (Fig. 2.26). This growth is so pronounced that it results in the tearing apart of nearby tissue and a lengthwise space develops in the xylem. The two large metaxylem vessels and the smaller metaxylem vessels between them develop very thick lignified walls with numerous pits as they differentiate.

The larger veins also become invested to varying degrees with a sheath of mechanical supporting tissue consisting of thick-walled, lignified, *sclerenchyma fibers*. These larger veins are continuous with the vascular bundles of the stem which follow the same general pattern of maturation and differentiation (Fig. 2.27).

Many small veins lie parallel to and between the major veins. These smaller veins are surrounded by parenchyma instead of sclerenchyma. In addition, there are many short cross-connecting veinlets

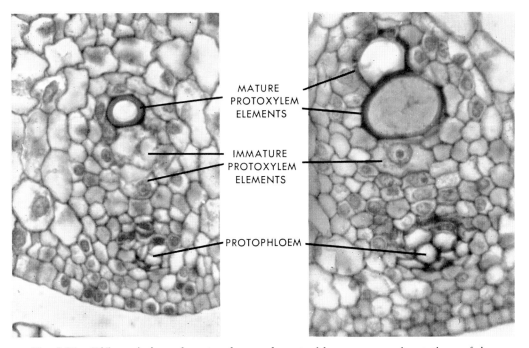

Fig. 2.25. Differentiation of protoxylem and protophloem as seen in sections of immature veins in very young corn leaves. The section at the left is the younger.

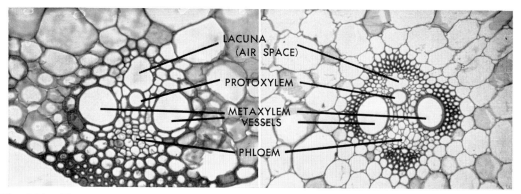

Fig. 2.26. Transverse section of a major vein in a nearly mature corn leaf. Note the 2 large metaxylem vessels and the lacuna (air space).

Fig. 2.27. Vascular bundle from a corn stem. Compare with Fig. 2.26.

occurring at right angles to the lengthwise veins (see Fig. 1.4).

## TISSUE SYSTEMS OF THE CORN STEM

Corn leaves have a clasping leaf base which is wrapped around the stem and attached to the stem at a joint or node. The veins are connected to and continuous with the vascular bundles of the stem. In the nodal region the vascular bundles are much twisted and form a complex network that is difficult to trace. In the internodes of the stem, however, the vascular bundles extend lengthwise and are essentially parallel.

When cross sections of corn stems are cut in an internodal region it may be observed that the vascular bundles are scattered throughout a background tissue of *ground parenchyma* which, in nontechnical language, is usually called *pith* (Fig. 2.28). (Such an arrangement is characteristic of many monocot stems.) It has been noted that the vascular bundles have much the same structure as the larger veins of the leaf. The phloem is oriented toward the outer surface while the xylem is oriented oppositely, toward the center of the stem. In mature stems especially the bundles are heavily invested with sclerenchyma fibers (Figs. 2.27, 2.28).

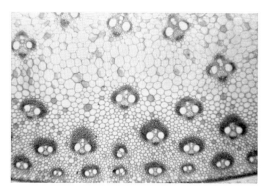

Fig. 2.28. Cross section of a corn stem showing the scattered distribution of vascular bundles in the ground parenchyma tissue.

## THE TISSUES OF A HERBACEOUS DICOT STEM

A herbaceous stem is one that completes its growth during a single season and does not continue growth in the next season. In herbaceous dicot stems the vascular bundles are often arranged in a ring at some distance inward from the epidermis, as opposed to the scattered arrangement of bundles in the corn stem (Figs. 2.29, 2.30). The tissue between the epidermis and the vascular bundles is the *cortex*. The tissue inside the vascular bundles is the *pith*. In cross sections of a young internode the recognizable tissues are epidermis, cortex, vascular bundles,

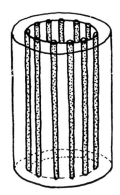

Fig. 2.29. Drawing showing the arrangement of vascular bundles in a ring as is characteristic of many herbaceous dicot stems.

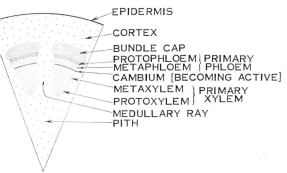

Fig. 2.30. Schematic drawing of a wedge-shaped section of a herbaceous dicot stem.

and pith. The parenchyma tissue between two bundles is sometimes referred to as a *medullary ray*. In corn, by way of comparison, there is no obvious distinction among pith, cortex, and rays; the background tissue is referred to as ground parenchyma.

Not all herbaceous dicot stems have separate vascular bundles; the procambial tissue may be in the form of a hollow cylinder separating the cortex from the pith.

The first-formed xylem (protoxylem) differentiates from the innermost cells of the procambium strands; differentiation of the xylem proceeds outward from these protoxylem points (rather than inward as in the protostele). The first-formed phloem (protophloem) develops from the outer cells of the procambium strands and differentiation proceeds inward. The xylem and phloem are thus opposite each other along a radial line through each bundle and differentiate toward each other (Figs. 2.30, 2.31).

Differentiation of xylem and phloem elements proceeds until most or all of the procambium cells are converted. The strand may then be called a *vascular bundle*. In some plants the procambium tissue becomes completely changed to xylem and phloem. In others the procambium cells lying between xylem and phloem may retain their meristematic abilities and give rise to a lateral meristem called the *vascular cambium*. This will be discussed later.

Most of the cells of the cortex are parenchyma. Frequently they contain chloroplasts and in young stems especially the cortex is a ventilated chlorenchyma. Many stems are angled in cross section. Supporting tissues (collenchyma or sclerenchyma or both) are found in the corners. Also the *bundle cap* outside each vascular bundle is made up of supporting tissue.

The pith tissue also consists mainly of parenchyma cells. In many mature stems the cells of the pith are dead or have been torn apart by stem enlargement. In young stems the turgidity of the pith cells (and the cortex cells) is largely responsible for mechanical support.

The *medullary rays* that connect the pith with the cortex consist of parenchyma cells. When the rays are broad, as they are when the vascular bundles are spaced far apart, there is no sharp separation among cortex, rays, and pith. When the bundles are close together the rays appear as radially and longitudinally disposed sheets of cells. Very frequently, rays occur in the bundles as well as between them. These are referred to as *vascular rays* or as *xylem and phloem rays*. The ray cells possibly accomplish radial translocation between xylem and phloem and are also important in many plants as food storage tissues.

A careful distinction should be made between the epidermis of young stems and the cork tissue in the bark of older stems. The epidermis is a primary tissue which is developed from cells derived directly from the apical meristem. It consists of a single layer of cells which is cutinized and has stomates. Cork tissue is a secondary tissue formed by a cork cambium,

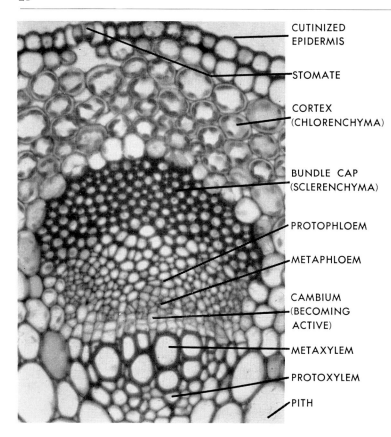

Fig. 2.31 Photomicrograph of a cross section of one vascular bundle from a herbaceous dicot stem, in this case Alsike clover.

as will be discussed later. It displaces the epidermis when the stem enlarges beyond a certain size.

In summarizing the basic structure of the stem it should be noted that all the tissues that have been discussed in detail are *primary tissues*. This means that they have been formed from cells laid down in the stem tip. The primary tissues include epidermis, cortex, protophloem and metaphloem, protoxylem and metaxylem, pith, and rays.

## CAMBIUM AND THE CONCEPT OF SECONDARY GROWTH

In the discussion of the maturation of tissues in the vascular bundle it was noted that in many plants a zone or layer of cells between the metaxylem and the metaphloem fails to differentiate and retains its meristematic nature. This layer is the *vascular cambium* and it is a lateral meristem. By division of its cells it gives rise to *secondary xylem* and *secondary phloem* (Figs. 2.32, 2.33). It should not be surprising that the cambium cells are able to grow and divide, since they have immediate access to soluble organic foods in the phloem and to water and inorganic chemicals in the xylem. Most of the divisions of the cambium are longitudinal and parallel to the outer surface of the plant. As a result the daughter cells are produced in radial rows. When the division rate is high the rows of undifferentiated daughter cells may be several cells in depth. Divisions may occur in more than one cell in each radial row but the greatest rate of division appears to occur in the middle of this cambial zone.

Differentiation of cells occurs at both the outer end and the inner end of each radial row in the cambial zone (Figs. 2.32, 2.33). The innermost cells become part of

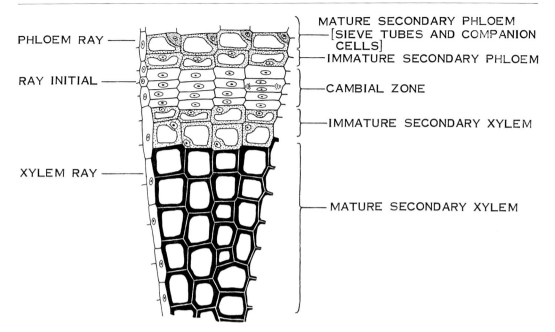

Fig. 2.32. Schematic drawing of a portion of a section of a woody stem showing relation of cambium to secondary xylem and phloem.

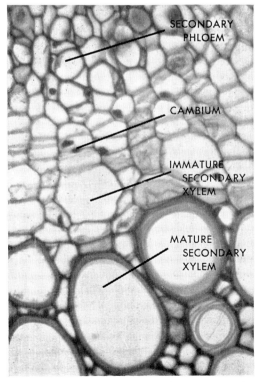

Fig. 2.33. Photomicrograph of a stem section showing cambium and immature xylem elements.

the xylem and the outermost cells become part of the phloem. The cambium may be a deep zone of tissue or may be reduced to a single layer when the differentiation rate catches up with the division rate.

In the above described manner the cambium is able to increase both the xylem and the phloem while remaining as a meristematic layer between them. The differentiating xylem cells increase in size considerably before maturing and this constantly forces both the cambium and the phloem outward. This process of expansion is resisted by the outer tissues of the stem and as a result cells in the cortex become greatly distorted. Also, the older nonfunctional phloem cells become crushed. The cambium itself is under stress due to the increase in circumference of the enlarging xylem mass inside it. This stretching is accommodated by periodic radial divisions of cambial initials which result in pairs of initials lying side by side where only single ones existed previously. Evidence of this can be seen in cross sections of stems by tracing a radial row of xylem cells outward to a point where it becomes two radial rows.

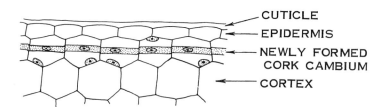

Fig. 2.34. Origin of the cork cambium in outer cells of the cortex.

## ORIGIN OF THE CORK CAMBIUM

With the increase in mass of secondary tissues the epidermis is subjected to such pressures that it is eventually broken. Before this happens another lateral meristem is formed that is capable of producing cork tissue. This is the *cork cambium* and it may form in the epidermis, in the cortex, or even in the parenchyma cells of the phloem (Fig. 2.34).

The divisions of the cork cambium are parallel to the surface of the stem. Radial rows of cells are formed in this way and most of them differentiate into *cork cells*. However, a limited number of cells within the cork cambium layer become *cork parenchyma*. The cork cells eventually die but before they do their walls become heavily impregnated with a fatty substance called *suberin* which gives cork its characteristic nature (Fig. 2.35). Such cells are the major component of the bark of woody stems.

## SECONDARY GROWTH IN HERBACEOUS STEMS

In some herbaceous dicot stems, as in almost all monocot stems, there is little or no cambial activity and consequently no secondary growth. In others there is a limited amount of secondary growth due to a cambium within each bundle. In a third category ray parenchyma cells between bundles revert to a meristematic condition and form an *interfascicular cambium* that is continuous with the cambium in the bundle. In this way a complete cylinder of cambium is formed. In those stems where the procambium tissue is a cylinder rather than a series of separate provascular strands the cambium develops from the beginning as a continuous cylinder.

Differentiation of secondary xylem may result in the formation of vessels, tracheids, fibers, and parenchyma. In the newly formed secondary phloem tissue, sieve

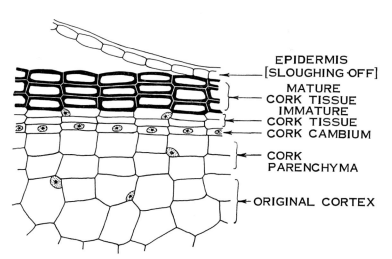

Fig. 2.35. Positional relationships of cork, cork cambium, cork parenchyma, epidermis, and cortex.

# VASCULAR PLANTS: GENERAL STRUCTURE

tubes, companion cells, fibers, and parenchyma may be differentiated.

The vascular rays are continuous from xylem to phloem across the cambium. They are extended in both directions by cells of the cambium called *ray initials* (Fig. 2.32). Periodically the number of rays is increased by conversion of cambial cells that had been producing vascular elements into ray initials. Observation of cross sections of stems shows many rays of various lengths with the shortest rays being the newest.

## SECONDARY GROWTH IN WOODY STEMS

The evolutionary history of seed plants indicates clearly that woody stems evolved before herbaceous stems. For purely pedagogical reasons, it has seemed preferable to reverse the natural order and discuss the apparently simple monocot stem and herbaceous dicot stem before the woody stem.

Very early in the first season's growth of a woody twig, before any significant amount of secondary growth has occurred, the distribution of tissues is very similar to that of a herbaceous dicot stem. As secondary growth proceeds, however, all the primary tissues outside the cambium become distorted or lost. The only primary tissues that are not displaced are the primary xylem and the pith.

A woody stem may be divided into three major zones: the *wood*, the *bark*, and the *cambium*. The various tissues in each

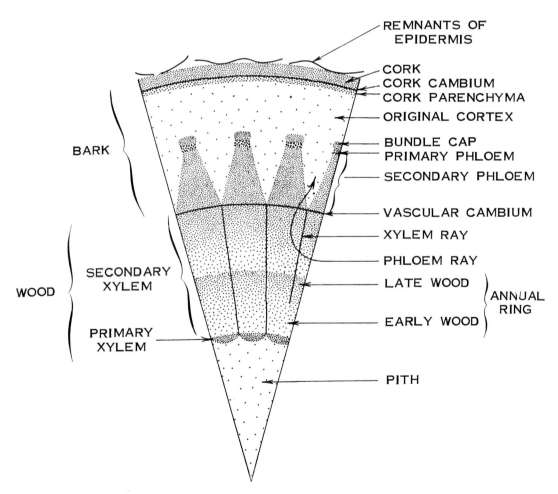

Fig. 2.36. Schematic diagram of a wedge-shaped section of a woody stem.

of these are listed below in order as they occur from the outside to the inside of a stem in which secondary growth has begun (Figs. 2.36, 2.37, 2.38):

BARK
- broken remains of old epidermis
- cork
- cork cambium
- cork parenchyma
- remains of original cortex
- bundle cap
- primary phloem (if not already crushed)
- older secondary phloem
- younger secondary phloem

CAMBIUM

WOOD
- younger secondary xylem
- older secondary xylem
- primary xylem (metaxylem and protoxylem)
- pith

Vascular rays extend from xylem to phloem across the cambium, but in the progressively older and outermost phloem they are stretched laterally as the stem increases its circumference. Thus they appear as wedges with the narrowest point of the wedge touching the cambium zone.

In many trees the bark is characterized by deep fissures that are natural results of the increasing girth of the stem. The bark is always under lateral tension due to internal expansion pressures; this tension is relieved periodically by the formation of lengthwise splits that extend deeply into the living tissues of the bark. Each time this happens a new cork cambium layer develops internally to the fissure.

Cross sections of basswood twigs (Figs. 2.37, 2.38) are often used to illustrate the tissues of woody plants. In the phloem of this plant small strands of sieve tubes and companion cells are surrounded by phloem fibers. This arrangement prevents much of the crushing of functional phloem that results from expansion pressures. An examination of freehand sections of living basswood twigs is of considerable value as an aid to the interpretation of the standard prepared slides because there is a sharp contrast between the brilliant white of the phloem fibers and the duller colors of the cortex and phloem rays.

In most trees of temperate regions there is a line of demarcation between successive annual increments of secondary wood and, as is widely known, the age of a tree can be determined by counting annual rings (Figs. 2.36, 2.38, 2.39, 2.40). With microscopic examination of wood sections it may be noted that the innermost cells of each annual ring are larger and thinner-walled than the outer cells of the same ring. The inner portion is the *early wood* that is formed when the available supplies of water are high and the transpiration rate is low. The outer portion of an annual ring is the *late wood* formed when the leaves are fully out and the transpiration rate is high enough to reduce the availability of water to growing cells. The contrast between the late wood of one year and the early wood of the following year accounts for the apparent demarcation between successive annual rings.

Early wood and late wood are commonly termed *spring wood* and *summer wood* but these terms are somewhat inappropriate since the factors which operate to induce the formation of smaller thicker-walled cells are in operation before the calendar start of summer.

As a tree grows older the central portion of the wood gradually becomes inactive in conduction. Cellular bubbles or *tyloses* may grow from the living xylem parenchyma cells into the lumens (openings) of the vessels and tracheids (Fig. 2.41). Air penetrates into the tissue and oxidation of organic waste materials turns the wood dark. This nonfunctional xylem is called *heartwood* in contrast to the functional xylem which is the *sapwood*. There is no constant conversion rate of sapwood into heartwood but it gradually increases from year to year (Fig. 2.40). The heartwood has no other function than its contribution to

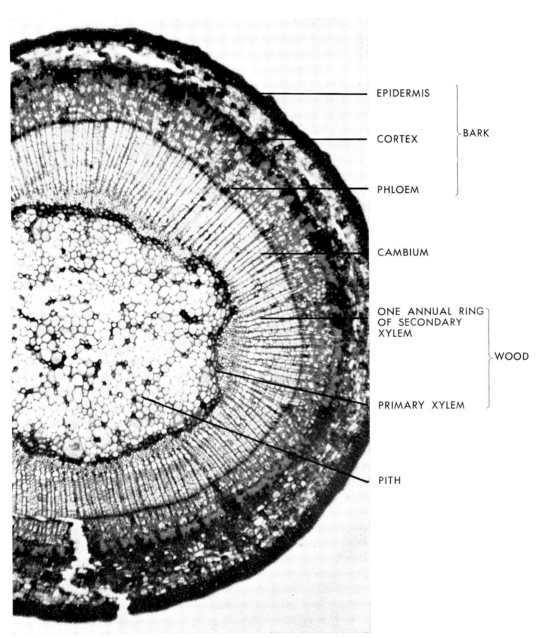

Fig. 2.37. Transverse section of a 1 year old basswood twig.

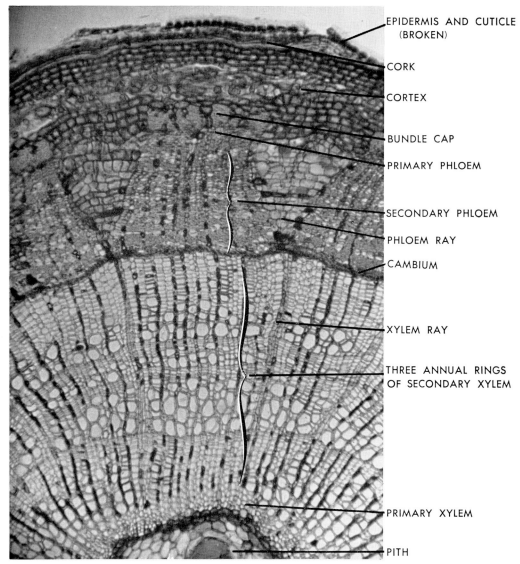

Fig. 2.38. Portion of a transverse section of a 3 year old basswood twig for comparison with Fig. 2.37.

the mechanical strength of a tree. In many trees it is highly prized as a source of wood for furniture.

## THE NATURE OF THE ROOT

The growth of the root system and the growth of the shoot system are mutually interdependent. The root system is nonphotosynthetic and must depend on the aerial parts of the plant for supplies of organic food. On the other hand, stems and leaves require water and inorganic chemicals from the soil which must be absorbed and transported by the roots. The extent of the root system is seldom realized; often it is more spread out in space than is the stem system. The number of branch roots is very large and the health of a plant depends a great deal on the continued active growth of these young roots. Such growth is primarily the result of two related activities:

# VASCULAR PLANTS: GENERAL STRUCTURE

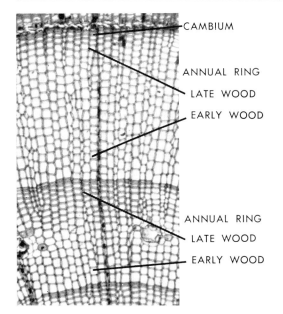

Fig. 2.39. Transverse section of a pine stem illustrating the contrast between late wood and early wood which makes the annual rings evident.

Fig. 2.40. Transverse section of a tree trunk showing annual rings, xylem rays, and the color contrast between heartwood (dark) and sapwood (light).

Fig. 2.41. Tyloses expanding into a xylem vessel.

(1) the increase in number of cells due to cell divisions in the root tip and (2) the enlargement of cells in the region immediately behind the root meristem.

The tip of a root is covered by a conical mass of cells that are held together very loosely. This is the *root cap* (Fig. 2.42) and portions of it may extend a considerable distance back along the root. New root cap cells are formed constantly in the inner portion of the cap. Meanwhile the walls of the outer root cap cells become gelatinous and slough off in the soil. This sacrifice results in a slippery layer that protects the root tip from mechanical injury as it is forced through the soil.

The tissues covered by the root cap are

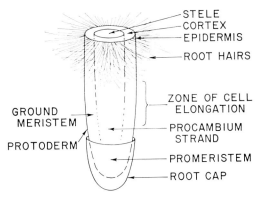

Fig. 2.42. Schematic representation of the nature of a root tip.

basically similar to those in the apical meristem of the stem, except that no leaf primordia are formed. A small cluster of cells that divide relatively slowly in more or less random planes of division constitutes the *promeristem*. From this are derived the *protoderm*, the *procambium*, and the *ground meristem*. As is discussed in more detail later in this chapter, these meristems give rise to the *epidermis, cortex,* and *stele,* respectively (see Fig. 2.45).

As the root develops, the cortex, which is made up almost entirely of parenchyma cells, becomes by far the largest of the root tissues. The planes of division in the ground meristem are almost always perpendicular to the long axis of the root and thus the root is made up largely of longitudinal rows of cells. As cells in the longitudinal rows become more and more remote from the tip the division rate slows down and eventually stops altogether. When cell division stops in these cells they continue to absorb water rapidly and increase in size. The major dimensional increase is lengthwise, as may be observed by tracing a column of cells back from the tip and noting the size changes. In roots this region of cell elongation is not as extensive as it is in stems, seldom being more than 2 or 3 mm long. The increase in length results in a pushing of the root tip through the soil since the older parts of the root are firmly anchored in the soil and are not easily displaced.

Once any cell has finished elongating, it begins to change in other ways until eventually it assumes its mature condition. Cell types vary in the time and method of differentiation; there is no sharp boundary between the region of differentiation and the preceding one.

An important differentiation process occurs in the epidermal cells. Some of them form long tubular outgrowths called *root hairs*. They begin as small swellings in the epidermal cell walls but may reach a final length of several millimeters. Each root hair is thus an outgrowth of the epidermal cell from which it arises (Figs. 2.43, 2.44).

ABSORPTION BY ROOTS

The number of root hairs is vast and they effectively increase the total surface area of root tissue in contact with the soil. This permits a more efficient intake of water and dissolved chemicals from the soil.

In order for all parts of the root, especially root hairs, to grow vigorously it is necessary that several environmental factors be properly balanced. The temperature of the soil must be suitable. The soil must

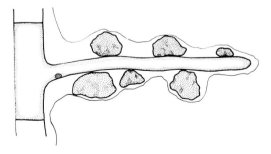

Fig. 2.43. Root hair showing epidermal cell from which it arose plus relation to soil particles and soil water.

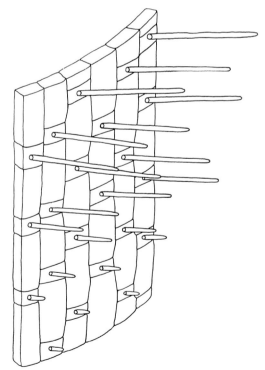

Fig. 2.44. Origin of root hairs from epidermal cells of a root.

be well aerated since the living cells of most roots require abundant supplies of oxygen for respiration. There must be adequate soil water but not enough to eliminate the air in the soil. Constant supplies of organic food must be made available by transport from the leaves.

The entrance of water into the root hairs is basically a diffusion phenomenon (osmosis). The relative water concentration inside the root hairs is less than that of the soil solution and the diffusion gradient is inward. The effect of the transpiration pull on the water column in the xylem possibly increases the steepness of the diffusion gradient and thus increases the rate of water uptake.

The entrance of chemical substances (other than water) from the soil into the roots is a more complex matter involving several related phenomena. Entrance by simple diffusion of molecules or ions in solution undoubtedly occurs but it is not considered to be of major importance by most authorities.

*Direct ion exchange* is the term applied to situations in which ions of equivalent electrical charges are traded between cells and soil particles. Root hairs grow in between soil particles and often become tightly adherent to them (Fig. 2.43). This makes it possible for ion exchanges to occur without the substances actually diffusing in solution. (Obviously this brief statement barely touches on a concept which has had a great influence on modern soil management practices. The successful use of anhydrous ammonia as a fertilizer is largely dependent on the principle of ion exchange.)

*Active solute absorption* is a term used to describe a situation in which certain chemical ions are accumulated inside cells in far greater concentration than they occur in the soil solution. Such accumulations indicate movement of the ions against the normal direction of the diffusion gradient. Undoubtedly this accumulation requires the expenditure of energy. Since it has been demonstrated that active solute absorption occurs only in healthy vigorous roots with high respiration rates, it has been postulated that some of the energy released by respiration is utilized in this way.

DIFFERENTIATION OF ROOT TISSUES

In the region of differentiation various cells begin to take on characteristics of their mature conditions. In cross sections through this region three concentric zones may be recognized (Fig. 2.45):

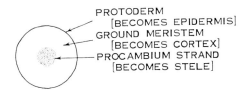

Fig. 2.45. Transverse section of a root tip before differentiation of vascular tissues in the procambium strand.

1. The outer layer of cells is the epidermis which matures from a protoderm (as in stems and leaves).
2. The inner core is made up of small elongated cells with dense protoplasm. This is destined to become the vascular cylinder or stele. At this stage it is a procambium strand.
3. The wide zone of more or less nonspecialized cells between the epidermis and the future stele is the cortex, which matures from the ground meristem.

Epidermal cells do not become cutinized as they do in stems and leaves. Furthermore many of them give rise to root hairs as noted previously. When the root hairs grow old and die they collapse and the epidermis ceases to be a functional tissue.

The cells of the root cortex remain relatively thin walled and have a considerable volume due to absorption of water. Often they accumulate reserve foods. In many roots they comprise the largest volume of the root. Inward moving water and dissolved substances diffuse from cell to cell in the cortex in their passage toward the stele.

The innermost layer of cells in the cortex, immediately in contact with the stele,

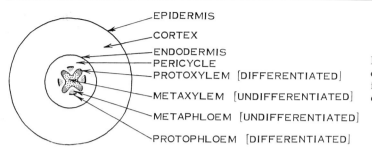

Fig. 2.46. Transverse section of a root in which some differentiation of tissues has occurred.

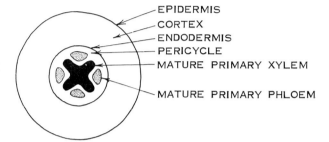

Fig. 2.47. Transverse section of a root after differentiation of all primary tissues has been completed.

is called the *endodermis*. Actually the endodermis is a single-layered cylinder of cells which separates the rest of the cortex from the stele (Figs. 2.46, 2.49, 2.50). A peripheral band in the radial walls of each endodermal cell becomes impregnated with a substance, possibly like suberin, that is relatively impermeable to water. This is called the *Casparian strip* (Fig. 2.51). Any substances that enter or leave the stele must pass through the living protoplasm of the

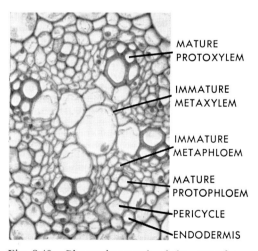

Fig. 2.48. Photomicrograph of the central portion of a buttercup root at a stage similar to Fig. 2.46. Note the differentiated protoxylem and the undifferentiated metaxylem.

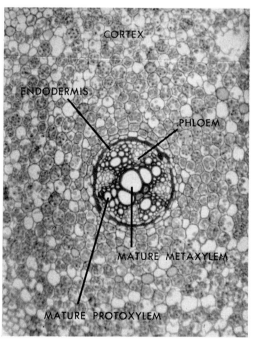

Fig. 2.49. Photomicrograph of a transverse section of a buttercup root at a stage similar to Fig. 2.47. Note especially the fully matured metaxylem, the endodermis, and the stored food in the cortex cells. The magnification is lower than that in Fig. 2.48.

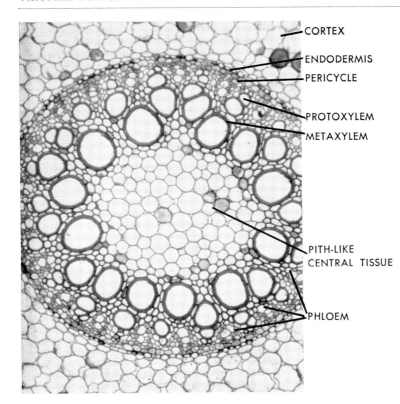

Fig. 2.50. Transverse section of the central portion of an asparagus root. Note the numerous protoxylem points and the well-defined endodermis.

endodermal cells since the Casparian strips bar intercellular radial movements. This insures that the whole of the endodermis functions as one continuous differentially permeable membrane surrounding the stele. In older parts of the root the endodermal cells may become completely thickened except for a few *passage cells* opposite the xylem.

The cells of the procambium strand grow in length but do not become laterally stretched as do cortex cells. They continue to be densely protoplasmic. At first all of these cells appear alike but as they begin to mature and differentiate into the vascular cylinder several different tissue systems become evident.

The xylem of mature roots appears in cross section as a star-shaped core of thick-walled empty cells. The differentiation of protoxylem cells begins at the points of the star and progresses inward. In sections of young roots mature protoxylem may be evident only at the points of the star, while the inner cells are still living and thin walled (Figs. 2.46, 2.48). Later all of the cells in the center become differentiated into mature metaxylem cells (Figs. 2.47, 2.49).

Exceptions exist in certain thick roots, such as corn prop roots and asparagus roots, in which a number of thin-walled nonxylem cells remain in the center. This

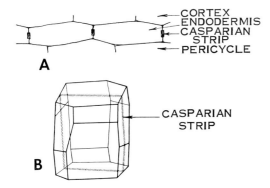

Fig. 2.51. Nature of the endodermis. A. Transverse section of an endodermis showing position of Casparian strips in radial walls. B. Three-dimensional view of a single cell of the endodermis showing the Casparian strip as a peripheral band in the radial walls.

tissue resembles the pith tissue of stems (Fig. 2.50).

Root xylem may contain both vessels and tracheids. Parenchyma tissue is more extensive in roots than in stems and usually there are no conspicuous amounts of xylem fibers.

Most of the cells that lie in small masses between the points of the xylem star and thus alternate with them become differentiated into phloem tissue. This arrangement of primary xylem and primary phloem in an alternating fashion is one of the distinguishing differences between root anatomy and the anatomy of stems. Cells in the phloem remain thin walled and do not stain as strikingly as do xylem cells (Figs. 2.47, 2.49).

The *pericycle* is the outermost layer or layers of cells of the vascular cylinder. It is in immediate contact with the endodermis to the outside and with the xylem and phloem on the inside. This seemingly insignificant tissue has three important functions:

1. The apical meristem of lateral roots forms in the pericycle opposite the points of the xylem star. When lateral roots begin to develop they force their way outward through the cortex (**Fig. 2.52**).
2. Part of the vascular cambium forms in the pericycle. This will be discussed further below.
3. The cork cambium also forms in the pericycle. This gives rise to a layer of cork cells similar to that in stems. When the cork is formed all the tissues exterior to it (endodermis, cortex, and epidermis) are sloughed off.

It should be noted that pericycle and endodermis may also occur in stems. However, they are very difficult to recognize and since they have no apparent special functions in stems they are generally ignored in elementary treatments of stem structure.

## Secondary Growth in Roots

All the tissues that grow and mature from cells laid down by meristematic ac-

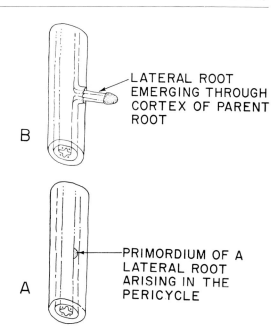

Fig. 2.52. Origin of lateral roots. A. Root primordium forming in the pericycle. B. Lateral root emerging through the cortex.

tivity in the apical meristem are called primary tissues or tissues of primary origin. These include the root cap, epidermis, cortex, endodermis, pericycle, primary xylem, and primary phloem. As in stems, secondary tissues of dicot roots are formed by the activity of lateral meristems. One of these is the vascular cambium which at first is a lobed cylinder lying inside the phloem and outside the protoxylem points (Fig. 2.53). Its formation is due to the fact that residual procambium cells between the xylem and phloem, as well as the pericycle cells outside the xylem points, retain an ability to become meristematic. Usually the successive divisions of cambial initials are parallel to the outside of the root, resulting in the formation of radial rows of cells. The innermost cells of the cambial zone become secondary xylem and the outermost ones become secondary phloem while the layers in between remain as the cambium. The original shape of the cambium is lobed but eventually it becomes more or less uniformly cylindrical (Fig. 2.54). The

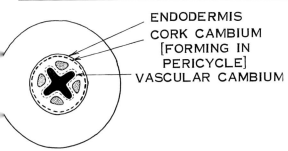

Fig. 2.53. Transverse section of a root showing position of newly formed cambium and cork cambium.

essential details of cambial activity have been discussed previously.

Xylem tissue is never displaced in space by cambial activity but the phloem is continually pushed outward. Primary phloem and the older secondary phloem are crushed by this expansion and disappear. As noted above, the endodermis, cortex, and epidermis are sloughed off when the cork tissue is formed (Fig. 2.54).

As a root enlarges it comes to be more and more like a stem. Only the organization of the primary xylem remains as a clue to the original differences between young roots and stems.

The vascular rays in older roots are broader than they are in stems and there may be a much more extensive development of cork parenchyma.

Many kinds of plants, including almost all monocots, do not have the ability to form cambial tissues and thus their roots consist of primary tissues only.

## SUMMARY OF ROOT FUNCTIONS

The main functions of the root system are considered to be *anchorage, storage of food, absorption,* and *conduction.*

The aerial parts of a plant are subjected to severe stresses due to the combined action of wind and gravity; a familiar example of such stress is the wild tossing of tree branches during a storm. Roots form a widespread network in the soil which effectively serves to anchor the plant so that the stem may grow erect.

The storage of food by roots has important economic aspects since the roots of carrots, turnips, sugar beets, garden beets, radishes, and sweet potatoes are widely used in the human diet. In plants the food stored in underground protected structures such as storage roots is utilized to promote rapid growth, flowering, and fruit development in a succeeding year.

A consideration of absorption by roots is a matter of prime economic as well as biological significance. Previously, in the discussion of root hairs, the entrance of water and inorganic chemicals into roots was outlined as was the importance of a healthy vigorously growing root system to normal growth of the shoot.

The maintenance of proper soil conditions for good root growth is one of the primary aims of agriculture. Good soil structure permits easy entrance of water yet allows for drainage of excess amounts. In addition to holding water, good soil structure must allow for air circulation. Soils must have sufficient chemical sub-

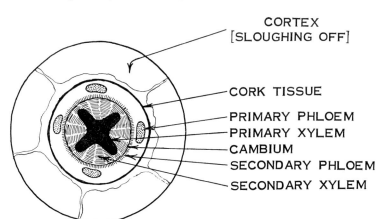

Fig. 2.54. Arrangement of primary and secondary tissues in a root after considerable cambial activity has taken place.

stances for normal growth but these should be released slowly so that they are not rapidly leached out of the soil in drainage water. They must have abundant microorganisms, which are vital to the maintenance of soil fertility, yet should not be heavily infested with root-destroying organisms. In other words, a good soil has a delicately balanced and exceedingly complex mixture of physical, chemical, and biological factors. At the elementary level these can be considered only in brief form.

SOIL NUTRIENTS

With the major exceptions of carbon dioxide and gaseous oxygen, all the essential chemicals for plant growth enter the plant through the roots from the soil. There are at least 14 chemical elements required for plant growth. They are listed below with some indication of their major functions in metabolism.

Nitrogen (N) is an essential part of such compounds as amino acids, nucleic acids, and chlorophyll.

Phosphorus (P) is an essential part of nucleic acids. It also plays an important role in many enzyme reactions and in energy transfer systems.

Sulfur (S) is part of certain essential amino acids.

Potassium (K) does not become bound in specific compounds. It remains mobile and enters into many essential reactions.

Calcium (Ca) forms part of the pectic compounds that cement plant cells together and also plays an important role in the functioning of cell membranes.

Magnesium (Mg) is an essential part of the chlorophyll molecule. It also bears an intricate relationship to calcium in membrane functions.

Iron (Fe) is not a part of the chlorophyll molecule but is indirectly essential to chlorophyll formation. Also, it plays an important role in oxidative enzyme systems and has a relationship to the availability of phosphorus.

Manganese (Mn), boron (B), copper (Cu), zinc (Zn), and molybdenum (Mo) are essential elements required in minute quantities. In many cases these micronutrients act catalytically as parts of enzyme systems. Other micronutrient elements may be added to this list as research proceeds in this field.

Carbon (C), hydrogen (H), and oxygen (O) enter into the activities of living matter in so many ways that no attempt to list them will be made.

Many nonessential chemical elements enter plants and accumulate. Some of them may have toxic effects on the plants themselves or, as in the case of selenium, on animals that feed on the plants. Others have no serious effects. People have learned to use certain plants as collectors of specific chemicals. For example, iodine has been recovered from the cells of seaweeds that accumulate the iodine from ocean waters.

The evident relationship between roots and the soil makes a brief consideration of soil composition necessary. Rock particles occur in various sizes. The smallest particles are of colloidal size and have a special significance because of their enormous total surface area. Much of the inorganic chemical substance in the soil is bound tightly to the soil particles by adhesive forces. As noted previously, root hairs come into contact with the soil particles and due to the large total surface areas involved rapid absorption is made possible by direct ion exchange. Careless management of soil may result in the destruction of good soil structure and the concomitant decrease in the total surface area available to the root hairs.

Good soil normally contains a large percentage of partially decayed organic matter called *humus* which is important in the maintenance of soil structure. In roots with secondary growth the loss of outer primary tissues adds significantly to the soil humus.

Microscopic examination of any fertile soil reveals a vast array of minute plants and animals that enter into complex relationships with each other, with the soil, and with plant roots. The source of food for most of them is the humus, although some of them obtain their energy in other ways. Some of them are destructive in cultivated land but most of them are beneficial, for they bring about a release of essential chemical elements bound up in the humus. It is apparent that biological processes in the soil are not separable from physical and chemical ones.

Carbon is locked up in organic matter by photosynthesis. When organic matter is burned or respired, carbon is returned to the air as carbon dioxide. The supply of this gas in the air is essential to the continuance of all familiar forms of life on this planet, since it is a raw material of photosynthesis. Although carbon dioxide is returned to the air in many ways, the most significant source is the respiration of the organisms of decay living on the dead organic matter of the soil.

Some of the carbon dioxide goes into solution to form a weak acid that has a weathering effect on soil particles. This is important in the gradual release of chemical elements in forms available to plant roots.

Some of the organic compounds in the soil, loosely referred to as *humic acids,* help to maintain other ions in solution. Iron often tends to form insoluble and thus unavailable chemical unions unless some of these complex organic acids are present in the soil.

*Mycorrhizal relationships* between soil fungi and roots are vital to the successful growth of some plants. In this type of relationship fungus threads penetrate into or between the cells of the root. They may obtain all or a part of their nutrition from the root and in return serve in a similar capacity to that of the root hairs. Mycorrhizae are found often in plants like orchids, heaths, and gymnosperms growing in acid soil.

## THE DISTRIBUTION OF ORGANIC NUTRIENTS IN PHLOEM

Since they have no chlorophyll, roots are unable to carry on photosynthesis. Transport of dissolved organic foods to roots from major sites of photosynthesis is an essential function of phloem. The phloem solution actually moves toward any site where organic foods are being used: stem tips, developing fruits, cambial zones, and so on. However, most plants have large numbers of roots and the demands for food by the root system are very great. Thus the major direction of movement in the phloem is usually downward from leaves to roots.

One important theory concerning movement in the phloem is called the *mass flow hypothesis.* Simple sugars such as glucose and fructose are able to pass readily through the living membranes of the sieve tube elements. These sugars originate primarily in the leaf chlorenchyma and pass through the vein sheath cells into the phloem. Once inside the sieve tubes the simple sugars are apparently converted to sucrose by enzyme action. This larger molecule sugar does not pass readily through the cell membranes and is retained in the sieve tubes. This procedure has several effects:

1. A *nutrient sink* is created inside the sieve tube. The simple sugars continue to diffuse inward because their concentrations do not build up as a result of the conversion to sucrose.
2. The increase in sucrose, which is soluble, changes the osmotic properties of the sieve tube elements; they tend to absorb more water as a result.
3. The tendency toward increasing volume in a nonexpandable system results in a mass flow of the solution in the sieve tubes.

Many ingenious experiments have been used to demonstrate the existence of high concentrations of sucrose in the phloem solution and there is little question concerning the basic premises of the mass flow hypothesis. However, it is a relatively slow process and seems inadequate to account for some of the more rapid flow rates which are recorded in research literature.

If the mass flow is considered as a pushing mechanism and the concept of a pulling mechanism is added (to create a push-pull system) then greater speeds of flow might be anticipated. Such a pulling mechanism may be generated as a result of the following factors acting in concert:

1. At the point of use (destination) of the organic food the sucrose molecules must be digested so that the simple sugars can move outward through the sieve tube membranes into surrounding cells.
2. The loss of dissolved molecules from the phloem solution means an immediate lowering of the osmotic attraction for water inside the phloem cells. In a sense many of the water molecules that had been held in the phloem due to the presence of sugar are freed to move elsewhere when the sugar is gone. This concept can be expanded further when it is realized that most of the simple sugars are completely removed from solution in surrounding cells by conversion to building products such as cellulose and proteins or to storage products such as starch.
3. The freed water molecules are taken up immediately by water-demanding systems, most notably the xylem that is adjacent to the phloem. (It should be recalled that the xylem solution is frequently under a stress condition which amounts to a negative pressure and any available water molecules would be incorporated immediately into this system.) Xylem tissues are continuous throughout the plant and operate as a sealed system. Thus water molecules freed from the phloem in the roots or anywhere else in the plant could be taken up at the site of release.
4. The combined losses of solutes and freed water molecules to surrounding cells could generate a cohesion-type pull on the phloem solution comparable to the transpiration pull on the xylem solution generated in the leaves.
5. This mechanism would work even if the xylem solution was not under tension due to transpiration (as in humid weather). The phloem solution in the leaf veins could pull water out of the xylem by osmosis and then give up water to the xylem in the root tissues, thereby creating a circulatory flow which would move both the xylem and the phloem solution.

DISCARDED

# THREE  THE PLANT BODY

IN the preceding chapters the basic structure of various vegetative parts of plants has been described and it is now possible to consider briefly the plant as a whole. There is a vertical axis with a growing point at the stem tip and another growing point at the root tip (Fig. 3.1). An orderly procession of emergences on the stem tip results in the formation and placement of leaves in characteristic patterns on the developing *shoot system* (a term often used for the stems and leaves together). In the axil of each leaf there appears a bud containing a branch meristem which is similar to that of the main axis in most respects. Flowers, when they appear, are extreme modifications of branches.

Each root tip of the *root system* is protected by a unique structure, the root cap, and is clothed with evanescent epidermal outgrowths, the root hairs. Lateral roots emerge above the level where root hairs begin to die, originating deep within the parent root tissue rather than as superficial swellings (as in the case of leaf primordia).

Each of the components of the plant body exhibits a wide variety of morphological expressions, not only among the large number of existing plants but also among individuals of the same kind grown under varying conditions. Familiarity with the terminology of such variations is important in making and understanding useful descriptions of plants. In the following paragraphs portions of this voluminous terminology are examined briefly.

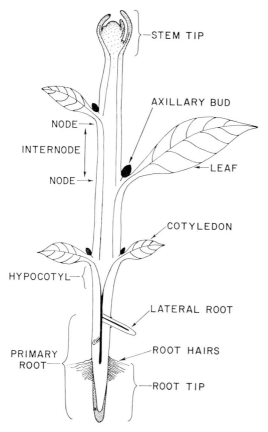

Fig. 3.1. Schematic diagram of an entire plant in the vegetative condition. (Compare Figs. 2.7, 2.42.)

## THE LEAF
### LEAF TYPES

Each leaf has an expanded *blade* portion and a leaf stalk or *petiole* (Fig. 3.2A). There may or may not be a pair of appendages at the base of the petiole. These are the *stipules*. In some plants they may be present when the buds open but fall off shortly thereafter, leaving only faint scars on the twig. In other plants such as the garden pea (see Fig. 3.10) the stipules are persistent and large enough to contribute significantly to the total photosynthetic product. In some plants the leaves are *ses-*

# Vascular Plants: Variations and Evolutionary Origins

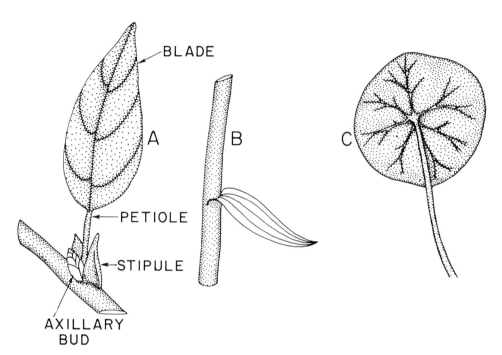

Fig. 3.2. A. Basic leaf type showing parts of leaf. B. Sessile leaf. C. Peltate leaf.

*sile;* they do not have petioles and the blades are attached directly to the stem (Fig. 3.2B). The leaves of a few plants like the common nasturtium (Fig. 3.2C) have the petiole attached toward the center of the blade instead of at its base. Such leaves are said to be *peltate*.

The blade may be undivided and *simple* as in Fig. 3.2A, or it may be subdivided one or more times into a set of *leaflets*. Subdivided leaves are said to be *compound*. If the leaflets are linearly arranged along the leaf axis the leaf is *pinnately compound* (Fig. 3.3A). The honey locust leaf (Fig. 3.3B) is twice pinnately compound. If the leaflets are attached at a common point, fanning outward like the fingers of a hand, the leaf is said to be *palmately compound* (Fig. 3.3C).

## Leaf Arrangement

As noted in Chapter 2, the joints of stems where leaves are attached are called nodes and the stem segments between them are internodes. There are three basic patterns of leaf arrangement on stems. If only one leaf is attached at each node the arrangement is *alternate* (Fig. 3.4A). If pairs of leaves occur at the nodes the arrangement is *opposite* (Fig. 3.4B). A special variation of this arrangement in which successive pairs are at right angles to each other along the stem is called *decussate* (Fig. 3.4C). When three or more leaves

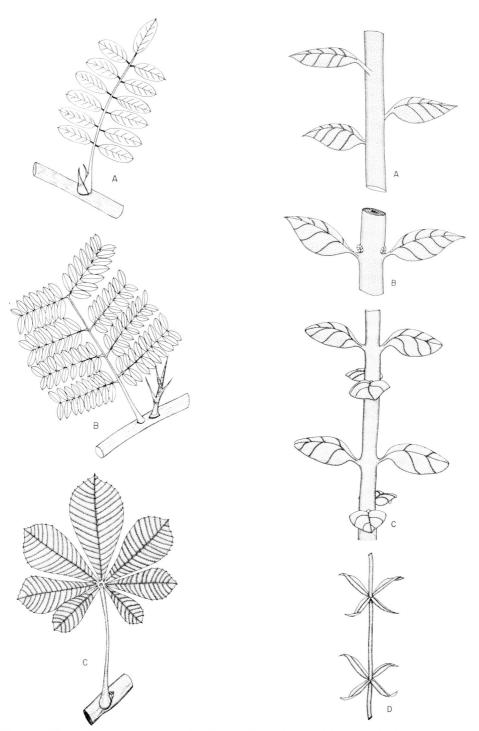

Fig. 3.3. A. Pinnately compound leaf. B. Twice pinnately compound leaf. C. Palmately compound leaf.

Fig. 3.4. A. Alternate leaf arrangement. B. Opposite leaf arrangement. C. Decussate leaf arrangement. D. Whorled leaf arrangement.

occur at each node the arrangement is *whorled* (Fig. 3.4D).

VENATION

The vein pattern *(venation)* is termed *parallel* when, as in most monocots, the several main veins are seen to be parallel (Fig. 3.5A). Or it may be *netted,* as in most dicots, where the visible vein system forms an evident network (Figs. 1.4, 3.6). If the major veins seem to originate from a common point the venation pattern is *palmate* (Fig. 3.6); if there is one major vein forming a lengthwise *midrib* in the blade with secondary veins joining the midrib at intervals the venation is said to be *pinnate* (Fig. 3.2A).

LEAF MARGINS

The nature of the leaf margin is also useful in a descriptive sense. Figure 3.2A shows a leaf with a smooth or *entire* margin; Fig. 3.5B shows a series of types of leaf margins. When the margin has small forward-pointing teeth it is said to be *serrate.* When the teeth are somewhat larger it is *dentate.* When the indentations extend halfway or more to the midrib the leaf is said to be *lobed.* The leaf may be *palmately lobed* (Fig. 3.6) or *pinnately lobed* (Fig. 3.5) depending on the pattern of the major veins.

## SPINES, THORNS, AND PRICKLES

Many plants have sharp projections of various types that can be injurious to ani-

Fig. 3.5. *Top:* Parallel venation in a monocot leaf. *Center:* A series of leaf margin types. *Bottom:* Pinnately lobed leaves (Gray).

Fig. 3.6. A palmately veined and palmately lobed leaf showing the characteristic netted veination of dicot leaves.

mals and thus may serve as deterrents to grazing. These are indiscriminately called spines or thorns in popular usage, but in conventional botanical usage these terms have different meanings.

*Spines* are modifications of leaves or leaf parts. In the Japanese barberry the fact that each spine subtends a bud is indicative of its origin as a modified leaf (Fig. 3.7D). In the black locust (Fig. 3.3A) the spines occur in pairs at the bases of leaves, indicating they are modified stipules.

*Thorns* are modified branches. In a shrub commonly known as buckthorn (Fig. 3.7A) the thorn is the entire terminal portion of a main branch which becomes hardened into a sharply pointed structure. The thorns of the hawthorn (Fig. 3.7B) occur in the axils of leaves and quite clearly are modified branches. The complex (and dangerous) thorns of the honey locust may be several times branched (Fig. 3.3B).

The *prickles* of roses (Fig. 3.7C), raspberries, blackberries, and so on are epidermal outgrowths and are not modifications of whole plant parts.

## TYPES OF BUDS

As noted previously, buds are developmental structures that contain stem-tip meristems and ultimately give rise to continuations of main stem axes or their branches. The following definitions give some indication of possible variations in structure and functions of buds (see Figs. 3.8 and 3.9):

1. *Active buds* are those that develop continuously into mature branches without an intermediate dormant period. (Examples: tomato, potato, pea, bean.)
2. *Dormant buds* are those that enter a period of dormancy prior to the winter season or other season unfavorable to growth. In most cases the delicate new plant parts inside the bud are enclosed within a series of tightly overlapping modified leaves, the *bud scales,* which serve the primary function of protection against transpiration losses.
3. *Naked buds* are those dormant buds that lack bud scales.
4. *Terminal buds,* as the name implies, occur at the tips of branches and each one includes an apical meristem.
5. *Lateral buds* or *axillary buds* are buds occurring on the sides of stems, which arise in the axils of leaves. In some plants such as elm and basswood the terminal bud dies each year and the uppermost lateral bud gives rise to the twig that continues the branch in the following year (Fig. 3.8A).
6. *Latent buds* are usually the lowermost buds formed on a twig during a year's growth. They do not normally develop into branches in the next season unless the new growth above them is killed, as by a late frost, insect or dis-

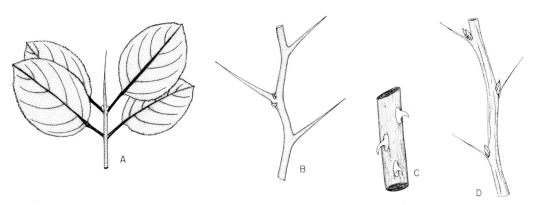

Fig. 3.7. A. Stem tip thorn of buckthorn. B. Modified branch thorn of hawthorn. C. Prickles on a rose stem. D. Modified leaf spines of Japanese barberry.

# VASCULAR PLANTS: VARIATIONS AND ORIGINS

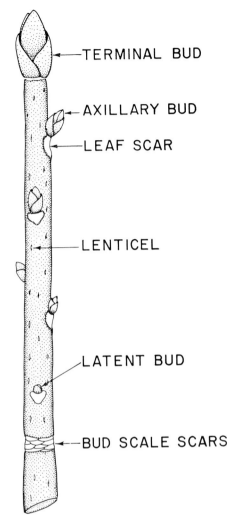

Fig. 3.8. One year twig.

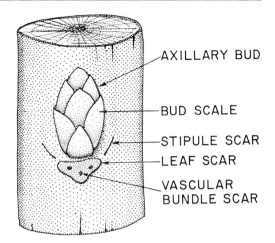

Fig. 3.9. Portion of twig showing one lateral bud with leaf scar below. Note the vascular bundle scars in the leaf scar and the paired stipule scars.

ease damage, or grazing by animals. They are latent due to inhibition by an excess of growth hormones originating in actively growing tissues above them. Death of these tissues eliminates the hormone source and the latent buds are freed of the inhibition. This procedure results in a type of insurance that allows woody plants to survive certain catastrophies by sprouting a new crop of leaves.

7. *Adventitious buds* are those not arising in normal positions such as at the tips of stems or the axils of leaves. The most common location for adventitious buds is on the callus tissue that often develops at the site of injury in many plants.
8. *Leaf buds* give rise to new branches that bear only leaves.
9. *Floral buds* give rise to branches that bear only flowers.
10. *Mixed buds* give rise to branches that bear both leaves and flowers.

## THE WOODY TWIG

The word *twig* is so widely used in common language that precise definition becomes difficult. In general, however, it is accepted as referring to the terminal portion of the branch of a woody stem (Fig. 3.8). The portion formed in a given year bears leaves and forms buds. When the leaves fall the scar is healed by a layer of cork. *Leaf scars* usually have shapes that are characteristic of the species and are useful in identification. Frequently within the leaf scar there are smaller scars, the *vascular bundle scars,* that indicate the positions of the broken vascular tissue. Pairs of *stipule scars* may or may not be evident alongside the leaf scars (Fig. 3.9).

When the buds open in spring the bud scales fall off, leaving behind on the twig a circle of closely arranged scars, the *bud scale scars.* These are usually recognizable for several years and may be used to determine the rate of growth of a branch. (The distance between any two successive sets of bud scale scars tells how much length growth the branch achieved in a given year.) Most woody twigs have numerous small but visible slits in their surfaces. These *lenticels* allow for gas exchanges between the living cells within and the external atmosphere. They more or less replace the stomates as the twig enlarges.

When a woody plant loses its leaves at the end of a growing season the plant is said to be *deciduous*. The term applies quite generally to most of the broad-leaved trees of the temperate zone. Trees that are continuously clothed with leaves throughout the year are called *evergreens*. It should be understood, however, that individual leaves do not survive for the entire life of the plant. Any set of leaves formed in a given year may survive two to several years but will eventually die and fall off.

## ADAPTATIONS FOR CLIMBING

An ability to grow tall is useful to plants in the sense that it gives them more living space for leaf functions. Yet the formation of a stout trunk with numerous branches to support the leaves utilizes enormous amounts of photosynthetic products in creating the cellulose and lignin needed to build such stems.

Some plants are able to utilize other plants or available inert structures to support their bodies instead of building stout stems. In this way they conserve their resources for other activities. There are several devices by which such plants climb over, twine about, or simply lean on their supports:

1. *Tendrils* are modified parts that are threadlike and touch sensitive. Growth on the touched side slows while growth on the opposite side continues at normal rates. This causes the tendril to grow in a coiled fashion which serves to wrap it around the touched object. In the garden pea the terminal leaflets are modified as tendrils (Fig. 3.10). In nasturtiums, *Clematis,* and some species of *Solanum* (Fig. 3.11) the petioles function as tendrils, while in grapes and in the passion flower (Fig. 3.12) the tendrils are modified branches.

   In some climbing plants such as the Virginia creeper branched tendrils arise at nodes of the stem. On contact with a supporting wall the tips of these tendrils become flattened into small discs which adhere firmly. The support provided by vast numbers of these tiny *holdfasts* is enough to hold the entire vine firmly in place (Fig. 3.13). In poison ivy and Boston ivy a similar type of support is provided by clusters of adventitious roots which arise at nodes and become firmly attached to the supporting structure.

2. *Twining Stems.* In some plants the stimulus to most rapid elongation of young internode cells seems to pass round the stem, spiraling upward. This

Fig. 3.10. Leaf of garden pea showing leaflets modified as tendrils.

Fig. 3.11. *Solanum* leaf with petiole tendril (Gray).

# VASCULAR PLANTS: VARIATIONS AND ORIGINS

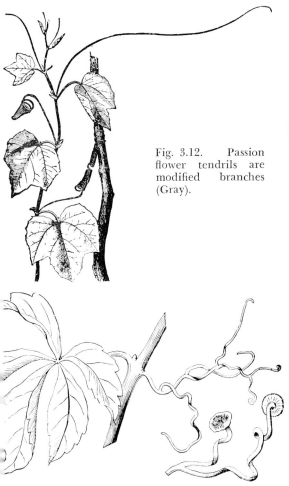

Fig. 3.12. Passion flower tendrils are modified branches (Gray).

Fig. 3.13. Virginia creeper tendrils (Gray).

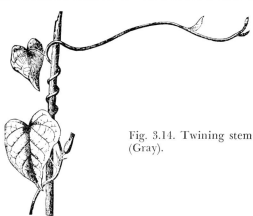

Fig. 3.14. Twining stem (Gray).

causes the whole apical region to swing around in a slow twining movement best visualized in time lapse motion pictures. In plants such as beans or morning glories this movement serves to wrap the stem around any available support such as the stems of other plants (Fig. 3.14). When beans are grown commercially or in the home garden it is a standard practice to provide supports about which the beans can twine easily.

## PLANT LONGEVITY

The longevity of plants may be indicated in a general way by three terms: *annuals*, *biennials,* and *perennials*.

1. *Annuals* are plants that go from seed to seed in a single growing season. The seed germinates, the plant grows, flowers develop giving rise to seeds and fruits, and then the plant dies—all in a single season. Many field crops (corn, soybeans, wheat), garden plants (peas, beans, tomatoes), and such ornamental flowering plants as zinnias, petunias, and marigolds are grown as annuals.
2. *Biennials* usually grow from seed germination to vegetative maturity in the first growing season and complete flower and seed production in the second season. Two common biennials are hollyhocks and carrots.
3. *Perennials* live for several to many years. If they are *herbaceous perennials* the aerial plant parts die at the end of each growing season but underground parts survive from one year to the next. Phlox and delphinium are common examples in the flower garden.

*Woody perennials* have persistent aerial stems that get larger year after year due to cambial action. *Shrubs* are woody perennials with several stems arising at ground level while *trees* have a single main trunk with branches arising above ground. *Vines* are climbing stems with various means of support, as discussed under adaptations for climbing. Many vines such as grapevines are woody perennials; others such as *Clematis* are herbaceous perennials; still others such as the scarlet runner bean are annuals.

## MODIFIED STEMS

A rather large and frequently confusing terminology exists for modified stems. It is hoped that the terms and definitions selected for inclusion here will not add to the confusion.

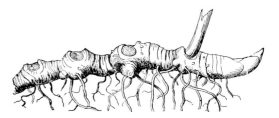

Fig. 3.15. Rhizome of Solomon's seal (Gray).

1. A *rhizome* is a prostrate stem above or below ground that is in itself the main axis of stem growth.
2. A *runner* is a prostrate stem arising as a branch at the base of an erect stem and spreading horizontally above ground.
3. A *stolon* is also a prostrate stem arising as a branch at the base of an erect stem but spreading horizontally below ground. Stolons are easily confused with roots at first glance, but the presence of internodes and nodes with small but clearly evident scale leaves should make the distinction certain.

    Examples: *Iris,* cattails, and Solomon's seal have rhizomes (Fig. 3.15). Strawberries and the airplane plant have runners (Fig. 3.16). Quackgrass and many mints have stolons (Fig. 3.17).
4. A *tuber* is an enlarged, fleshy, underground stem that develops as an enlargement of a short stolon and that functions primarily as a food storage organ (Fig. 3.18). The "eyes" of the potato tuber are axillary buds and the "eyebrows" are scars of vestigial leaves.
5. A *bulb* is a fleshy underground storage organ consisting of a short vertical stem with nodes and internodes condensed which is surrounded by closely overlapping modified leaf bases (Fig. 3.19). The leaf bases are scalelike and fleshy, serving as food storage organs. Some common examples are the bulbs of tulips, lilies, and onions.
6. A *corm* is another type of underground stem that serves as a food storage organ. It is vertically compressed but expanded horizontally. Nodes and internodes are condensed and only dead, dry, scaly remains of leaves are attached (Fig. 3.20). Some common examples of corms are those of gladiolus and crocus.
7. Most cactus plants as well as some specialized members of the lily and spurge

Fig. 3.16. Runners of airplane plant.

# VASCULAR PLANTS: VARIATIONS AND ORIGINS

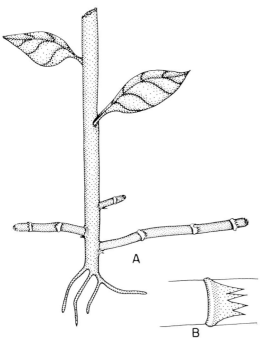

Fig. 3.17. Stolons emerging from base of an erect stem (A) and detail showing scalelike leaf at a node of a stolon (B).

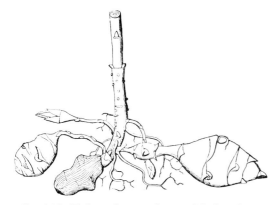

Fig. 3.18. Tuber of Jerusalem artichoke (Gray).

Fig. 3.19. Bulb (Gray).

Fig. 3.20. Corm (Gray).

families have aboveground fleshy *succulent stems* that are able to absorb large amounts of water during the brief rainy seasons of desert areas. The main photosynthetic tissues of such plants are usually in the stems. Leaves when produced at all are commonly small and evanescent, falling off the plant shortly after their formation.

(Note: In some other families the *leaves* rather than the stems are thick, fleshy, and *succulent*. Some examples well known to gardeners are the stonecrops and hen and chickens (Fig. 3.21).

Fig. 3.21. Succulent leaves of the hen and chickens plant (Gray).

## TYPES OF ROOT SYSTEMS

When the primary root develops strongly and dominates the root system it is referred to as a *tap root* (Fig. 3.22). Many trees such as the oaks have tap roots; carrots, beets, and turnips are examples of economically important tap roots in which large quantities of food are stored. The tap roots of dandelions are well known to most home gardeners.

If the primary root dies or does not become the dominant root several other roots may develop, resulting in a *fibrous root system* (Fig. 3.23). This type of root system is well illustrated by corn, the cereal grains, and many other grass plants. In some cases, as in dahlias and sweet potatoes, portions of the fibrous root system become large *fleshy storage roots* (Fig. 3.24).

Root systems do not always arise through branching of the primary root. The *seminal roots* (seed roots) of corn (Fig. 3.25) develop from the *scutellar node* of the embryo axis. The *prop roots* of corn (Fig. 3.25) arise from the lowermost aerial nodes of the stem. Many plants may be propagated by placing cuttings or slips in the right environment. Such plants develop *adventitious roots* at the base of the cut stem. In the tropics particularly many plants grow on other plants (as *epiphytes*); some of them have special *aerial roots* (Fig. 3.26) that seem to be able to absorb water from a humid atmosphere.

## SOME GENERAL ENVIRONMENTAL MODIFICATIONS

Although the modifications of plants to fit specific environments are so numerous that consideration of them is a course in itself, a few examples may be looked at briefly, particularly those related to the availability of water.

*Xerophytes* are plants adapted to environments where water is scarce, as in deserts, or where water is present but difficult to absorb, as in some types of bogs. Many

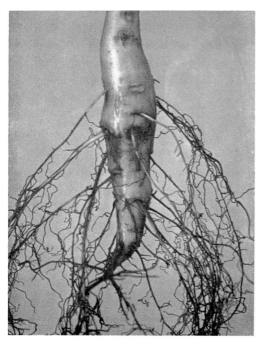

Fig. 3.22. A tap root system.

Fig. 3.23. A fibrous root system.

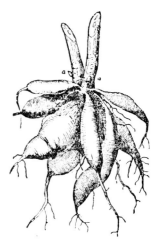

Fig. 3.24. The fleshy storage roots of dahlia (Gray).

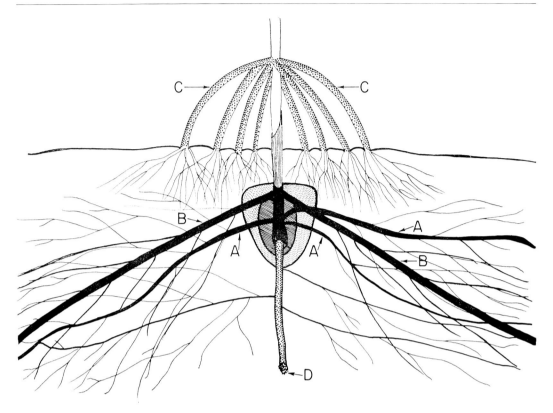

Fig. 3.25. Seedling from germinated corn grain. A. Seminal roots arising from the scutellar node. B. Adventitious roots arising from the coleoptile node. C. Prop roots (which are also adventitious) arising from the next node upward. D. The tip of the primary root has died as often happens under field conditions.

coniferous evergreens, for example, exist in essentially xeric conditions during winter when the ground is frozen and water cannot enter their roots. Since they keep their leaves throughout the year winter transpiration losses become serious. Pine needles are leaves with greatly reduced surface area in relation to volume; the intercellular spaces are small; the epidermis is heavily cutinized; and the stomates are sunken (Figs. 3.27, 3.28). All these devices help reduce transpiration losses.

The household rubber plant which is widely grown in college greenhouses is a tropical plant that grows naturally in areas where water is generally available but transpiration rates may temporarily exceed the capacity of the water transport system to replace the losses. The leaves of this plant have adaptations to conserve water and reduce transpiration losses: thick cuticles, sunken stomates, no stomates on the upper epidermis, and a special water storage tissue (Fig. 3.29).

Many desert plants, as noted in a discussion of cactus stems, drop their leaves at the end of the rainy season and depend on water stored in the stems to survive during the dry season.

A rather common modification of leaves when exposed to xeric conditions is to become rolled up in such a way that a minimum surface is exposed to the dry atmosphere. Corn leaves, for instance, have sets of special cells called *bulliform cells* located in the upper epidermis. These cells are responsible for this rolling effect. (This is illustrated in Chapter 1, Fig. 1.6.)

The oleander leaf has pits filled with hairs in the lower epidermis and curiously

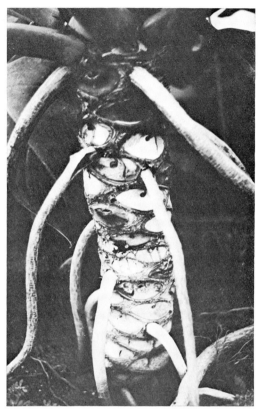

Fig. 3.26. Aerial roots of a tropical epiphyte.

Fig. 3.27. Twig of pine with needles in pairs, arising from short spur shoots on the main branch.

Fig. 3.28. Cross section of a needle of Austrian pine showing thick cuticle, sunken stomates, compact chlorenchyma.

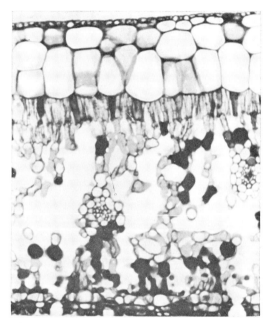

Fig. 3.29. Cross section of household rubber plant leaf showing thick cuticle, sunken stomates, and the water-storage tissue lying just below the upper epidermis.

the guard cells of the stomates are raised on stalklike epidermal cells in the pits (Fig. 3.30).

*Hydrophytes,* as the name implies, are plants adapted to growing in wet soils or in water. Water lilies, for instance, have floating leaf blades. Examination of these blades shows that the stomates are located only in the upper epidermis. The leaf tissues are loosely arranged with large amounts of internal air spaces, allowing maximum photosynthetic efficiency (Fig.

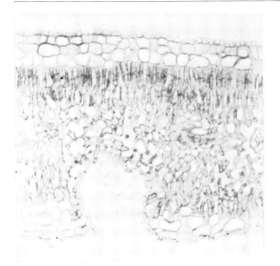

Fig. 3.30. Section of an oleander *(Nerium)* leaf showing one of the hairy pits in the lower epidermis.

"pitcher" is formed which becomes filled with rainwater (Fig. 3.32). This plant grows in acid peat bogs where normal root functions are considerably inhibited. The pitchers provide water to the plant but in addition they serve to trap small insects which are unable to escape. When the insects die and decompose the recycled chemical elements from their bodies are absorbed and used in the metabolism of the plant. *Nepenthes,* a relative of the pitcher plant that grows in humid tropical forests, has a similar pitcher partially supported by a tendril (Fig. 3.33).

3.31). In such leaves there are often large, many-lobed, sclerenchyma cells called *sclereids*.

The leaves of the pitcher plant are folded and sealed in such a way that a

Fig. 3.32. Leaves of a common bog pitcher plant (Gray).

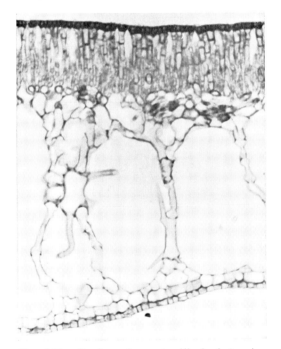

Fig. 3.31. Section of a water lily leaf showing the characteristically abundant internal air spaces and portions of a sclereid.

Fig. 3.33. Leaf of *Nepenthes,* a tropical relative of the pitcher plant (Gray).

## THE ESCAPE FROM THE SEA

It is evident that the leaf is a highly modified structure. To understand its possible evolutionary origin it is necessary to consider some of the factors affecting the survival of land plants in general.

The molecular oxygen in the upper layers of the present atmosphere of the earth exists in the form of ozone that provides a protective shield for land-dwelling organisms against excessive concentrations of ultraviolet radiation from the sun. It is generally considered, though, that the earth's atmosphere was originally free of any significant amounts of molecular oxygen and thus the ultraviolet radiation reaching the earth's surface would have been a serious hazard for any organisms that emerged from the sea.

Organisms growing in water are protected to a large extent from such injury due to absorption of ultraviolet light in the surface layers of water bodies and it follows that living organisms must have had a long evolutionary history in aquatic environments before the oxygen released by algal photosynthesis accumulated in the atmosphere to a concentration capable of providing an effective ultraviolet screen.

The emergence of both animals and plants from aquatic to aerial environments captivates the imagination. Unfortunately our direct knowledge concerning the emergence of plants is meager and much of what can be said of the process is admittedly speculative.

It is generally accepted that heritable changes occur constantly in living organisms. Environmental influences then affect natural selection among progeny and eventually new types of organisms develop that possess characteristics enabling them to exist in environments that were untenable by their ancestors.

This argument is often reversed in what has been termed teleological thinking. One might state or imply, for instance, that primitive plants developed a cuticle in order to withstand desiccation in an aerial environment. This statement is teleological. If it were rephrased to avoid the teleological implications one could state more correctly that the development of the cuticle enabled plants to withstand the desiccating effects of an aerial environment. In any discussion of cumulative evolutionary changes the danger of teleological expression is ever present and must be guarded against.

When living forms of green algae are considered as a whole it will be seen that they have many of the attributes of the higher green plants. Among these are similarities in pigmentation and an ability to form starch.

The fossil record has given us evidence of the general nature of primitive land plants but unfortunately not much evidence of the transitional stages. Thus we can deal only with speculations concerning the events that led to the conquest of the land. One popular theory is based on an interpretation of plants shaped like certain of our common seaweeds that have dichotomously branching (Y-shaped) plant bodies. In a simple expression of this theory one of the branches remained in the mud and evolved into the root system while the other turned upward and evolved into the shoot system.

One criticism of this theory is that the living seaweeds of this general nature are more apt to be among the brown or red algae while most authorities agree that the land plants evolved from the green algae. However, some species of green algae have plant bodies that are made of densely intertwined branching filaments held together in a tough mucilage. These structures often show irregular branching that approaches dichotomy at times and might serve as examples of green algae to fit the above theory (Fig. 3.34).

### The Hypothetical Emergents

There may have been an invasion of the atmosphere before there was an invasion of the land. The hypothetical plant to be considered had a complex, erectly growing axis of some sort (Fig. 3.35) in which the component cells were firmly

Fig. 3.34. A green alga with a complex branching thallus composed of intertwined branching filaments embedded in firm mucilage. Schematic but based on specimens of *Chaetophora incrassata*.

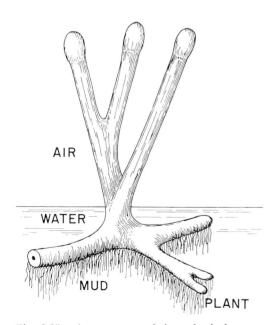

Fig. 3.35. Appearance of hypothetical emergent plant demonstrating evolutionary transition stages between the green algae and the land plants.

bound together by a cementing substance derived from pectic compounds.

Cell division became more or less limited to the terminal portion of the axis, a feature not unknown in the algae where many branching forms undergo cell division only in the younger cells near the tips of the branches.

A plant body of this type would have had a certain rigidity due partly to the firmness of the gelatinous substances and partly to mutual pressures of adjoined cells. It might have grown upward out of the water or have been exposed periodically by the lowering of the water level in shallow ponds or tidal seas. In any case, the out-of-water branch tips would have been exposed to the drying action of the air. The external gelatinous covering would have dried and hardened but in so doing would have protected the living cells somewhat from direct water loss. In this way the stage may have been set for the evolution of the cuticle. However, a complete seal over the surface would have interfered with the gas exchanges necessary to both photosynthesis and respiration. Thus it is probable that the cuticle did not become a completely successful innovation until stomates evolved.

The threadlike outgrowths shown in Fig. 3.35 are anchoring *rhizoids* while the swollen tips of the emergent branches contain reproductive cells called *spores*. (The significance of spores in the life cycle is discussed in Chapter 4.) It is noted in passing that the terminal position of the *sporangia* (spore containers) was characteristic of primitive land plants.

The Wind-Disseminated Spore

Stresses and strains due to drying eventually caused the sporangium wall to tear open exposing the mature spores inside to the atmosphere. The gelatinous walls of the spores hardened and the protoplasm became partially dehydrated. The ability to exist in such a condition was not entirely new since many green algae had evolved resting spores resistant to desiccation previously. The spores in this dehydrated condition became subject to wind dispersal as soon as the walls hardened enough so that one cell did not stick to the next.

The evolution of the wind-disseminated spore may or may not have occurred in this way but the fact that it did occur meant a tremendous advantage to the species concerned in the struggle for survival. Windspread spores are disseminated over far greater distances than swimming spores and the dissemination of progeny is one of the more important selective factors in evolution.

It follows that the higher the sporangia were pushed above the water level the farther the spores were spread and the height of the axis above water thus became an important selective factor in evolution. However, mechanical factors in the support of the aerial tissue may have limited this growth and additionally there are physical limits to the rate at which simple cell-to-cell diffusion of water can replace that unavoidably lost by transpiration. Furthermore, the fact that the cuticle was not easily wettable meant that external water movements due to capillarity, previously of considerable significance, were no longer operative. Quite evidently the evolution of a more efficient water transport system in such plants would have provided definite advantages in the struggle for survival of species.

PRIMITIVE VASCULAR PLANTS

The development of a central core of elongated thick-walled cells effectively increased the mechanical strength of the aerial branches and certain similarities between these supporting cells and the water-conducting cells of xylem suggest an evolutionary relationship. It is equally possible, however, that the conducting cells of the xylem had an independent origin. Nonetheless, once a water-conducting tissue did become functional the aerial branches were able to attain greater heights without being desiccated.

(It is important to recall here that xylem cells, despite their complex wall structure, are merely dead cells, empty of protoplasm, that provide a passageway for water movement; the water movement is generated by transpiration losses from living green cells exposed to the internal atmosphere of the photosynthetic tissue system. The addition of *lignin* to the walls of xylem cells gives them the structural strength to resist collapse when negative pressures generated by transpiration exist in the water column.)

The hypothetical vascular invader of the atmosphere would thus have been a plant with a slender axis consisting of parallel rows of cells bound together by derivatives of pectin. It would have had a terminal sporangium and a central core of xylem. Portions of the plant would have been submerged and attached to the muddy bottom.

The question of whether plants such as these ever existed was answered dramatically by the discovery of fossilized plants which had been living in a Devonian swamp (approximately 350 million years old). The plants of this Devonian swamp are known to us as *psilophytes* (Fig. 3.36). Their stems were slender and several cm in height. They did not have true leaves. The terminal sporangia consisted of layers of sterile wall cells enclosing masses of spores. The epidermis had a rather thick cuticle and possessed stomates. There was a central core of xylem and around it was a layer of phloem. The tissue inside the epidermis consisted of chlorenchyma, which was the functional photosynthetic tissue. Because the cells of this tissue had air spaces between them, allowing both circulation of an internal atmosphere and connections to the external atmosphere through stomates, it may be referred to as a *ventilated chlorenchyma*. (Compare Fig. 3.39.)

These plants did not have roots but some of the branches were prostrate and anchored to the mud with rhizoids. One of the prominent features of the aerial system was that the stem branched dichotomously. This feature has significance in the origin of the fern-type leaf, as will be seen.

The discovery of the psilophytes led

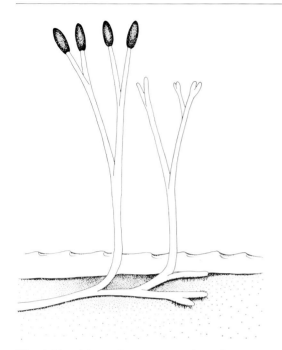

Fig. 3.36. Schematic representation of a type of psilophyte that occurred in a Devonian age swamp. Some such plants were 10–15 cm tall.

Fig. 3.37. Aerial branches of *Psilotum*. Note dichotomous branching and position of sporangia.

to the solution of a puzzle concerning the classification of two living plants that seemed to have no real niche in the then-existing classification schemes. One of these plants, *Psilotum*, grows in the tropics and subtropics the world around, while the other, *Tmesipteris*, is restricted to Australasia.

A COMPARISON OF *Psilotum* WITH THE FOSSIL PSILOPHYTES

*Psilotum* bears a striking resemblance to the fossil psilophytes. Its stem is dichotomously branched (Fig. 3.37) and lacks true leaves. It has a central core of xylem surrounded by phloem and a cutinized epidermis with stomates (Fig. 3.38). Its outer cortex is green and photosynthetic and by definition is a ventilated chlorenchyma. It has a subterranean branching rhizome but does not have true roots. The only significant point of difference is that the sporangium of *Psilotum* is three parted and attached laterally (Fig. 3.39). Even this feature can be interpreted as an extreme reduction of a short branch system bearing three terminal sporangia.

It can be said that the vascularized stem was a major breakthrough in the evolution of plants. Most of what has happened to the higher plants since then may be interpreted as an elaboration of the basic pattern set at least 400 million years ago.

(So far the fossil record has not shown us any evidence of the gametophyte generation of the ancient psilophytes. However, the gametophyte generation of *Psilotum* is well known. It consists of an axis much like a short segment of a sporophyte rhizome and grows buried in the soil. It is associated with a fungus that invades its outer tissue. The fungus obtains its basic organic foods from the humus in the soil and presumably makes them available to its host. It is possible that the *Psilotum* gametophyte pays for its keep by creating certain essential organic chemical substances for which the fungus lacks the necessary enzyme systems.)

During the millions of years that the

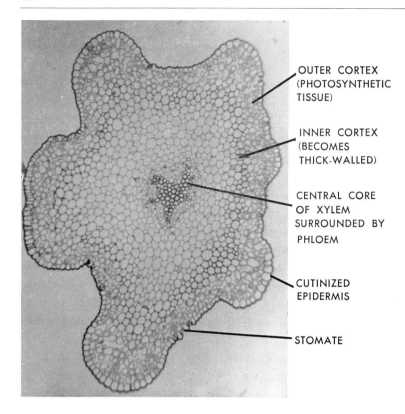

Fig. 3.38. Transverse section of a stem of *Psilotum*. The photosynthetic tissue in the outer cortex is a ventilated chlorenchyma.

primitive vascular plants existed many different types were developed. Some of them are known to us rather completely as fossils but others left no more than a fragmentary record.

From the ancient stock of primitive vascular plants many divergent lines emerged, prospered, and vanished. Of these the following major groups still occur:

1. remnants of the ancient psilophyte stock
2. club mosses and quillworts
3. horsetails
4. ferns
5. gymnosperms
6. flowering plants

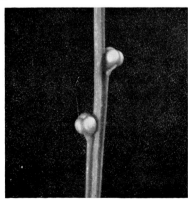

Fig 3.39 Portion of a stem of *Psilotum* showing details of the 3-part sporangium.

## CLUB MOSSES AND QUILLWORTS

*Selaginella* (Fig. 3.40) is one of the two main genera of living club mosses.

A more familiar genus in this group (to residents of the temperate zones, at least) is *Lycopodium* (Fig. 3.41) which includes plants known commonly as running pine, ground pine, and ground cedar. (*Selaginella* does grow in the temperate zones but the species that do so are inconspicuous and usually known only to those botanists particularly interested in this group of plants.) The tropical and subtropical species are larger and more conspicuous. Many of them are cultivated, particularly in the greenhouses of colleges and botanical gardens. Species of *Selagi-*

Fig. 3.40. Branches and vegetative leaves of a species of *Selaginella*.

*nella* provide some very interesting clues as to how the seed habit may have developed but their actual relationships to modern seed plants are extremely remote.

All club mosses have true stems and true roots but their leaves have a basically different nature than the leaves of ferns, which will be discussed later. The name club moss is appropriate since many of them look like overgrown mosses, with clublike structures at the tips of their stems. Each of the clublike structures is a *cone* or *strobilus* (Fig. 3.41). It consists of a group of sporophylls attached to a central axis. The sporophylls of most species are separated from the vegetative leaves in this manner. In *Selaginella* the strobilus consists of four rows of sporophylls. Externally all of the sporophylls look alike; they are very little different in appearance from the vegetative leaves.

The major features of the life cycles of the club mosses are discussed in Chapter 4. Here the significance of leaf structure in this group will be considered. Microscopic examination of a leaf of *Selaginella* shows that it has a cutinized epidermis with stomates (Figs. 3.42, 3.43), chlorenchyma tissue, and an internal atmosphere. In these ways it is like the leaf of a flowering plant.

A peculiar feature is the presence of a minute tongue-shaped flap, the *ligule*, at the base of each leaf of *Selaginella* (Fig. 3.44). Ligules are absent, however, from the leaves of *Lycopodium*. The significance of this difference and the function of the ligule when present are not clearly understood.

Another distinction between *Selaginella* and *Lycopodium* is dealt with in Chapter 4. (*Lycopodium* produces only one kind of spore and is said to be *homosporous*

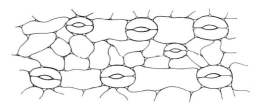

Fig. 3.42. Stomates in the epidermis of a leaf of *Selaginella*.

Fig. 3.41. The branching stem system, vegetative leaves, and terminal cones of a species of *Lycopodium*.

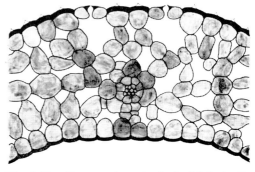

Fig. 3.43. Cross section of a leaf of *Selaginella*.

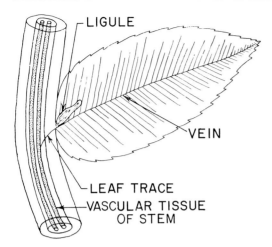

Fig. 3.44. Vegetative leaf of *Selaginella* drawn to show the nature of a microphyll. Note the single vein, the single leaf trace, and the absence of a leaf gap in the vascular tissues of the stem.

while *Selaginella* produces two kinds of spores and is therefore *heterosporous*).

The club moss leaf possesses only one vein and this is unbranched. If this single vein is traced to its connection with the vascular tissue of the stem (Fig. 3.44), it can be seen that there is no *leaf gap* or break in the stele at the point of union. These two characteristics mark this type of leaf as being fundamentally different from the fern-type leaf which is discussed later in this chapter.

The club moss leaf is said to be a *microphyll* and two theories exist as to its possible origin.

1. It could have resulted from the enlargement, flattening, and vascularization of spinelike outgrowths *(enations)* that occurred on the stems of many psilophytes. *Psilotum* (Fig. 3.39) still has such outgrowths.
2. It could have resulted from a flattening process affecting one member of a dichotomous branch pair. The other member of the pair would have continued as a stem segment.

Whether or not either theory is correct, the microphyllous nature of the club moss leaves distinguishes them as a group.

It should be noted that fossil records uncovered in recent years indicate that the ancestors of the club mosses coexisted with the psilophytes; thus the possibility exists that they may have had separate origins.

Many of the fossil club mosses such as *Lepidodendron* and *Sigillaria* were giant treelike forms existing in swamp forests which would be strange sights to our modern eyes (Fig. 3.45). In some of them the base of the tree was branched several times, dichotomously, to form a broad supporting base for the aerial stem. These bases are called *stigmarian appendages* and from them the much smaller roots emerged. Some species of *Selaginella* have leafless branches called *rhizophores* from which the roots emerge when the tip touches the ground.

The evolutionary origins of roots are still somewhat of a mystery. All roots, even those of flowering plants, have the primitive arrangement of vascular tissues; this suggests that they evolved as modifications of the mud-inhabiting portions of the psilophyte branch system. However, the positive geotropism of roots, their lack of chlorophyll, the presence of the characteristic root cap, and the mode of origin of lateral roots all point to their being special organs. The evolutionary development of phloem tissue to the point where it could transport large quantities of food rapidly was a likely prelude to the evolution of roots.

The *quillworts* are a small group of plants in the genus *Isoetes* that are essentially unknown to the general public even though they range from tropical to subarctic habitats. They grow in wet places for the most part; many of them are actually submerged.

Their leaves are considered to be microphylls but they bear little superficial resemblance to club moss leaves. They are somewhat widened near the base and are usually narrowed toward the tip with a quilllike appearance, hence the name. The leaves are attached to a short fleshy stem which somewhat resembles a corm (Fig. 3.46). Details of the life cycle are not discussed in this text but it should be noted

Fig. 3.45. Reconstruction of *Lepidodendron* (left) and *Sigillaria* (right) from fossil remains in a Coal Age swamp forest (CNHM).

that *Isoetes* is heterosporous and that its sporangia are among the largest found in living vascular plants.

## HORSETAILS

The genus *Equisetum* consists of a number of species characterized by jointed hollow stems bearing scale leaves at the nodes (Fig. 3.47). The leaves are nonfunctional and occur in definite whorls. Photosynthesis is carried on in the green stems. The stems are lined with vertical ridges which are harsh to the touch because they are covered with small tubercules of a siliceous substance. *Equisetum* is often called scouring rush because pioneers used these stems for scouring pots and pans.

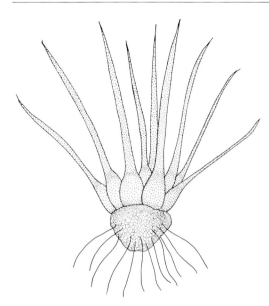

Fig. 3.46. Habit sketch of *Isoetes*, one of the quillworts.

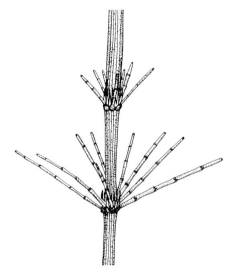

Fig. 3.47. Portion of a branching species of *Equisetum*.

Some species branch freely (Fig. 3.47) while others do not branch at all. Many of them have underground rhizomes that spread the plant very effectively by vegetative means. True roots are attached to the rhizomes.

*Equisetum* is the only living genus in this group but it had many relatives in the Coal Age swamp forests. Some of them, like the giant *Calamites* (Fig. 3.48), were treelike and had secondary growth. *Sphenophyllum* was smaller and vinelike with wedge-shaped functional leaves (Fig. 3.49).

FERNS

The living ferns of today compose a group of several thousand species growing in many diverse habitats. Some float in water and a few have become adapted to very dry habitats but most of them grow in shady places where abundant moisture is available in the soil.

Details of the basic fern life cycle are given in Chapter 4; it is important here to understand that ferns do not produce seeds. The main method of dispersal of species is by spores which are transported in air. Spores that land in suitably moist places germinate to produce small, independent, green gametophytes.

Leaves of the sporophyte generation of ferns are much like those of higher plants in that they have a branching system of veins, a cutinized epidermis with stomates, and an internal atmosphere associated with the photosynthetic tissue. In other words, fern leaves have a ventilated chlorenchyma.

In many fern leaves the veins are branched from one to several times in a distinctly dichotomous fashion. When this vein system is compared with the dichotomously branching stem system of a plant like *Psilotum* or one of the fossil psilophytes one is led to a basic hypothesis that the *fern leaf evolved through a flattening of a branch system and the development of tissue between the branches*. This type of leaf is called a *macrophyll* in contrast to the microphyll of a club moss. The dichotomy of veins in mature secondary leaves of many ferns is not always obvious but it is illustrated in classical fashion by the leaflets of the common maidenhair fern (Fig. 3.50).

Most fern leaves are called *fronds*. They may be very large and are often several times compounded. The stems of many of the modern ferns are *rhizomes* from which many short roots originate.

Fig. 3.48. Reconstruction of a Coal Age swamp forest. The large tree in the right foreground is *Calamites* (CNHM).

The rhizome also gives rise to new leaves near its tip (Fig. 3.51). In the bud stage each leaf is curled up like a fiddlehead. As the leaf matures it seems to unroll toward the tip (Fig. 3.52).

The vascular connections between the main vascular tissue of a fern rhizome and the base of the petiole are called *leaf traces*. Just above the point where the leaf traces join with the main vascular supply of the stem there is an apparent break in the vascular continuity of the stem. This is called a *leaf gap* (Fig. 3.53). Such gaps are common in the stems of plants that bear macrophylls but, as was noted in the discussion of the club mosses, they do not occur in stems bearing microphylls. The vascular tissue at the level of a leaf gap is not actually interrupted, however. Instead it is deflected slightly to both sides of a small mass of parenchyma tissue, much as the current of a river is deflected by an island (Figs. 3.53, 3.54).

From the highly stylized and simpli-

Fig. 3.49. Reconstruction of *Sphenophyllum* from fossil remains of a Coal Age swamp forest (CNHM).

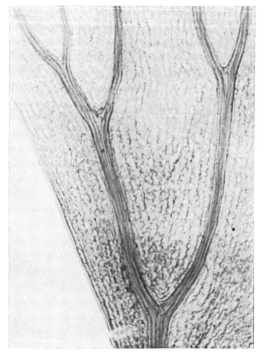

Fig. 3.50. Cleared leaflet of the maidenhair fern showing the dichotomous vein pattern.

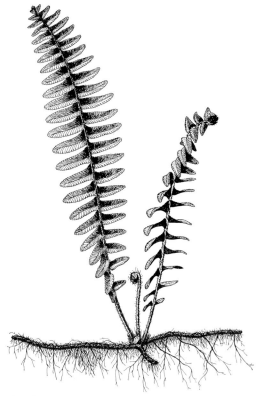

Fig. 3.51. Mature fern sporophyte with rhizome, roots, and large compound leaves.

fied sketch in Fig. 3.55 it may be seen that a stem bearing macrophylls might well have arisen from the fusion of leaf bases. This would account for certain observable facts such as (1) the continuity of leaf veins with vascular bundles in the stem and (2) the inevitability of leaf gaps in the stem, particularly when leaves are arranged in spiral patterns so that one leaf is not directly above the next in the series.

Many of the ferns of temperate regions tend to have stems that are more or less inconspicuous rhizomes. Their leaves, on the other hand, are the conspicuous parts. The strong main axis of the leaf is often mistaken for a stem.

Many of the ferns that lived in the Coal Age swamp forests were large trees; there are a few species of living ferns, mostly tropical, that have erect stems which may grow to several feet in height. Their trunks are somewhat barrel-shaped and

# Vascular Plants: Variations and Origins

Fig. 3.52. Fiddleheads (young fern leaves).

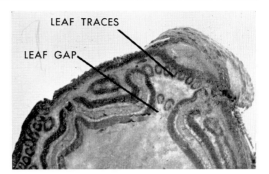

Fig. 3.54. Appearance of leaf traces and leaf gaps as seen in a transverse section through the nodal region of the stem of a tree fern.

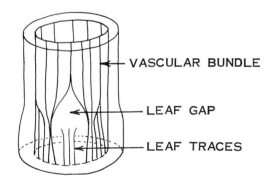

Fig. 3.53. Appearance of a leaf gap as seen in longitudinal perspective following removal of leaf base and axillary bud. Based on a dissection of the nodal region of a stem of nasturtium.

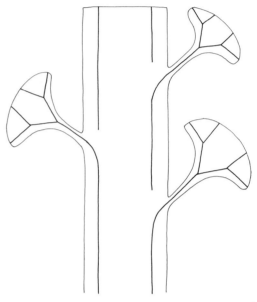

Fig. 3.55. Schematic longitudinal section of stem illustrating hypothetical concept that stems with leaf gaps arose from a fusion of leaf bases.

are clothed with the stubby remains of dead leaf bases and roots. Usually the fronds (leaves) of such ferns are very large and compound.

Not so many years ago the classification of plants that was used in most standard textbooks divided the plant kingdom into the following four divisions: the *Thallophytes* included all of the algae and the fungi; the *Bryophytes* included the mosses, liverworts, and hornworts; the *Pteridophytes* included the ferns and their allies, the club mosses and horsetails; and the *Spermatophytes* (seed plants) included the gymnosperms and the flowering plants. A comparison of this outdated classification system with those in current use is given in Chapter 11.

During this century scientists have realized that ferns are more closely related to the modern seed plants than they are to the club mosses or the horsetails. Moreover, the ability to produce seeds is no longer the primary criterion by which a major group of plants is separated from all others. As has been indicated elsewhere in this chapter this change in point of view was bolstered by the discovery and appreciation of fossilized primitive vascular plants.

### The Seed Plants

The plants that dominate the land surfaces of the earth today are the seed plants, which are separable into two broad groups: the *gymnosperms* and the *angiosperms*. The latter group includes all the flowering plants while the former includes the cone-bearing trees that make up the evergreen forests.

While the relationships between these two groups are not close, they have many features in common. Discussions of their basic structure have been undertaken in Chapter 5.

The fossil record clearly shows that an ancient group of fernlike plants, now extinct, actually bore seeds on leaves that somewhat resembled those of certain living tree ferns. These fossil plants have long been known as the *seed ferns* but modern reevaluations of their possible relationships indicate they should be considered as primitive gymnosperms.

# FOUR

ALTHOUGH a number of algal species are adapted to life on land, the term *land plants* generally refers to such plants as mosses, ferns, pine trees, and flowering plants. Primarily for purposes of organizing discussions, they are sometimes divided into the *higher green land plants* and the *lower green land plants*. The *bryophytes*, which include the *mosses, liverworts*, and *hornworts*, form a major segment of the lower group.

## A GENERALIZED MOSS LIFE CYCLE

The basic outline of plant life cycles was mentioned in Chapter 1 and the moss cycle has been chosen here as the first example to be discussed in depth. The reason for this choice is the relative ease with which the beginning student can understand the concept of *alternation of generations* from a study of moss.

The spores of mosses are the products of meiosis and therefore may be termed *meiospores*. Each spore has a cutinized wall and its protoplasm is partially dehydrated. These dormancy modifications permit a long survival in environments unsuited to active metabolism. At maturity moss spores are released into dry air and are distributed in air currents. Of minute size and light weight, they may be distributed for hundreds of miles from their points of origin. Experiments have shown that the spores of many species retain their viability for several years, if kept in dry places. Normally they germinate readily when transferred to moist places such as damp soil, crevices in the bark of trees, or continuously wet rock surfaces.

Upon germination the spore wall cracks open and the protoplasmic contents emerge as a cell containing numerous lens-shaped chloroplasts. This cell divides again and again in parallel planes to form a row or filament of cells which is distinctly algalike in appearance (Fig. 4.1). It soon begins to branch and does so until a green tangled web is formed on the surface of a suitably moist habitat. The whole web is referred to as a *protonema* and at first it consists of green branched filaments. Ultimately, however, some of the branches fail to develop chloroplasts and their walls become dark in color. These branches of the protonemal system are called *rhizoids*. An unexplained feature of rhizoids is that their cross walls are oblique rather than at right angles as are the cross walls of the green branches.

After a period of growth in this algalike phase certain cells of the protonema give rise to small masses of cells called *buds*. These may be dormant, in which case they look like small dark potatoes un-

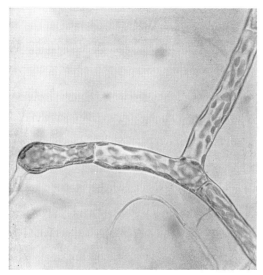

Fig. 4.1. Protonema of moss originating from germination of a moss spore.

# Reproduction in the Lower Green Land Plants

der the microscope, or they may be green and active (Fig. 4.2). At its apex each bud has a single meristematic cell which is shaped somewhat like an inverted pyramid with three planes of cell division. When a particular bud germinates it gives rise to one of the leafy moss plants (Fig. 4.3) which are familiar to everyone. In the initial phases of this growth a moss plant may be seen to have three rows of leaves, but this arrangement is soon distorted.

A given protonema may eventually cover several square inches of its habitat and may form several hundred or even several thousand buds. Since all the buds tend to germinate simultaneously it is common for moss plants to grow in dense clumps.

The leaves and stems of mosses are not true leaves and stems since they lack highly specialized conducting tissue. Their nature will be discussed subsequently.

When the leafy moss plant reaches maturity it is capable of producing *gametangia* (gamete cases). These are of two kinds: *antheridia* and *archegonia*. An antheridium is a cigar-shaped structure that

Fig. 4.3. Young leafy moss plant arising from a protonema.

consists of an outer jacket of sterile cells and an inner mass of cells, each one of which eventually gives rise to a *sperm* (Fig. 4.4A).

The archegonium is a flask-shaped structure having a long *neck* and a swollen basal portion called the *venter*. One of the cells within the venter, at first no different from any other cell, becomes larger and better nourished than its neighbors. This cell differentiates into the female gamete or *egg* (Fig. 4.4B).

The neck of the archegonium consists of an outer cylinder of cells and a single row of cells in the center of the cylinder. These *neck canal cells* are actually part of the same row of cells as the egg. When the egg is mature the canal cells disintegrate and the disorganized contents dissolve in water, leaving a free canal to the egg (Fig. 4.4C).

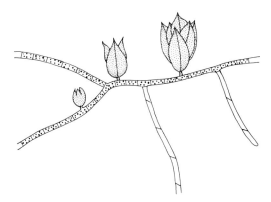

Fig. 4.2. Moss protonema with green branches, rhizoids, and buds.

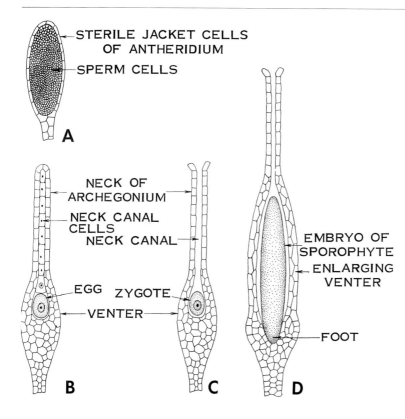

Fig. 4.4. Gametangia of a moss. A. Antheridium. B. Archegonium with neck canal cells present. C. Archegonium with neck canal cells dissolved and neck canal open. D. Archegonium with embryo.

The gametangia are formed at the tips of the main branches in some mosses (Fig. 4.5) and at the tips of short side branches in others. Usually there are series of protective leaves which overarch the branch tip and there may be a number of sterile hairs *(paraphyses)* between the gametangia.

Fig. 4.5. Leafy moss plant with terminal rosette of leaves partially dissected to show location of gametangia and paraphyses.

The total number of gametangia is often large (Fig. 4.6).

Archegonia and antheridia may occur together in the same branch tip, on different branches of the same plant, or on different individuals entirely. When they occur on the same branch they may develop at the same time or one type may mature later than the other. It follows from the above that it is a difficult matter to determine whether or not a moss species has gametophytes that are sexually different. Even if the gametangia appear on different leafy plants, both plants might have come from the same protonema. The point would be settled by growing the species from single spores under controlled conditions.

Fertilization depends on the presence of free surface water films on the plants. Dew and light rain create favorable conditions. When mature antheridia absorb water they swell and burst open at the tips. This discharges the sperms into the surface water films where they become ac-are many archegonia per stem tip only one sporophyte will develop per tip. Apparently the growth of the first successful embryo inhibits the potential growth of any other zygotes that might be formed (Fig. 4.7).

*The division of the zygote nucleus and all succeeding divisions of nuclei in the embryo are mitotic and the embryo is thus a diploid structure.*

The moss embryo is cigar-shaped and each of the two ends is a growing point (Fig. 4.4D). The basal portion is called the *foot;* it burrows down into the haploid tissue of the stem tip. The foot is an absorbing organ, taking water as well as inorganic and organic nutrients from the gametophyte for the nourishment of the embryo. At this stage the embryo sporophyte is a parasite, completely dependent for its nutrition on the gametophyte.

The middle portion between the tips of the embryo develops into the *stalk* or

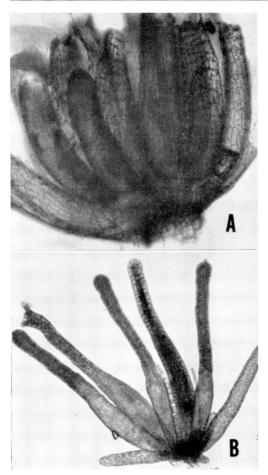

Fig. 4.6. Gametangia of mosses. A. Cluster of living antheridia dissected from a leafy moss plant. B. Cluster of living archegonia dissected from a leafy moss plant.

tively motile. Frequently sperms are transferred from plant to plant by splashing rain drops. Mature archegonia secrete a chemical substance that stimulates the sperms to swim in the right direction and sooner or later a sperm finds its way down the neck canal of an archegonium. There it fuses with the egg to form the *zygote.* Only one sperm fuses with each egg.

The archegonia usually do not all mature at once and this is, in a sense, a safety device that spreads the chances of fertilization over a period of time.

Once the zygote is formed it begins to grow immediately into the *embryo* of the sporophyte generation. Even though there

Fig. 4.7. Stem tip of a moss with leaves removed to show enlarged venter of an archegonium enclosing an embryo. (Compare with Fig. 4.4D.)

*seta* of the sporophyte while the upper tip portion eventually develops into the *capsule*.

As the embryo gets longer the tissue of the venter of the archegonium grows also. In this way a continuous protective layer is maintained over the embryo (Fig. 4.4D). The neck of the archegonium is not affected by the enlargement. Usually it turns brown and withers. Eventually the stalk portion of the embryo begins to grow so rapidly that the enlarged venter is torn away from its attachment to the stem tip. It is then carried upward as a membranous cap over the tip of the developing capsule of the sporophyte (Figs. 4.8, 4.9, 4.10). At this stage the remnant of the archegonium becomes known as the *calyptra*.

The young sporophyte of a moss usually becomes green and is capable of carrying on photosynthesis and manufac-

Fig. 4.8. Group of moss plants showing the relation of sporophytes to gametophytes. The capsules of the sporophytes are in various stages of development and each capsule is covered by a calyptra.

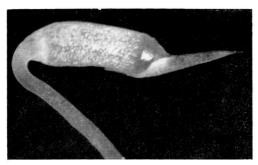

Fig. 4.9. Capsule of a species of *Mnium* with calyptra in place.

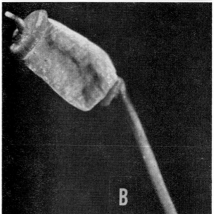

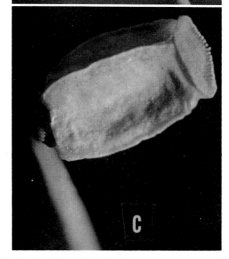

Fig. 4.10. Capsules of the hair-cap moss *(Polytrichum)*. A. With calyptra ("hair-cap") in place. B. With calyptra removed. C. With operculum removed to show peristome teeth.

turing its own organic nutrients. However, the sporophyte still is dependent on the gametophyte for most of its water and the inorganic nutrients that come from the soil.

The capsule that develops at the upper end of the enlarging sporophyte is a container in which spores are produced; it has essentially a cylindrical organization (Fig. 4.11). There is a central core of sterile tissue surrounded by a narrow cylinder of spore-producing tissue. Outside this is a wider cylinder of green cells that function in photosynthesis. The outermost cylinder is a cutinized epidermis that often contains functional stomates. Thus the green tissues of a moss capsule make up a ventilated chlorenchyma.

The open end of the capsule is covered by a circular lid called the *operculum* (Figs. 4.10B, 4.11). This structure falls off at maturity (or may be picked off with a dissecting needle) revealing a ring of toothlike structures *(peristome teeth)* around the opening (Fig. 4.12).

The spore-producing tissue consists at first of a large number of diploid *spore mother cells* (Figs. 4.11A, 4.13A). Each one of these undergoes meiosis to produce a *tetrad of haploid spores (meiospores)* (Fig. 4.13B). Thus the possible total number of spores is four times the number of spore mother cells. As the spores mature each one develops an outer layer of waxy material like cutin.

When the spores are mature there is a natural sequence of events by which the capsules are opened. The chief agency in this process is dry air and it is interesting to contrast the requirement of dry air for spore dispersal with the requirement of free surface water for fertilization.

In species of the common moss genus *Mnium*, for example, the calyptra shrivels and falls off, exposing the capsule which then begins to dry. As it dries it shrinks; this causes the operculum to be popped off. The teeth around the capsule straighten out as they dry (Fig. 4.12) and this exposes the spores. The teeth are very sensitive to changes in atmospheric humidity. They tend to close the opening during damp

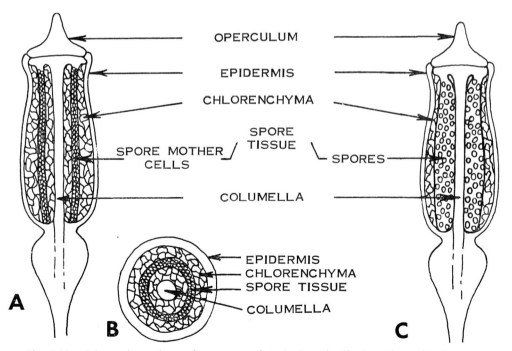

Fig. 4.11. Schematic sections of moss capsules. A. Longitudinal section with immature spore tissue. B. Median transverse section of immature capsule. C. Longitudinal section after maturation of spores.

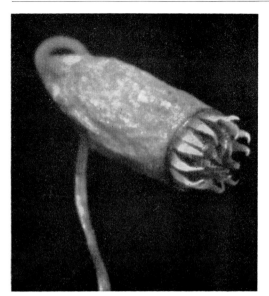

Fig. 4.12. Capsule of a species of *Mnium* with operculum removed to show the peristome teeth.

weather and their movements as they dry out may help to loosen the packed mass of spores within. Once the spores are out of the capsule they are picked up by air currents and may be carried long distances. The presence of the cutinized wall plus their naturally dehydrated protoplasm protects them from damage by desiccation during this voyage. When a spore lodges in a suitably moist place the cell contents swell and cause the spore wall to crack open. This is the beginning of the protonema of the next generation.

ADAPTATIONS OF MOSSES TO A
TERRESTRIAL EXISTENCE

Moss gametophytes have shown much diversification in their adaptations to life on land, particularly in the matter of being able to endure desiccation. It is a common observation that moss gametophytes so dry and shrivelled that they appear dead will become bright green and look fresh within minutes after being wetted. Moss leaves in many species are completely free of a cutinized surface. Thus they are easily wettable and water moves on their surfaces by capillary action. Also when moss leaves are

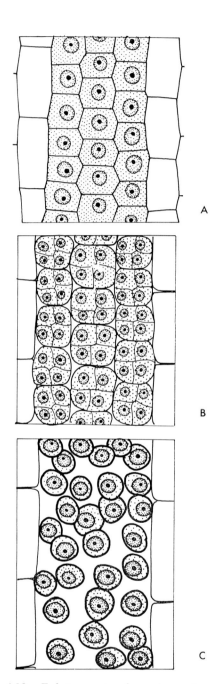

Fig. 4.13. Enlargements of portions of spore tissue of a moss capsule at three stages of development. A. Spore mother cells before meiosis. B. Tetrads of spores (meiospores) after meiosis. C. Separated and maturing spores.

wet they are totally exposed for absorption of carbon dioxide from the air.

Resistance to water loss is not entirely lacking in the mosses. Some species have erect sheets of green cells on the upper surface of the leaves. These are spread apart and exposed to the air when the plant is moist. But under dry conditions the rest of the leaf curls up over these photosynthetic layers in a protective fashion. Most moss leaves are only one cell in thickness except in the central portion. The thickened midrib is called a *costa* and is not a true vein since it lacks vascular tissue (Fig. 4.14).

The moss sporophyte is essentially nonvascular although in a number of species some of the cells in the stalk bear resemblances to certain types of conducting elements. The sporophyte has a single axis of growth with the capsule at the upper end and the foot at the lower. The foot is an absorbing organ imbedded in the gametophyte. The stalk is tough and wiry. Most immature moss capsules contain chlorenchyma and manufacture their own organic food. Thus the sporophyte is only partially dependent on the gametophyte.

Moss capsules have a cutinized epidermis and the basal portions of the capsules of many mosses have stomates surrounded by pairs of functional guard cells (Fig. 4.15). The cutinized epidermis enables the sporophyte to reduce the transpiration rate; it rarely exhibits evidences of wilting as does the gametophyte.

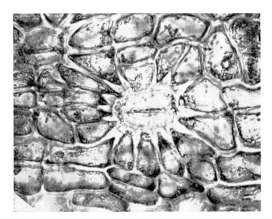

Fig. 4.15. Stomate from the basal region of a moss capsule.

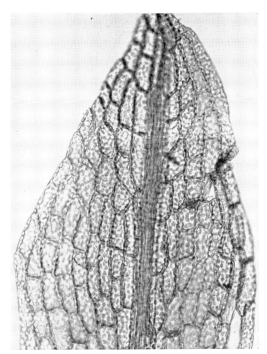

Fig. 4.14. Leaf of a moss as seen with the compound microscope.

As noted above, the main survival mechanism of the moss gametophyte is an ability to remain alive in the desiccated condition and to recover quickly when water becomes available. The major water movements are capillary in nature and occur in the surface films that cover the plant body. There is thus a striking contrast between the two generations of mosses with respect to the mechanisms by which they exist in the aerial environment.

The gametophyte of the peat mosses (*Sphagnum*) is one of the few moss types that has economic significance. In either the living or the dead condition these plants have a tremendous capacity for absorbing and holding water and this makes them valuable as mulches in garden soil and in the packing of live plants for shipment.

*Sphagnum* grows in many parts of the world but achieves its greatest development in the cooler regions, particularly in the areas where retreating glaciers have created numerous small lakes and ponds with no drainage outlets. In such conditions the waters become acid and this favors the growth of *Sphagnum*.

The stems of this plant are densely covered with overlapping leaves. Some of the branches hang downward and twine about the central axis thus serving as effective wicks in moving water upwards. The leaves are only one cell in thickness and have two kinds of cells: narrow, green, photosynthetic cells and large, dead, water storage cells which are spaced between the green cells (Fig. 4.16). The space occupied by the water storage cells is much greater than that of the green cells, so the color of the leaf is pale green.

Under the conditions of growth in highly acid waters the older parts of the plants do not decay after death. They sink to the bottom and gradually accumulate to form deposits of *peat* that in time fill up the lake completely.

The sporophyte of *Sphagnum* consists of a foot and a capsule connected by a short stalk that never elongates. When the spores are mature the tip of the gametophyte branch in which the foot of the sporophyte is imbedded begins to grow upward and carries the whole sporophyte up into the air (Fig. 4.17).

The capsule is dark in color and heavily cutinized. It is possible to recognize pairs of cells arranged like guard cells in the capsule wall but stomates do not develop between them. The heat of the sun causes gases within the sealed capsule to expand and eventually it explodes and discharges the spores into the air violently.

THE PLANT BODY OF *Marchantia*

One of the common representatives of the *liverworts* is *Marchantia*, a plant with many remarkable features including the development in the gametophyte generation of a cutinized epidermis with perforations like stomates. The gametophyte of *Marchantia* (Fig. 4.18) is a flat ribbon-like thallus that is held close to the soil by numerous rhizoids and branches periodically in a Y-shaped fashion. It is of fairly common occurrence on soil along stream banks, in moist woods, and on shaded cliffs. Sometimes it becomes extremely abundant in burned over forest areas provided moisture is plentiful.

Another common liverwort, somewhat

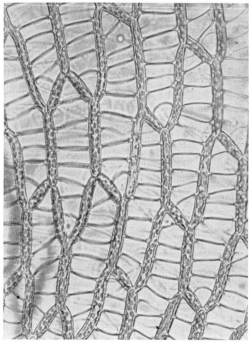

Fig. 4.16. *Sphagnum* leaf showing arrangement of chlorenchyma cells and water storage cells.

Fig. 4.17. *Sphagnum* gametophytes with attached sporophytes. The capsules shown are actually somewhat smaller than BB shot.

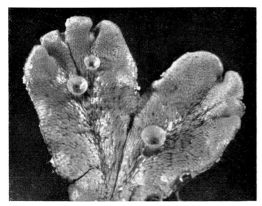

Fig. 4.18. *Marchantia* thallus with gemmae cups. (Compare with Fig. 4.21.)

closely related to *Marchantia*, is *Ricciocarpus* (Fig. 4.19), which may be found floating on the water in ponds or anchored firmly by rhizoids in the mud of stream banks.

In *Marchantia* the upper epidermis appears to be divided into rhomboidal areas that are actually the outlines of air chambers below. Each air chamber is separated from the next one by a wall of cells and the walls of all of the chambers support the epidermis as walls of the rooms in a house support the roof (Fig. 4.20).

The epidermis is cutinized and in the center of the portion of the epidermis over each air chamber there is a round air pore surrounded by a "chimney" consisting of four tiers of four cells each. This opening functions as a stomate and it is possible that the lowermost tier of cells in the chimney may be able to regulate the size of the opening. This arrangement of cells around an opening, when compared to the stomates and guard cells of the sporophytes of many plants, is certainly an outstanding example of parallel evolution.

From the floor of each air chamber in the thallus there arise numerous algalike filaments of deeply green cells that comprise the main photosynthetic tissue of *Marchantia*.

*Marchantia* has an especially effective method of asexual reproduction. Numerous, wafer-shaped, vegetative buds called *gemmae* are formed in open cups on the surface of the thallus (Fig. 4.21). When detached the gemmae (Fig. 4.22) quickly give rise to new plants.

When sexual reproduction occurs in *Marchantia* it becomes evident that there are separate male and female plants (Fig. 4.23). Prior to this time, however, this distinction cannot be made since both plants look alike in the vegetative condition.

The male plants bear antheridia (Figs. 4.24, 4.25) imbedded in disclike segments of the thallus that are raised into the air on slender stalks (Fig. 4.23). When mature the sperms are discharged into films of water on the surfaces of the discs that act as splash platforms; i.e., the splattering of rain drops when they fall on the discs causes droplets containing sperms to be spread to nearby plants, some of which are female.

The archegonia (Fig. 4.26) are to be found on the undersurfaces of similar discs that have several fingerlike projections extending outward (Fig. 4.23). When sperms are transferred to the female plants they swim toward the archegonia in surface water films. Commonly fertilization occurs in several archegonia and embryos begin to form in each of them.

The mature sporophytes that develop from the embryos resemble moss sporophytes in that each one consists of a *foot*, a *stalk*, and a *capsule*. However, they are smaller and less complex structurally. Also, they are more completely dependent on the gametophyte than are the sporophytes of mosses.

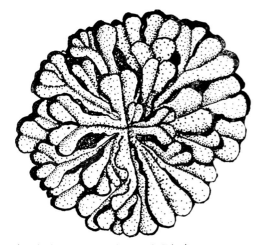

Fig. 4.19. Gametophyte of *Ricciocarpus*.

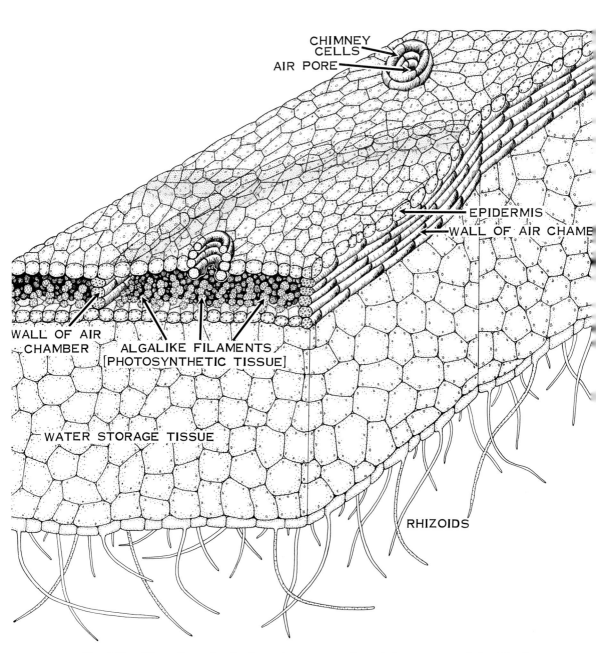

Fig. 4.20. Three-dimensional schematic representation of the structure of the *Marchantia* thallus.

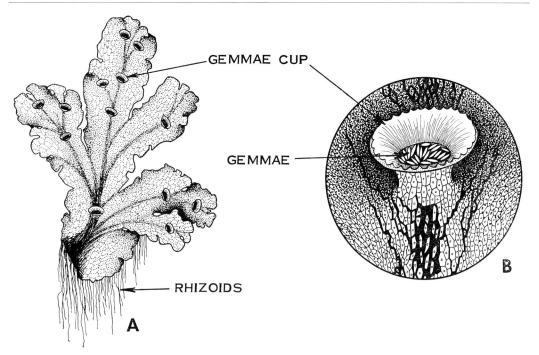

Fig. 4.21. A. Thallus of *Marchantia* with gemmae cups. B. Enlargement of a gemmae cup.

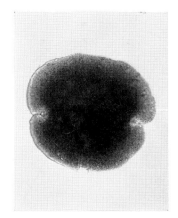

Fig. 4.22. Photomicrograph of a single gemma of *Marchantia*.

The wall of the capsule is only one layer thick and has no stomates. The whole sporophyte, including the capsule, is retained within the enlarged archegonium until the spores are fully mature. At this time the stalk elongates rapidly, pushing the capsule downward until it emerges from the protective sheaths that had surrounded it (Fig. 4.27).

Dry air then causes a loss of water and the resultant shrinkage tensions tear open the capsule wall, allowing the spores to be released.

Mixed with the spores are numerous special structures called *elaters* (Figs. 4.28, 4.29). These are long, pointed, dead cells with spirally thickened walls. As they dry they change shape constantly and their movements help loosen the spore mass so that individual spores can be carried away in air currents.

Experiments with large numbers of plants grown from spores have shown that half of the spores give rise to female plants and the other half to male plants. On the other hand, when plants of a given sex are propagated vegetatively (as by gemmae) they always give rise to plants of the same sex. From such experiments it has been concluded that sexuality in *Marchantia* is controlled by genetic factors and that these are separated during meiosis.

This contention was strengthened by experiments with another liverwort *(Sphaerocarpos)* in which tetrads of spores

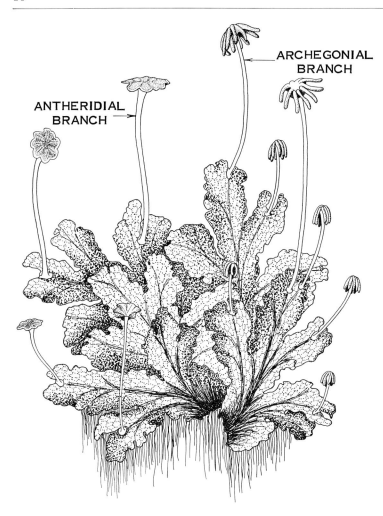

Figs. 4.23. A group of male and female plants of *Marchantia* with antheridial branches and archegonial branches.

were isolated before they broke apart. When the individual spores of each tetrad were then separated and allowed to grow into mature plants it was found that two of these plants were female and the other two male. The research that led to this discovery is of exceptional interest to science since it further demonstrated that the presence or absence of a particular chromosome was related to the actual expression of sexuality.

LEAFY LIVERWORTS

The gametophytes of the leafy liverworts are small and delicate plants much like some mosses. Their leaves are one cell thick and lack the midribs (costae) that are characteristic of many moss leaves. Most of the common leafy liverworts are prostrate and appear to have only two rows of leaves (Fig. 4.30). A third row of leaves may be reduced in size or missing entirely but there is little doubt that the primitive members of this group had three rows of leaves and grew erect. In some species the leaves are finely dissected, appearing like branched filaments of algae. Frequently the leaves overlap; this increases the efficiency of capillary water movements. Members of this group are like mosses in that they are able to endure desiccation. They are not, apparently, as aggressive in temperate regions as mosses and do not occur in as many diverse habitats. Certainly they

# Reproduction, Lower Green Land Plants

Fig. 4.24. Photographs of antheridial branch (left) and archegonial branch (right) of *Marchantia*.

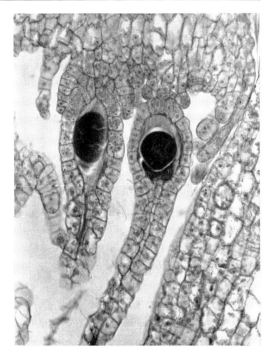

Fig. 4.26. Partial section of an archegonial branch of *Marchantia* with archegonia in position.

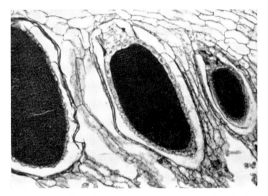

Fig. 4.25. Partial section of an antheridial disc of *Marchantia* with imbedded antheridia.

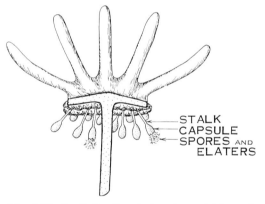

Fig. 4.27. Partially dissected archegonial branch of *Marchantia* showing the attached mature sporophytes.

are unknown to the vast majority of people while mosses are familiar to almost everyone.

The gametangia of leafy liverworts are borne at the tops of the main stem or on short side branches. They differ from the gametangia of mosses in minor respects. The sporophytes are like those of the mosses in that they have a single axis of growth with a foot, a stalk, and a capsule in which the spores are formed. They are, however, much simpler structures and are more completely dependent on the gametophyte generation than are moss sporophytes. The stalk does not elongate until the spores are mature. When it does grow it elongates rapidly. This thrusts the capsule up into the air and the drying action of the air causes it to rupture immediately. Prior to this stage the capsule is immersed in protective leaves and membranes at the tip of

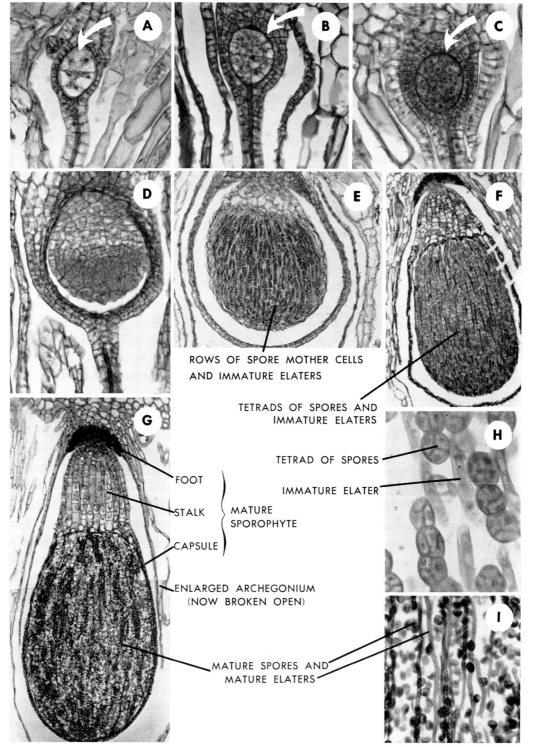

Fig. 4.28. Series of photomicrographs showing development of the *Marchantia* sporophyte.
A,B,C. Arrows point to early stages in the development of the embryo within the enlarging venter of the archegonium.
D. The spherical embryo shows beginnings of differentiation into foot, stalk, and capsule.
E. The cells in linear series within the capsule are spore mother cells. The elongated cells will become elaters.
F. The spore mother cells have undergone meiosis to form tetrads of spores. (See also the enlargement in H.)
G. Mature sporophyte just at the time the stalk begins the rapid elongation that pushes the capsule out of its protective coverings.
H. Tetrads of spores and immature elaters enlarged from F.
I. Mature spores and elaters enlarged from G.

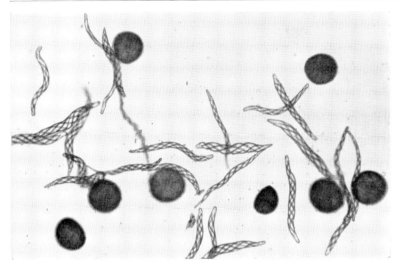

Fig. 4.29. Spores and elaters of *Conocephalum* (a relative of *Marchantia*).

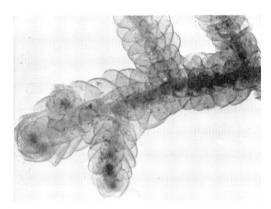

Fig. 4.30. Photomicrograph of a species of the leafy liverwort *Porella*.

the gametophyte. The capsule does not open by a lid as does the moss capsule. Instead it splits open lengthwise into four valves (Fig. 4.31).

## Hornworts

The most common plant of this group is *Anthoceros;* it holds a special place in the thinking of many authorities on plant evolution because of its unique sporophyte.

The gametophyte of *Anthoceros* is a small, thin, dark green thallus that lacks the internal differentiation found in *Marchantia*. The gametangia are borne on the

Fig. 4.31. Leafy liverwort with attached mature sporophytes.

upper surface and are imbedded in the thallus.

The sporophyte becomes a slender, erect, green spire, 3 cm or more in height (Fig. 4.32). The epidermis is cutinized and has stomates surrounded by pairs of functional guard cells. The central core of tissue, the *columella*, becomes sclerenchyma-like. This is surrounded by a cylinder of spore-bearing tissue (Fig. 4.33). Outside this is the green photosynthetic tissue. At the base of the sporophyte is a foot imbedded in the gametophyte tissue (Fig. 4.34). Between the foot and the capsule is a zone of meristematic cells that adds new tissues to the base of the capsule. The youngest spore mother cells are at the base of the capsule adjacent to the meristematic region. Above them the spore

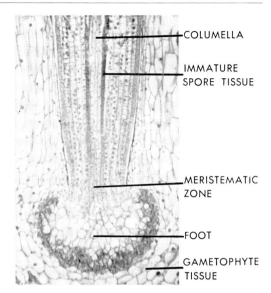

Fig. 4.34. Longitudinal section of the basal portion of an *Anthoceros* sporophyte showing the foot and the intercalary meristem.

Fig. 4.32. Gametophytes of *Anthoceros* with attached sporophytes.

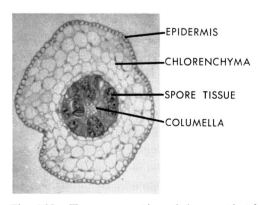

Fig. 4.33. Transverse section of the capsule of *Anthoceros*.

mother cells are in the early stages of meiosis. By following the sporogenous tissue upward in the capsule one can often observe all stages in meiosis from spore mother cells at the base to mature spores at the tip (Fig. 4.35). The tip of the capsule splits into two valves and permits the mature spores to be blown away.

Evolutionists often take appraising looks at the sporophyte of *Anthoceros* because it is so close to being an independent plant like a simple psilophyte. If the columella were to evolve a little further

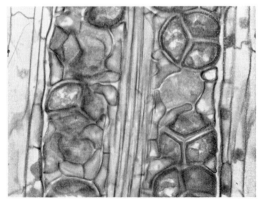

Fig. 4.35. Enlarged portion of the *Anthoceros* sporophyte showing tetrads of spores.

toward being xylem and if the foot could be stimulated to grow rhizoids into the soil instead of parasitizing the gametophyte, then the resulting plant would indeed be a psilophyte. However, this should be recognized as being only an interesting speculation since there are several serious objections to using *Anthoceros* as an ancestor to the vascular plants.

## FERNS AND THE SPORE-BEARING LEAF

Ferns are plants that have true roots, true leaves, and true stems. They do not produce seeds nor do they have flowers. However, their spores are produced on leaves.

The general appearance of the fern sporophyte and the vegetative nature of the fern leaf are discussed in Chapter 3 (see Fig. 3.51). The details of structures and reproductive mechanisms which follow apply mainly to a group called modern ferns that includes many of the familiar ferns found in temperate regions.

### The Fern Gametophyte

As it was with the mosses, the spore is a convenient point of entry into the life cycle of ferns. The fern spore is able to exist in dry air for considerable periods of time. When it is transported to a suitably moist habitat the spore germinates to form a short algalike filament (Fig. 4.36). One or more cells of this protonemal stage may give rise to tubular rhizoids that are not separated from the parent cell by cross walls. The protonemal stage is of short duration; the terminal cell soon begins to divide by oblique cell walls to form a roughly triangular apical cell. By dividing alternately to the right and to the left the

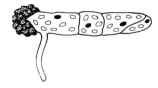

Fig. 4.36 Germination of a fern spore resulting in an abbreviated, algalike, protonemal stage of the gametophyte.

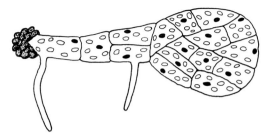

Fig. 4.37. Young fern prothallus (gametophyte) shortly after the apical cell has become active.

apical cell causes a flat membrane to be formed (Figs. 4.37, 4.38). The daughter cells formed by this activity continue to divide for some time. The enlargement of all of the cells formed by these divisions causes the membrane to take on the shape of a heart with the apical cell in the notch. This gametophyte of a fern is often called a *prothallus*. As it grows larger it may become somewhat thickened in the middle due to an occasional division of the apical cell in a third plane. Many rhizoids devel-

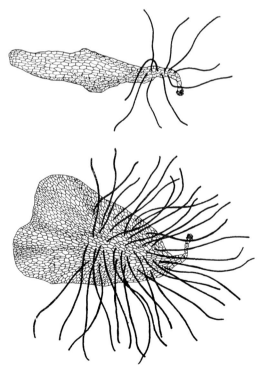

Fig. 4.38. Immature fern prothalli as seen from the ventral surface.

op from the lower surface of the central midrib region. Their primary function is the anchorage of the prothallus to the ground so that it is in contact with capillary soil water. Although it is difficult to prove, the rhizoids may have additional significance in the absorption of mineral nutrients and water from the soil.

The prothallus is green and is completely independent in its nutrition. It is a small structure, rarely reaching the size of a fingernail. All of the nuclei are haploid and all of the divisions are mitotic.

Most ferns produce both antheridia and archegonia on the same gametophyte (Fig. 4.39). The antheridia begin to be

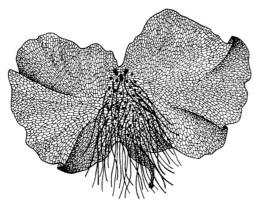

Fig. 4.39. Mature fern prothallus with antheridia, archegonia, and rhizoids (ventral surface).

formed earlier than the archegonia. They are spherical bodies attached to the lower surface of the prothallus (Fig. 4.40). When the sperms are mature the contents of the antheridium swell and force the outermost wall cell to open like a lid, allowing the sperms to escape. Once in the external film of water the multiciliate sperms begin to swim rapidly.

Fig. 4.40. Sectional view of a fern prothallus showing structure of an antheridium with maturing sperms.

The archegonia are not formed until the central portion of the gametophyte becomes more than one layer thick and the venter of the archegonium is sunken in the gametophyte tissue (Fig. 4.41). Only the neck protrudes and it has a decided curvature. The neck canal cells disintegrate and dissolve out of the neck canal when the egg is mature. Sperms are attracted to the archegonia and eventually one enters the canal and reaches the egg (Fig. 4.41).

THE FERN EMBRYO

The zygote that results from the fusion of the egg and sperm begins to divide shortly by mitosis and gives rise to a more or less spherical mass of diploid cells (Fig. 4.42). This is the embryo of the sporophyte generation. Like the moss embryo it is nurtured and protected by the surrounding gametophyte tissue. However, it differs from the moss embryo in that it is spherical rather than cigar-shaped. Furthermore, it develops four regions of meristematic activity rather than two as in mosses (Fig. 4.42C).

One of these regions is a foot that serves to absorb nutrients and water from the gametophyte tissue. Of the rest, one region gives rise to a *primary root* and another to a *primary leaf* (Fig. 4.43). These two structures develop rapidly. Each one has a strand of specialized conductive tissue that is continuous from one to the

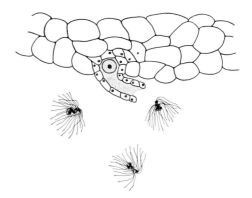

Fig. 4.41. Sectional view of a fern prothallus showing the structure of a mature archegonium. Note the sunken venter with a large egg, the curved neck with the open neck canal, and the motile sperms swarming about the archegonium.

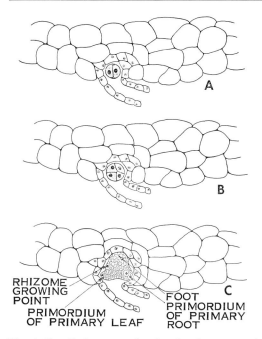

Fig. 4.42. Early stages in the development of the embryo of a fern sporophyte within the archegonium.
A. First division of the zygote.
B. Second division of the zygote.
C. After many divisions the embryo becomes 4-lobed. The lobes give rise to a primary leaf, a primary root, a foot, and a rhizome growing point.

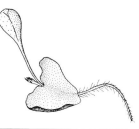

Fig. 4.44. Fern prothallus with attached young sporophyte. The primary root and primary leaf are well developed and the rhizome has begun to grow.

other. The primary root is a true root and functions as such. The primary leaf pushes up into the air, usually through the apical notch of the gametophyte. It becomes green and is a photosynthetic structure. The primary root and the primary leaf are soon able to take care of all of the nutritional needs of the young sporophyte and it becomes completely independent of the gametophyte even though the two generations remain attached for some time (Figs. 4.43, 4.44).

The fourth lobe of the embryo develops slowly to this time. Then it becomes active as a stem-tip growing point and gives rise to the type of prostrate stem known as a *rhizome*. As the rhizome grows it gives rise to more leaves and more roots (Fig. 4.45). These secondary leaves are much bigger than the primary leaf and have a different shape. By the end of the first growing season the primary leaf and the primary root are dead. The only surviving structures are those formed by the rhizome growing point that developed from the fourth lobe of the embryo. All of the succeeding roots and leaves of a mature fern plant come from this structure.

It is not known whether the fern gametophyte has been reduced from a more complex structure or has merely failed to develop beyond the present level in which its algal ancestry is apparent. Note, however, that the sporophyte generation can begin only where the gametophyte generation is able to grow. Presumably fern spores are scattered into all sorts of environments, but they develop into mature gametophytes only where supplies of water near the surface of the soil are reasonably constant.

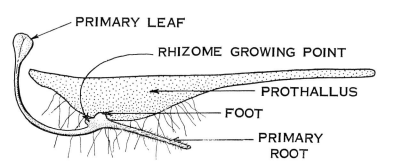

Fig. 4.43. Section of a fern prothallus and attached young sporophyte at a comparable stage to the illustration in Fig. 4.44.

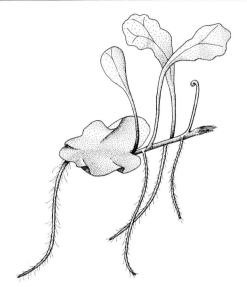

Fig. 4.45. Somewhat older fern sporophyte, still attached to the prothallus but no longer dependent on it. The rhizome has begun to grow rapidly giving rise to secondary leaves and adventitious roots.

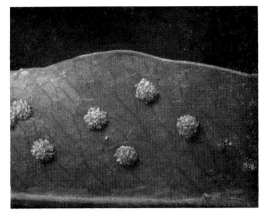

Fig. 4.47. Undersurface of a fern leaf showing sori without indusia (naked sori).

## THE FERN SPOROPHYLL

When the leaves of many ferns are mature they develop a large number of rusty-looking spots on their undersurfaces (Figs. 4.46, 4.47, 4.48, 4.49). Each of these spots is called a *sorus* (pl. *sori*) and each sorus consists of a cluster of *sporangia* (spore cases).

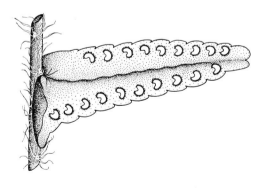

Fig. 4.46. Undersurface of a fern leaf with sori.

Fig. 4.48. Undersurface of a fern leaf showing sori with indusia.

The sorus may be naked (Fig. 4.47) or covered by a protective membrane called an *indusium* (Fig. 4.48). Figure 4.49 illustrates a cross section of a fern leaf with a sorus having an indusium.

The appropriate term to be applied to the spore-bearing leaf is *sporophyll* and it should be noted that the development of sporophylls was one of the major events in

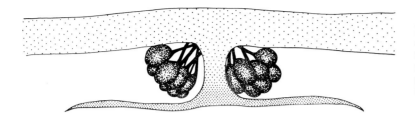

Fig. 4.49. Schematic cross section of a fern sporophyll showing the sorus as a cluster of sporangia, here covered by an indusium.

plant evolution. (Actually it was not one event since sporophylls undoubtedly evolved in several independent lines.)

THE FERN SPORANGIUM

Each sporangium is somewhat flattened and has a slender stalk by which it is attached to the leaf. The sporangium walls are 1 cell layer in thickness and are transparent except for 1 row of conspicuous cells extending partially around the edge. This row of cells, which looks something like a zipper, is the *annulus*. Careful examination reveals that all of its walls, except the walls facing outward, are very thick and rigid (Fig. 4.50).

In many of the modern ferns each sporangium contains 16 spore mother cells. Each of them is diploid as are all of the rest of the cells of the sporophyte. Each spore mother cell undergoes meiosis to form 4 spores and thus the potential total number of spores in a sporangium is 64. Each of the spores is, of course, haploid. As the spores mature they separate and develop heavily cutinized walls.

Dry air plays an important part in the release of the mature spores. First the indusium shrivels up and exposes the sporangia. Then the cells of the annulus begin to lose water rapidly through evaporation from their thin outer walls. Their radial walls are pulled toward each other, shortening the outer circumference of the annulus. This induces tensions in the lateral walls of the sporangium that eventually cause it to rip open and fold back on itself (Fig. 4.51). The process is very interesting to watch under the microscope. With a little imagination one can see resemblances between an opening sporangium and a baseball player getting ready to throw a ball. When the tension in the annulus cells reaches a peak it is released suddenly and the whole sporangium snaps shut violently. This hurls the spores out into the air where they are distributed by air currents. Each spore is capable of producing a new gametophyte.

The opening of spore cases during dry weather is a consistent feature of land plants. It is obvious that spores will travel

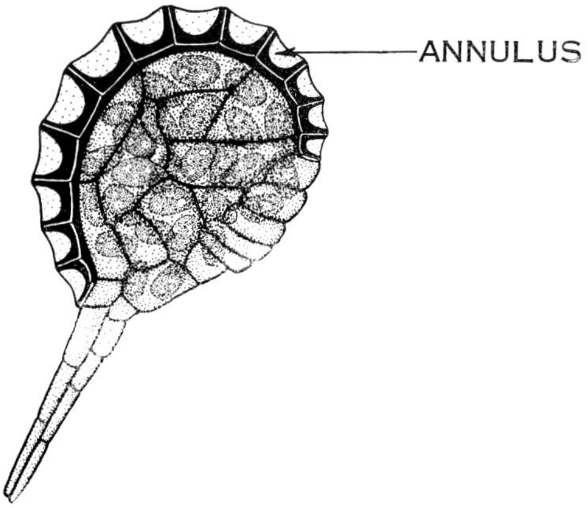

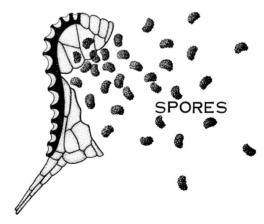

Fig. 4.50. Mature fern sporangium with enclosed spores.

Fig. 4.51. Fern sporangium discharging spores.

further when it is dry than when it is wet and they will not be induced to germinate prematurely in an environment that is only temporarily moist.

### THE HETEROSPOROUS WATER FERNS

In the ferns that have been discussed all the spores produced by a given species have the same appearance and give rise to gametophytes which are basically similar. These ferns are said to be *homosporous*.

In a few ferns and in most members of the other groups of plants to be considered the spores produced are of two kinds, *microspores* and *megaspores*. The sporangia which produce them may have different appearances as may also the sporophylls themselves. Such plants are said to be *heterosporous*. The water ferns *Salvinia*, *Azolla*, and *Marsilea* (Fig. 4.52) have highly advanced heterospory.

### THE HORSETAILS

As noted in Chapter 3, all the living horsetails belong in a single genus, *Equisetum*. They have jointed, hollow, green stems with scalelike nonfunctional leaves at the nodes. Some species are unbranched while others branch freely.

The life cycle is similar to that of the ferns. The conspicuous generation is the sporophyte; this alternates with a minute, green, independent gametophyte (Fig. 4.53) that is lobed or even shrubby in appearance when viewed under the microscope. The gametangia are partially imbedded in the gametophyte. It is known that many species produce antheridia and archegonia on the same gametophyte while others seem to have separate male and female gametophytes.

Spore production occurs in terminal strobili (Figs. 4.54, 4.55); each consists of a central axis to which are attached numerous *sporangiophores,* each bearing several sporangia. The sporangiophore is not called a sporophyll because it is interpreted as a reduced branch system that never went through the leaf stage. The sporangia are terminal on portions of this branch system that have become fused into the form of

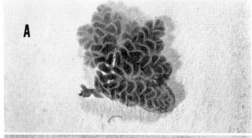

Fig. 4.52. Water ferns. *Azolla* (A) and *Salvinia* (B) have a floating habit. The leaves of *Marsilea* (C) are aerial but are attached to a rhizome which is mud inhabiting.

hexagonal plates with the sporangia pointing inward (Fig. 4.56).

The spores have two ribbonlike hygroscopic appendages called *elaters* (Fig. 4.57). The elaters straighten as they dry, loosening the spore mass and helping the spores to float more readily in the air while being dispersed. The spores are not especially thick-walled and apparently are not long-lived as are the spores of mosses and ferns.

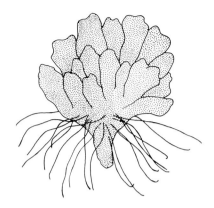

Fig. 4.53. Gametophyte of *Equisetum* as seen in the compound microscope.

## SELAGINELLA—AN INTRODUCTION TO THE CONCEPT OF HETEROSPORY

In Chapter 3, *Selaginella* was noted as being one of the two main living genera of club mosses. The life cycle of this plant is used here to provide an introduction to the concept of *heterospory*, one of the most significant evolutionary changes to occur among land plants.

In this plant the sporophylls are arranged in four lengthwise rows along a central axis. Each sporophyll closely overlaps the one above it, and the whole structure takes on the appearance of a club. It is called a *cone* or *strobilus* (Fig. 4.58). (It should be noted that the sporophylls are much like the vegetative leaves, although the latter are not compactly arranged.)

At the base of each sporophyll there is a single more or less egg-shaped sporangium (see Fig. 4.61). Microscopic examination shows that some of the sporangia contain a large number of small reddish-colored spores. These are called *microspores;* the sporangia containing them are called *microsporangia* (Fig. 4.59). The other

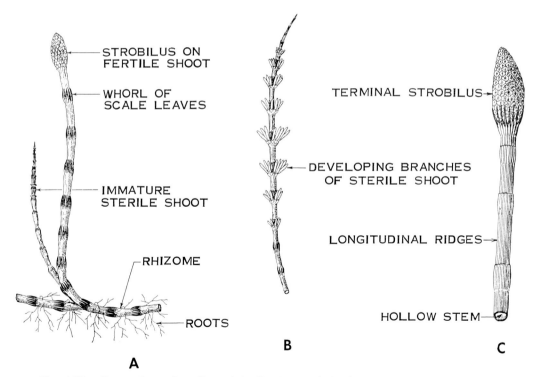

Fig. 4.54. Comparison of sterile and fertile shoots of *Equisetum*. A. Plant of *E. arvense* showing rhizome with roots, mature fertile shoot, and immature sterile shoot. B. Sterile vegetative shoot of the same species at a later stage of development, showing the whorls of branches arising from the nodes. C. Terminal portion of the stem of *E. hyemale*, a tall unbranched species, with mature strobilus.

Fig. 4.55. *Equisetum arvense*. The photographs A, B, and C, show changes in the mature strobilus as the sporangiophores separate allowing spore dispersal, while D is the terminal portion of a sterile shoot for comparison with Fig. 4.54B.

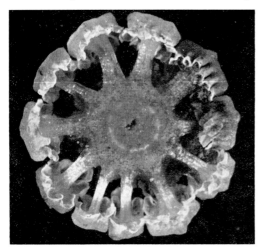

Fig. 4.56. A single whorl of sporangiophores dissected from a strobilus of *Equisetum*. The several sporangia on each sporangiophore are evident.

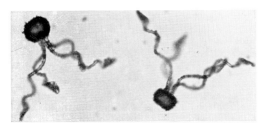

Fig. 4.57. Spores of *Equisetum* with extended elaters.

Fig. 4.58. Branches of *Selaginella* with terminal cones.

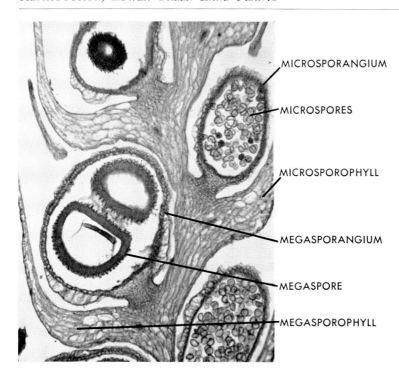

Fig. 4.59. Longitudinal section of a cone of *Selaginella* showing microspores in a microsporangium and megaspores in a megasporangium.

sporangia contain smaller numbers of light-colored spores which are as much as 1,000 times larger in volume than the microspores. They are called *megaspores* and the megaspores in each *megasporangium* look like golf balls in a plastic bag when viewed with appropriate magnification (Figs. 4.59, 4.61).

The condition described above in which two kinds of spores are produced by a given species is called *heterospory*. As will be seen later, the development of heterospory was an important step in the evolution of seed plants. Apparently this condition arose independently in several evolutionary series.

Why do some sporangia contain microspores while others on the same plant contain megaspores? *Selaginella* provides us with some information in answer to this question.

When the sporangia of *Selaginella* are very young they are all alike; each one contains a large number of spore mother cells (Fig. 4.60). In some of the sporangia all or most of the spore mother cells undergo meiosis to produce spores. This results in the large number of microspores.

In other sporangia most of the spore mother cells disintegrate before meiosis. In many species there is only one surviving spore mother cell and it undergoes meiosis to form four spores. These have the same amount of space in which to develop and the same food supply as do the larger numbers of microspores in neighboring sporangia; they grow accordingly. Without understanding the basic reason for the disintegration of all but one of the spore mother cells, we can accredit the differences between microspores and megaspores to *differences in their nutrition*.

The sporophylls that bear microsporangia are called *microsporophylls* while those that bear megasporangia are called *megasporophylls* (Fig. 4.61). There are several hundred species of *Selaginella* and among them there is considerable variation in the distribution of the two kinds of sporophylls in the strobilus. The microsporophylls may be segregated from the megasporophylls in one of several patterns or the two types may intermingle indiscriminately.

When the spores are mature both kinds of sporangia split open violently due

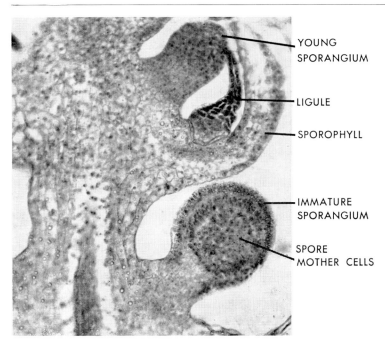

Fig. 4.60. Longitudinal section of a cone of *Selaginella* showing the young sporangia before they become differentiated into either megasporangia or microsporangia.

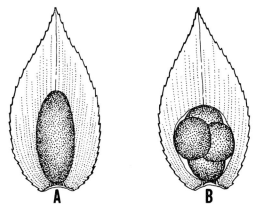

Fig. 4.61. Individual sporophylls from cones of *Selaginella*. A. Microsporophyll with attached microsporangium. B. Megasporophyll with attached megasporangium in which the outlines of the 4 large megaspores are evident.

Each microspore develops into a gametophyte that produces sperms only. This male gametophyte is very much reduced in size and is formed completely inside the microspore wall (Fig. 4.62). It consists of a single vegetative cell and one antheridium. When the sperms are mature the microspore wall breaks open and the sperms are relased into surface water films. This entire development is dependent on food stored in the microspore since there is no photosynthetic tissue in the male gametophyte.

The megaspore gives rise to a much larger gametophyte than does the microspore. A mass of haploid cells is formed inside the megaspore wall causing it to crack to tensions caused by the drying action of air. In some species this process hurls the spores out and away from the plant. They are relatively heavy and are not spread far before they fall to the ground. In other species they never quite escape from the strobilus. In such cases they may lodge in between the sporophylls or remain in the open sporangia, completing the life cycle there.

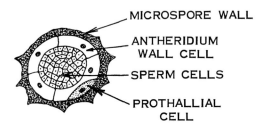

Fig. 4.62. Microspore of *Selaginella* with the enclosed male gametophyte (many times enlarged).

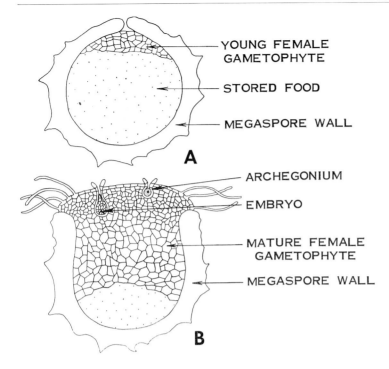

Fig. 4.63. Stages in the development of the female gametophyte of *Selaginella*. A. Early stage within the megaspore wall. B. Mature gametophyte partially protruding from the megaspore.

open (Fig. 4.63). The mass of cells then expands somewhat. In favorable situations it may become faintly green and manufacture a limited amount of food. However, it is dependent for the most part on food stored in the megaspore, food that was produced originally by photosynthesis in the sporophyte.

One to several archegonia are formed in this female gametophyte. The venter of each archegonium is sunken in gametophyte tissue and only the very short neck protrudes (Fig. 4.63). Fertilization occurs when a swimming sperm finds its way into one of the archegonia. The zygote develops into an embryo that eventually becomes the mature sporophyte. The embryo is nourished by the gametophyte (Fig. 4.64) but, as we have seen, the gametophyte itself is largely dependent on food stored in the megaspore by the preceding sporophyte generation.

The following three significant points emerge from this discussion of the life story of *Selaginella*:

1. Genetically similar spore-producing tissue may be affected by nutrition in such a way that two different kinds of spores are produced.

Fig. 4.64. Young sporophyte of *Selaginella*. The foot is embedded in the gametophyte, which is still enclosed in the megaspore wall.

2. The microspores, which are produced in large numbers, are small and have only a limited food supply. They produce very much reduced male gametophytes that develop entirely within the microspore wall.

3. The megaspores are fewer in number and are provided with larger quantities of stored food. This nourishes the female gametophytes and through them the embryos of the next sporophyte generation. This partial adoption of the female gametophyte by the sporophyte

of *Selaginella* is to be compared with a complete adoption in plants that produce seeds.

*Lycopodium* (Fig. 3.41) is a homosporous club moss. The strobili are basically similar to those of *Selaginella* but contain only one kind of sporophyll and produce only one kind of spore.

The gametophyte generation of *Lycopodium* shows a wide range of expression from being green and independent to being nongreen and associated with fungi that invade its tissues. These gametophytes are subterranean and so hard to find that it has only been in relatively modern times that they have been known at all.

The spores of *Lycopodium* are very heavily cutinized and in the past they were an item of commerce sold as "*Lycopodium* powder." One use was the coating of pills to make them easier to swallow. A more spectacular use, as a flash powder, resulted from their ability to ignite explosively when mixed with air and exposed to a flame.

# FIVE  REPRODUCTION

## THE LIFE CYCLE OF PINE—A FIRST LOOK AT THE SEED HABIT

Pines belong to a large group of plants called *gymnosperms*. The name means naked seeds and implies that the seeds are developed in an exposed position. This condition contrasts with that of the flowering plants or *angiosperms* in which the seeds develop within a protective tissue. Pines are not closely related to the flowering plants even though plants in both groups produce seeds. Nor are pines closely related to *Selaginella* even though both are heterosporous.

Pines are classified in a group of gymnosperms popularly known as *conifers* (cone-bearers) (Fig. 5.1). The two kinds of spore-bearing structures are segregated in separate cones. These are often called male and female cones but the terms are somewhat misleading. The technical terms *microsporangiate* and *megasporangiate cones* are awkward to use and it has been suggested that *pollen cones* and *seed cones* might serve satisfactorily in general usage. Both occur on the same tree and may even be found on the same branch.

### Seed Cones

The larger and more familiar cones are the mature seed cones (Fig. 5.2). Each one consists of a series of overlapping scale-like structures which have in the past been interpreted as being megasporophylls. However, each of these scales has a small bract associated with it in such a way that the bract might well be considered to be the modified leaf rather than the scale. This fact has been responsible for the modern interpretation that the seed cone is not a simple strobilus. This is a problem for advanced discussions; a simple solution is to call the parts on which the seeds are borne

Fig. 5.1. Growth habit of a mature coniferous tree.

*seed-bearing scales*. Two seeds are borne at the base of each of these scales (Fig. 5.3).

Depending on growth conditions and the particular species of pine, immature stages of both kinds of cones may or may not exist together in the overwintering buds. In spring when the buds swell and the protective bud scales fall away the first parts to emerge are the new pollen cones,

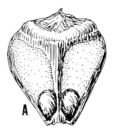 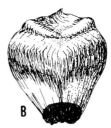

Fig. 5.3. The seed-bearing scale of pine. A. Dorsal surface of the scale with 2 winged seeds. B. Ventral surface of the scale with small subtending bract.

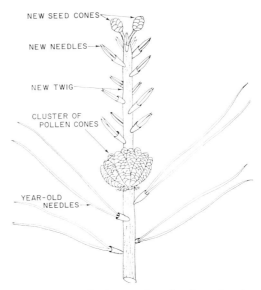

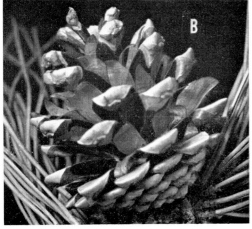

Fig. 5.2. Mature seed cones of pine at the end of the second year's growth. A. Before the scales open. B. After the scales open to allow seed dispersal.

Fig. 5.4. Developing twig of pine showing various parts as they might appear in late spring.

which occur in a cluster at the base of the new twig. As the twig continues to elongate the new needles become evident. Finally in late spring the new seed cones appear at the tip of the new twig (Fig. 5.4). They are barely visible when first evident

(Fig. 5.5) and in pines they take two years to reach maturity. They achieve only the size of small bird eggs by the end of the first summer; the major volume increase occurs during the growing season of the second year.

Fig. 5.5. Tip of pine twig in late spring showing newly apparent young seed cones.

POLLEN CONES AND DEVELOPMENT

Pollen cones (Fig. 5.6) differ from seed cones in many respects including numbers and location. They are borne at the base of the new year's growth of a twig rather than at its tip. There are usually many of them in a tight cluster rather than two or three. Both types of cones may appear on the same branch but in general seed cones occur only on the more vigorous branches while pollen cones can often be found on poorly developed twigs.

Pollen cones are simpler structures than seed cones and there is little question that the scales of the pollen cone are microsporophylls. Each one bears two parallel microsporangia attached to the lower surface, one on each side of the midrib (Fig. 5.7). The tip of each microsporophyll is membranous and curves upward to overlap the one above. By this method of overlapping scales all the microsporangia of a pollen cone are protected against premature drying.

Each microsporangium contains a large number of microspore mother cells and

Fig. 5.6. Pollen cones of pine. A. Opening of terminal bud in spring exposing young pollen cones at base of the new year's growth. B. Mature cones just before pollen dispersal. C. Cones become somewhat elongated at the time of pollen dispersal. D. Pollen cone of pine cut lengthwise to show arrangement of the microsporophylls.

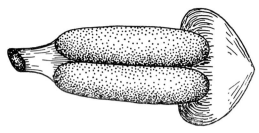

Fig. 5.7. Microsporophyll of pine with 2 microsporangia. (Dissected from a pollen cone.)

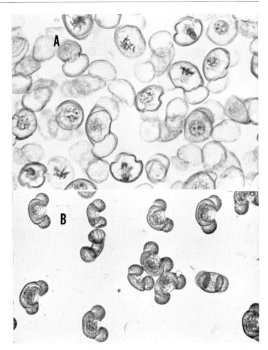

Fig. 5.9. Pollen formation in pine. A. Microspores in the process of becoming pollen grains. (The nuclei shown are haploid and mitosis is evident in some of them.) B. Mature pollen grains (whole mount).

each microspore mother cell undergoes meiosis to produce four microspores. These are arranged in the form of a spherical tetrad (Fig. 5.8) rather than a linear tetrad as are the megaspores. Soon all the microspores separate, become rounded in shape, and begin to form a thickened wall that has at least two layers.

At two points in the wall of each microspore the inner and outer walls separate in such a way that the outer wall becomes inflated, forming two balloonlike appendages. This adaptation increases the total surface area without altering the weight and makes possible a more efficient dispersal in air.

The contents of the microspore divide several times by mitosis (Fig. 5.9). Two of the cells formed are small and functionless. They are called *prothallial cells* and are thought to be vestigial remnants of the plant body of the male gametophyte. After these divisions the whole structure can no longer be called a microspore. The proper and more familiar name for it is *pollen grain* (Figs. 5.9B, 5.10). When the pollen grain has completed these divisions and developed its cutinized outer wall it is ready for dissemination.

At this time the main axis of the pollen cone begins to elongate and the micro-

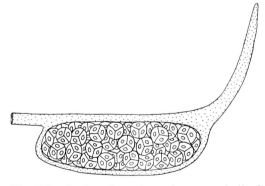

Fig. 5.8. Section through a microsporophyll of pine showing tetrads of microspores formed by meiosis from microspore mother cells in the microsporangium.

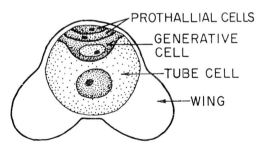

Fig. 5.10. Structure of a mature pollen grain of pine.

sporophylls are separated (Fig. 5.6C). In this way the microsporangia are exposed to dry air. Drying causes them to split open; exposed pollen grains are blown away in air currents. This process occurs in late spring; when it happens clouds of yellow pollen dust may be seen drifting among the branches of a pine tree.

OVULE AND SEED DEVELOPMENT

In its early stages of development a seed is called an *ovule* and when immature seed cones (Fig. 5.11) are carefully dis-

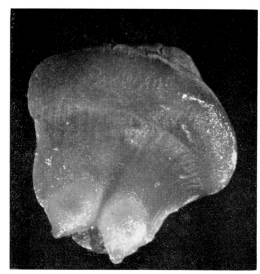

Fig. 5.12. Photograph of a young seed-bearing scale dissected from a young seed cone. Note the 2 ovules.

Fig. 5.11. Newly emerged seed cones at tip of developing pine twig (about 5 times actual length).

sected it may be seen that the ovules occur in pairs at the base of the upper surface of the seed-bearing scales. Each ovule begins as a small swelling with an oval outline (Fig. 5.12). The nature of this swelling indicates that it is a *megasporangium* (Fig. 5.13); this term is gradually replacing an older term, *nucellus*, which was applied to this tissue long before its true nature was understood. As the megasporangium grows it develops a ringlike zone of meristematic tissue some distance from its tip. Because of active cell divisions in this zone a collar-like band of tissue begins to overgrow and enclose the megasporangium. This is the *integument*. It serves to protect the inner tissue from excessive evaporation and against some predators. In mature seeds the integument forms the major part of the seed coat. The integument does not completely seal over the megasporangium, how-

ever. A small cavity is left inside and a small opening, the *micropyle*, extends from this cavity through the integument to the outside.

The ovule thus consists of two major parts: the megasporangium and the integument (Fig. 5.13C).

Deep within the megasporangium, one of the cells, at first no different from any of its neighbors, begins to enlarge considerably. This is the *megaspore mother cell*. When it achieves a certain stage of maturity the nucleus of this cell undergoes meiosis, giving rise to four haploid nuclei. Cell walls develop between the nuclei and four *megaspores* are formed. They are positioned in a row called a *linear tetrad* (Fig. 5.13D).

Next three of the four megaspores disintegrate and gradually disappear. The one surviving megaspore enlarges rapidly and synthesizes a large mass of protoplasm. Then the haploid nucleus of the megaspore begins to divide by mitosis. After several divisions a large number of haploid nuclei lie free in the cytoplasm without any new cell walls to separate them. This is the *free nuclear stage* of the female gametophyte. Finally cell walls begin to appear between nuclei and the mass is partitioned neatly

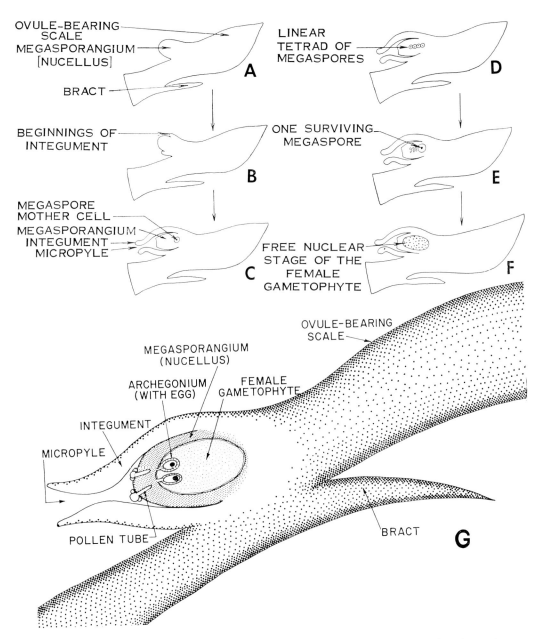

Fig. 5.13. Schematic sections of stages in the development of a pine ovule. A. Early appearance of the megasporangium (nucellus) on the surface of the scale, before the formation of the integument. B. Integument forms as a collar around the megasporangium. C. Integument fully developed and micropyle evident; megaspore mother cell in prophase of meiosis. D. Linear tetrad of megaspores in the megasporangium. E. Degeneration of 3 megaspores and growth of the 1 surviving megaspore. F. Free nuclear stage of the female gametophyte. G. Pine ovule at time of fertilization in spring of second year.

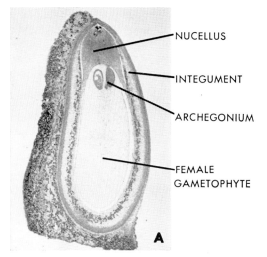
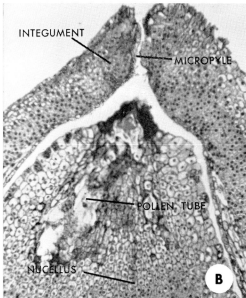
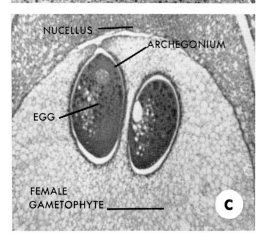

Fig. 5.14. Sections of a pine ovule containing a mature female gametophyte with archegonia. A. Entire section. B. Partial section showing micropyle and irregular growth pattern of the pollen tubes growing through the megasporangium (nucellus). C. Partial section showing details of the archegonia.

into an organized tissue. This tissue is quite large and it persists into the mature seed. In that portion of the mass closest to the micropyle there are developed anywhere from one to several archegonia, each one containing a very large egg cell (Figs. 5.13G, 5.14C).

The tissue just described is considered to be a female gametophyte by the following traits:

1. It developed from a haploid cell, the megaspore.
2. It consists only of cells with haploid nuclei.
3. At maturity it produces female gametes.

Note that the gametophyte is completely enclosed in sporophyte tissue and that the megasporangium does not shed the megaspores as in *Selaginella*. Not only has the female gametophyte become completely dependent for its nutrition on the sporophyte generation but it is no longer separated from the source.

In pines the development of the ovule proceeds to the point of egg formation during the first year but fertilization does not occur until the spring of the second year. In most other gymnosperms the ovule completes its entire development in a single growing season.

POLLINATION IN PINE

The act of transfer of pollen from its place of origin in the pollen cone to its destination in the seed cone is called *pollination*. It occurs in late spring when the pollen is mature. At the time of this event the ovules are still immature and in pine the female gametes in the ovules will not be ready for fertilization until spring of the second year.

At the time of pollination the young seed cones stand straight up at the tips of the branches to which they are attached (Fig. 5.11). The ovule-bearing scales are slightly separated so that a space exists between each one and the next. The air is full of pollen and some of the pollen grains drift down between the scales. Each ovule secretes a small drop of sticky fluid through the micropyle (Fig. 5.15). Pollen grains be-

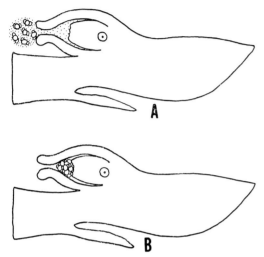

Fig. 5.15. The pollination mechanism in pine. A. The pollination fluid is extruded as a droplet from the micropyle. Note the adherent pollen grains. B. Reabsorption of the pollination fluid draws the pollen grains through the micropyle into the pollen chamber.

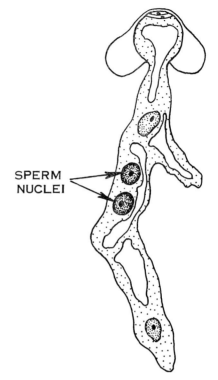

Fig. 5.16. Mature pollen tube of pine (male gametophyte).

come trapped in this fluid and are drawn through the micropyle into the pollen chamber when the fluid is reabsorbed. Once pollination is completed the stalk of the cone bends so that it hangs downward and growth of the scales eliminates the spaces between them, effectively sealing the cone.

## The Male Gametophyte

Each pollen grain gives rise to a tubelike outgrowth that slowly digests its way through the megasporangium tissue toward the developing female gametophyte (Fig. 5.14B). This structure is called the *pollen tube*. Several mitotic divisions occur during the growth of the pollen tube but only the last division is important since this produces two male gametes (Fig. 5.16). These are commonly called sperms, yet they lack flagella and cannot swim. The growth of the pollen tube carries them to an archegonium and deposits them in close approximation to the egg.

The first cell in the developmental series leading to pollen tube formation was the microspore. Furthermore, the mature pollen tube contains several haploid nuclei and produces male gametes. Therefore it may be considered properly as a male gametophyte in which the plant body of the gametophyte has been diminished to the two insignificant prothallial cells.

## Fertilization

In most pines the egg has been formed and the pollen tube has reached the archegonium by late spring of the second year. The sperm are discharged from the tip of the pollen tube; one of them creeps by amoeboid movements across the few remaining microns of distance separating it from the egg. The resultant zygote promptly begins to divide by mitosis to form the

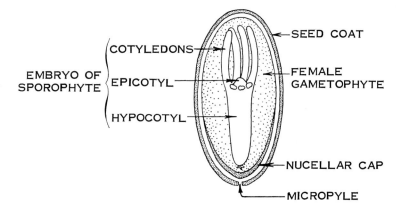

Fig. 5.17. Mature seed of pine dissected to show the embryo lying in the massive female gametophyte.

diploid mass of cells that is the embryo stage of the sporophyte. The early stages of embryo formation are complex and will not be discussed here. When the embryo is mature it consists of a lengthwise axis with a root tip growing point at one end and a stem tip growing point at the other. Surrounding the stem tip there is a whorl of leaflike appendages called *cotyledons* (Fig. 5.17).

The whole embryo lies lengthwise in the massive tissue of the female gametophyte. This tissue has accumulated much reserve food that is used later by the embryo during germination. The superficial resemblance of the female gametophyte tissue to the endosperm of such seeds as corn was recognized long ago and it has been called endosperm (improperly) for many years.

THE PINE SEED

By the time the embryo is mature the integument of the ovule has become much thicker and very hard. The female gametophyte has expanded enormously. The megasporangium is gone except for a papery membrane at the micropyle end which is called the *nucellar cap*. This matured ovule is now a seed. It can remain alive under dry conditions for long periods of time and then germinate when placed in a suitable environment.

In most pines some of the surface tissue of the ovule-bearing scale remains attached to the seed as a wing (Fig. 5.18). When the scales separate and the seeds fall out of the cone the wings slow down the rate of fall so wind movements can spread the seeds away from the base of the parent tree.

Fig. 5.18. Mature seeds of pine with attached wings.

## THE LIFE CYCLE OF FLOWERING PLANTS

The preceding brief discussions of the life cycles of lower plants serve as a background for the consideration of the life cycle of flowering plants.

One definition of a flower is that it is a modified branch bearing modified leaves. The leaves at the base of the flower are sterile while the ones near the tip are fertile. The fertile leaves or sporophylls are of

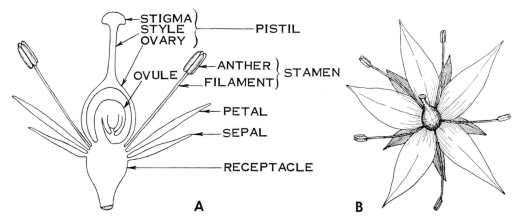

Fig. 5.19. Schematic representations of flower structure. A. Longitudinal section. B. Entire.

two types which are known as *stamens* and *pistils*. The sterile leaves in primitive flowers may have been all alike but in many modern flowers the lowermost ones are differentiated as *sepals* and those above as *petals*.

RECEPTACLE AND PERIANTH

The *receptacle* is the part of the flower (Fig. 5.19) to which the other parts are attached. Probably it represents the axis of the conelike primitive flower which no longer becomes elongated. If there is but one flower arising in the axil of a leaf then the stalk of this flower may be called a *peduncle*. If the flowers occur in definite clusters or *inflorescences* then the stalk of each individual flower is called a *pedicel*.

*Sepals* are the leaflike parts that are attached below (outside) all of the others on the receptacle. Often they are smaller than the petals; they are usually green. Normally they enclose the rest of the flower parts during the bud stage. It should be noted that in some plants such as lilies and tulips the sepals and petals are similar in size and color. In some other plants they may be lacking entirely. Collectively the sepals of a flower are referred to as the *calyx*.

*Petals* are the leaflike parts that are attached immediately above (and thus appearing inside) the sepals. They are often brightly colored. In many plants they may be small and inconspicuous or even lacking. Flowers with showy petals often attract insects that come in search of nectar and pollen. In so doing they may do the plant a service by transferring pollen from one flower to another. When referred to collectively the petals of one flower are termed the *corolla*.

The term *perianth* refers to all of the leaflike parts of the flower and thus includes both sepals and petals. Sometimes it is difficult to distinguish between sepals and petals. At other times one of these two parts is missing and it is difficult to decide which one is present. Under such circumstances it is helpful to refer to the structures as perianth parts.

STAMENS

*Stamens* are commonly thought of as being the male parts of the flower. In terms used to discuss previous plants, however, stamens are microsporophylls. Their leaflike nature has become obscured by modification. All that are left normally are a slender stalk or *filament* and the *anther*. The anther consists of four anther sacs with some connective tissue between them (Fig. 5.20).

Each anther sac contains a microsporangium that when immature is filled with a great many microspore mother cells (mi-

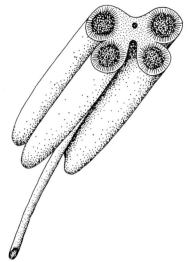

Fig. 5.20. Stamen with anther cut crosswise.

crosporocytes) (Fig. 5.21). Each one of the microspore mother cells undergoes meiosis to form microspores that are, of course, haploid (Fig. 5.22). The nucleus of each microspore then undergoes one mitotic di-

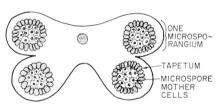

Fig. 5.21. Transverse section of a young anther showing the 4 microsporangia. Note presence of tapetum and numerous microspore mother cells.

vision to form two haploid nuclei. At this stage the microspore becomes a pollen grain. The wall of the pollen grain becomes thick and cutinized and has characteristic markings (Figs. 5.23, 5.24).

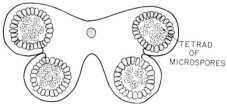

Fig. 5.22. Section of anther after meiosis has occurred. Note presence of tetrads of microspores.

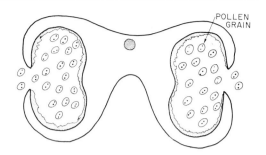

Fig. 5.23. Section of mature anther at time of dehiscence when the pollen is being released.

During this development the anther increases in size. The inner layer of wall cells of the anther sac, called the *tapetum*, breaks down to form a viscous fluid in which the microspore mother cells seem to float. This fluid may have an important effect on microspore development. When the pollen grains are mature the wall between adjacent pairs of anther sacs breaks down so that two large chambers are formed, one on each side of the anther. Finally a lengthwise slit develops down each side of the anther and the mature pollen grains are exposed to the air (see Fig. 5.30B).

### Pistils

*Pistils* are commonly thought of as the female parts of the flower. Each one is composed of three major segments. The *stigma* is the terminal portion of a pistil adapted

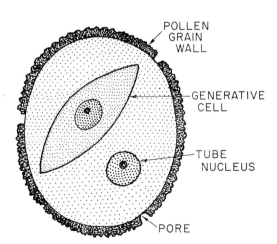

Fig. 5.24. Mature pollen grain as seen in section in a compound microscope.

primarily to receive pollen grains. The *style* is the portion that connects the stigma with the ovary. The *ovary* is the basal portion of a pistil and contains the ovules (Fig. 5.19).

Actually a simple pistil is a highly modified megasporophyll. In order to explain this concept it becomes necessary to discuss something of the theoretical origins of the pistil. According to a widely accepted theory the pistil evolved from a flattened megasporophyll with ovules on the under surface. This structure failed to unfold from the folded embryonic condition. The joined edges became sealed and thus the ovules were enclosed. As a result the ovules have an added degree of protection against desiccation and to some degree against predation by other organisms (see Figs. 5.25, 5.26 and further discussions of this topic in Chapter 7).

Many plants have flowers with compound pistils that have resulted from the fusion of two or more simple pistils. The

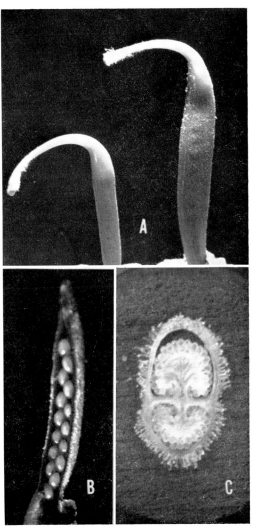

Fig. 5.26. Simple pistils and compound pistils. A. Simple pistils dissected from flowers of the perennial sweet pea. B. Longitudinal section of one of the simple pistils in A showing the double row of ovules. (Compare with the transverse section in Fig. 5.25C.) C. Transverse section of the compound ovary of a snapdragon flower showing 2 fused carpels with numerous ovules in each.

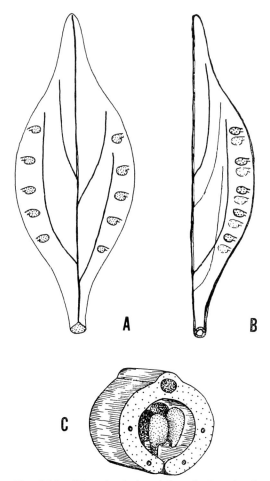

Fig. 5.25. Hypothetical origin of the simple pistil. A. Megasporophyll of a preangiosperm flower with ovules along the margin. B. Megasporophyll folded along the midrib, thus enclosing the ovules. C. Transverse section of the hypothetical simple pistil.

term *carpel* is used to designate each portion of a compound pistil that is recognizable as being one of the fused segments (Fig. 5.26C).

THE OVULE

The ovule (Fig. 5.27) is the structure that eventually becomes the seed. Depending on the kind of plant there may be from one to several thousand ovules in each ovary.

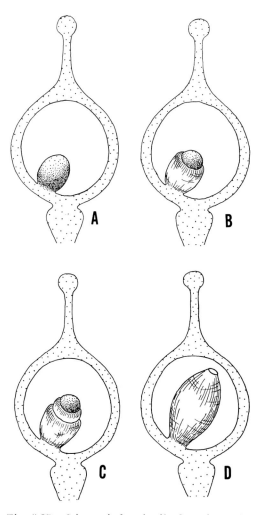

Fig. 5.27. Schematic longitudinal sections of a pistil illustrating stages in the development of an ovule. A. Origin of the megasporangium (nucellus) as a swelling on the inner ovary wall. B. Origin of the inner integument. C. Origin of the outer integument. D. Integuments completely enclosing the megasporangium except for the micropyle.

Each ovule begins as a small dome-shaped swelling on the inner wall of the ovary. As it grows, two ringlike layers of tissue form around its base, one inside the other. These two layers are integuments. It will be recalled that the pine ovule has but one integument. The inner dome-shaped mass is the nucellus or megasporangium. The integuments grow completely over the megasporangium except for a tiny hole in the center. This is the micropyle. (Note: In sections of the lily ovary which are widely used for teaching purposes the ovules are not straight. Instead they become curved during their growth in such a way that the micropyle end is next to the stalk [Fig. 5.28]).

One cell deep within the megasporangium enlarges enormously. The nucleus of this cell also increases greatly in size. This cell is a megaspore mother cell (megasporocyte) and like any other spore mother cell its nucleus undergoes meiosis to form four haploid nuclei. The cytoplasm of the megaspore mother cell may become subdivided to complete the formation of four megaspores (Fig. 5.28) but in some plants the megaspore nuclei do not become parts of cellular units.

FEMALE GAMETOPHYTE (EMBRYO SAC)

In many plants only one of the four megaspores formed after meiosis survives while the other three disintegrate and eventually disappear. The surviving megaspore grows rapidly; its nucleus divides by mitosis to form two haploid nuclei. These divide again to form four and again to form a total of eight haploid nuclei which lie free in the cytoplasm of the megaspore. This structure was termed the *embryo sac* long before its nature as a *female gametophyte* was recognized. Both names are now in common usage. Since all eight nuclei were formed by mitosis from the megaspore nuclei they are identical with respect to the numbers and kinds of chromosomes.

Commonly the nuclei are in two groups of four, one group at each pole of the embryo sac. One nucleus from each group migrates to the center; this pair of

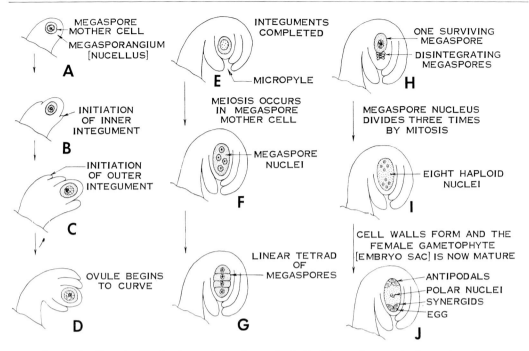

Fig. 5.28. Longitudinal schematic sections of pistils illustrating ovule formation, differentiation of the megaspore mother cell, meiosis, and embryo sac formation. Based in part on sections of lily ovules up to the point of megaspore formation.

nuclei is referred to as the *polar nuclei*. The remaining six nuclei and the cytoplasm around them become organized into six cellular units, three at one pole and three at the other. The three cells farthest away from the micropyle are called *antipodal cells*. Of the cells at the micropylar end of the embryo sac, one becomes the *egg*. The other two are called *synergids* (Fig. 5.28J).

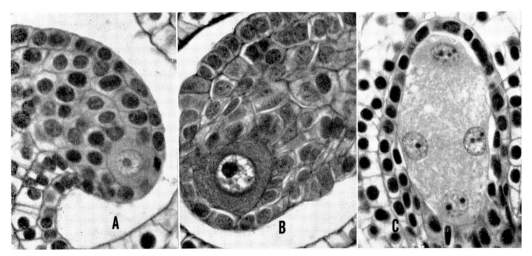

Fig. 5.29. Photomicrographs of sections of lily ovules for comparison with the early stages shown in Fig. 5.28. A. Compare with Fig. 5.28A. B. Compare with Fig. 5.28B. C. Compare with Fig. 5.28F.

Thus the mature embryo sac has seven cells, three at one end and three at the other, with a large binucleate cell in the middle. Logically it conforms to the definition of a gametophyte since it develops from a spore and produces a gamete. The three antipodal cells may be interpreted as the vestigial remains of the plant body of the female gametophyte. Whether or not the synergids are remnants of an archegonium is a question that cannot now be answered.

Students who have previously studied gamete formation in animals should again make a careful comparison with gamete formation in plants. The interpolated gamete-producing generation is lacking in almost all animals and gamete formation follows immediately after meiosis. The cells that become microspores in plants are, in a sense, analogous to the cells that become sperms in animals while the cells that become megaspores in plants are analogous to those that become eggs in animals. An awareness of this difference will aid considerably in the understanding of problems in inheritance since genetics courses are often taught with examples from both the plant and animal worlds.

## Pollination

At about the time when the embryo sac is mature the pollen grains in the anther are also mature. The anthers open and the pollen is dispersed. The transfer of pollen from anther to stigma is properly called *pollination;* i.e., it is incorrect to refer to pollen transfer as fertilization (Fig. 5.30).

The two chief agencies of pollination are wind and insects. Wind-pollinated flowers tend to have elaborate, sometimes feathery stigmas that trap pollen from the air. Insect-pollinated flowers tend to have sticky stigmas as well as devices that attract insects and cause them to be dusted with pollen. As the insects crawl around in the flower or visit other flowers of the same kind the pollen on their bodies becomes transferred to the stigma. Among the attractive devices are a general showiness of the flower and the secretion of nectar. Many flowers also have devices that tend to discourage the visitations of smaller insects that could rob the nectar without affecting pollination.

## Male Gametophyte (Pollen Tube)

Many pollen grains have thin places or pores in their outer cell walls. When an indvidual pollen grain lands on a stigma one of these pores comes into contact with the surface of the stigma. Through this pore a tubelike outgrowth (the pollen tube) emerges and penetrates the surface of the stigma. As was pointed out above, the pollen grain has two haploid nuclei. One of these, the *tube nucleus,* stays near

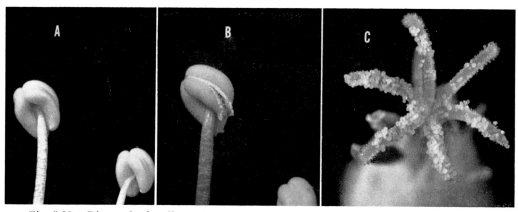

Fig. 5.30. Dispersal of pollen. A. Mature anther before pollen discharge. B. Mature anther with lengthwise split exposing the pollen for dispersal. C. Stigmas covered with pollen after pollination.

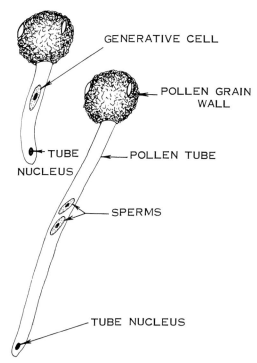

Fig. 5.31. Growth of the pollen tube. A. Emergence of tube from the pollen grain. B. Mature pollen tube.

the tip of the pollen tube and apparently controls the growth of the tube (Fig. 5.31).

The other nucleus has a layer of cytoplasm around it and forms a rather vaguely organized cell, the *generative cell*. During its journey down the pollen tube the generative cell divides once by mitosis to form two cells that function as male gametes. Thus the mature pollen tube contains a total of three haploid nuclei. Since all three have important functions it is evident that the male gametophyte has been reduced to an absolute minimum.

Many pollen grains are transferred to a stigma at the time of pollination. Most of them begin growth and many more pollen tubes enter the cavity of the ovary than there are ovules present. There seems to be a chemical secretion from the ovules that directs the growth of the pollen tubes toward the micropyles, but it is unuusal for more than one pollen tube to enter a single micropyle (Fig. 5.32).

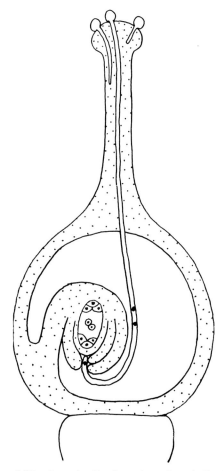

Fig. 5.32. Longitudinal section of a pistil with pollen tubes growing down style into cavity of the ovary. One tube is entering the micropyle of an ovule.

FERTILIZATION

The pollen tube, after entering the micropyle, grows through the nucellus and bursts into the embryo sac where most of its contents are discharged. The tube nucleus has no further function and begins to disintegrate. One of the male gametes unites with the egg cell to form the zygote. The other one unites with the two polar nuclei. This process, which involves two separate nuclear fusions, is called *double fertilization* (Fig. 5.33).

GROWTH OF THE ENDOSPERM

The mass of cytoplasm enclosing the triple fusion nucleus is now referred to

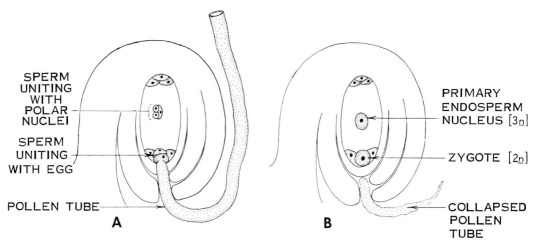

Fig. 5.33. Double fertilization. A. The sperms have been discharged from the pollen tube into the embryo sac. B. Union of one sperm with the egg forms the zygote, and union of the other sperm with the 2 polar nuclei forms the nucleus of the primary endosperm cell.

as the *primary endosperm cell* (Fig. 5.33B). The triploid nucleus with its three sets of chromosomes begins to divide rapidly by mitosis to form a large number of $3n$ nuclei. At first they lie free in the enlarging cytoplasmic mass of the endosperm cell (Fig. 5.34). This *free nuclear stage* of the endosperm is followed by a cellular stage when the cytoplasm becomes divided into cellular units, each containing one nucleus. Large masses of reserve food are accumulated in the *endosperm*, a unique tissue occurring only in the seeds of flowering plants (Fig. 5.34). This food is eventually used by the plant as nourishment for the embryo but many other organisms, including people, have come to rely on endosperm as a source of food. Many cultivated crop plants are important because of the food stored in the endosperm and much of the history of civilization is involved with the need for controlling land where such plants can be grown successfully.

Divisions in the endosperm occur rapidly but soon the zygote begins to divide

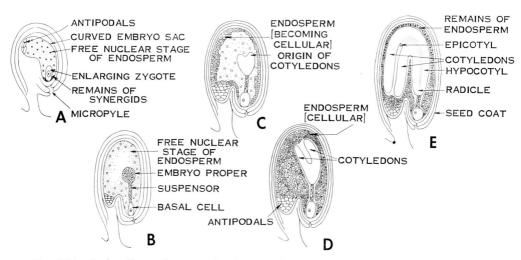

Fig. 5.34. Series illustrating growth of the endosperm and embryo based on *Capsella*.

by mitosis also. A mass of diploid cells is formed which is the embryo of the sporophyte (Fig. 5.34). As noted above the embryo obtains its nutrition from the food reserves in the endosperm. In some plants (such as corn) the seed becomes mature and enters a resting stage while the embryo is still small in comparison to the endosperm. In other plants (such as beans) the endosperm is completely used up before the seeds are mature. In such cases the reserve foods are transferred to the cotyledons of the embryo. Reserve foods, whether stored in the endosperm or in the cotyledons, are used during the germination period to provide young seedlings with metabolic materials until they are able to manufacture their own.

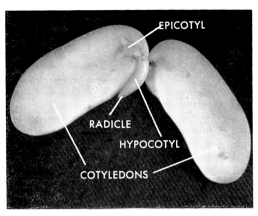

Fig. 5.35. Embryo of a bean seed removed from seed coat and spread apart to show the 2 massive cotyledons, the epicotyl, and the hypocotyl axis.

THE DEVELOPMENT OF A DICOT EMBRYO

In many plants, after the first division of the zygote the *basal cell* of the resulting cell pair enlarges considerably and does not divide further. The other member of this first cell pair divides several times in parallel planes to form a filament of cells which is called the *suspensor*. Eventually the terminal cell of the suspensor begins to divide in several planes to form a small spherical mass of cells which is the *embryo proper*. The function of the suspensor seems to be pushing the embryo proper into the endosperm tissue. It does not become a part of the mature embryo (Fig. 5.34).

The embryo proper soon loses its spherical shape and becomes somewhat elongated. One end remains rounded but the other end becomes flat-topped so that the whole structure is crudely triangular. Later the corners of the flattened side begin to grow rapidly, forming two lateral earlike appendages which are essentially parallel to each other. These appendages are the *cotyledons*. It should be remembered that they are parts of the embryo. In bean seeds they become so much larger than the other parts of the embryo that it is easy to lose track of this point (Fig. 5.35).

The small mass of meristematic cells that arises between and at the base of the two cotyledons is the growing point or apical meristem of the shoot system. Very commonly this growing point becomes active for a time before the seed matures. As a result several stem segments and young leaves may exist in miniature form between the cotyledons. The terms *plumule* and *epicotyl* are used interchangeably to describe this future shoot system in the embryo.

The major portion of the embryo axis below the cotyledons is called the *hypocotyl*. The root growing point at the basal end of the hypocotyl is the *radicle*. In the early phases of seed germination the radicle becomes active and gives rise to the primary root (Fig. 5.35).

THE MONOCOT EMBRYO

The flowering plants are divided into two large groups: *monocotyledons* and *dicotyledons*. One basic point of separation is an embryo difference: dicot embryos have two cotyledons while monocot embryos have but one. By and large the development of the monocot embryo is similar to that of the dicot embryo, which has already been described.

Traditionally, the embryo of corn or one of the cereal grains is used to illustrate the monocot embryo (Fig. 5.36). The corn embryo has a single appendage much larger than the embryo axis and attached laterally to it. Because it is shaped somewhat like a shield it has long been called a *scutellum* (from a Latin word meaning shield). A long-standing controversy exists concerning the exact nature of the scutellum and students should avoid the inference that it is automatically a cotyledon because of its position. All of the embryo axis below the attachment point of the scutellum is the radicle. Technically it might be considered as being a hypocotyl but the organization of its tissues is so entirely rootlike that the term radicle is more appropriate. The radicle is covered by a special tissue known as the *coleorhiza*.

Above the attachment point of the scutellum there is a leaflike membrane wrapped around the epicotyl as a protective sheath. This is the *coleoptile*. When the seed is mature the *epicotyl* consists of several segments of the future stem plus a corresponding number of undeveloped leaves and the apical meristem.

The mature grain of corn (Fig. 5.36) is actually a one-seeded fruit in which the seed coat and the ovary wall have become completely fused. The major portion of the corn grain is filled with endosperm tissue which may be white and flourlike or yellow and flinty. The outermost layer of cells in the endosperm contains considerable amounts of stored protein. It is referred to as the *aleurone layer*.

THE FRUIT

As discussed above, ovules are developed inside the ovary of the pistil; when mature they become seeds. Meanwhile the pistil enlarges, particularly in the region of the ovary, to become the fruit. The mature fruit may consist entirely of ovary tissue or of ovary plus accessory tissues.

A *simple fruit* matures from an individual pistil that may be simple or compound.

An *aggregate fruit* is formed when several pistils in the same flower mature as a part of one compound structure.

A *multiple fruit* involves parts of several flowers that mature as one compound structure.

Each of the above can be classified further as a *true fruit* in which only the pistil is involved or as an *accessory fruit* in which other parts of the flower in addition to the pistil are involved.

At maturity a given fruit may be *dry* or *fleshy*. Furthermore dry fruits may be *dehiscent*, meaning that they open at maturity; or they may be *indehiscent*.

These terms are useful in describing specific examples of fruits. In addition many types of fruits have been given special names. A few examples of special fruit types are discussed below.

An *achene* is a dry indehiscent fruit that contains a single seed; this seed is

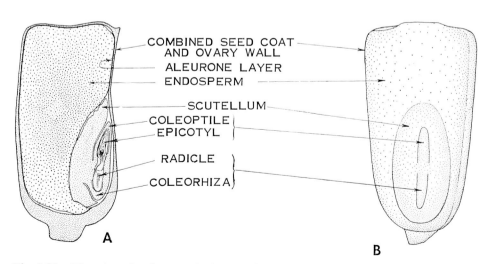

Fig. 5.36. Mature grain of corn. A. As seen in a section. B. As seen from the surface.

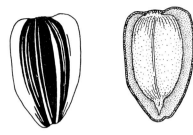

Fig. 5.37. Achene of sunflower entire and dissected.

united with the ovary wall only by its stalk. Sunflower "seeds" are actually achenes (Fig. 5.37).

A grain or *caryopsis* is also a dry indehiscent fruit containing a single seed. It differs from an achene in that the seed coat is completely and permanently united with the ovary wall (Fig. 5.36). The combined layer is often referred to as a *pericarp*.

A *nut* is another example of a dry indehiscent fruit that in many cases contains only a single seed. Nuts are usually larger than achenes and some portions of their ovary walls become very hard and thick. Hazlenuts and hickory nuts (Fig. 5.38) are good examples of the fruit condition while Brazil nuts are seeds rather than true nuts.

A *follicle* is a dry dehiscent fruit that splits open along one edge at maturity.

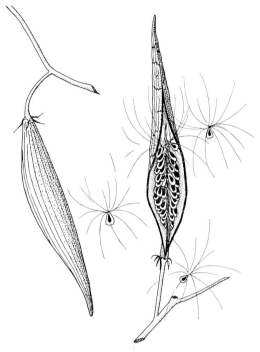

Fig. 5.39. Closed and open follicle of a milkweed.

The fruit of a milkweed plant is a familiar example of a follicle (Fig. 5.39).

A *legume* is the characteristic fruit of many members of the legume family, which includes peas, beans, afalfa, and clover. It

Fig. 5.38. Hickory nut with the enclosing husk split apart.

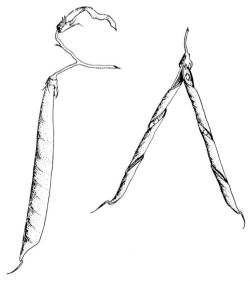

Fig. 5.40. Closed and open legume fruit based on the pod of the perennial sweet pea.

is a dry dehiscent fruit that splits open along both edges at maturity (Figs. 5.40).

A *capsule* is a dry dehiscent fruit formed from a compound pistil. Capsules vary considerably in the exact manner in which the individual carpels open at maturity to release the seeds. Irises, poppies, and lilies are examples of common plants that form capsules (Fig. 5.41).

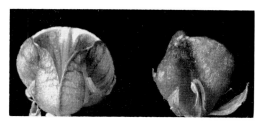

Fig. 5.41. Open and closed capsules of a member of the lily family *(Ornithogalum)*.

A *drupe* is a fleshy fruit in which the outer zones of the ovary wall become fleshy while the inner zone becomes hard and stony. Cherries, plums, and peaches are classified as drupes (Fig. 5.42).

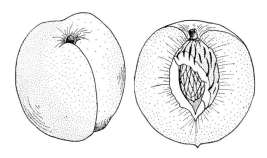

Fig. 5.42. Entire and dissected drupe of peach.

In true *berries,* the entire ovary wall becomes fleshy. Tomatoes (Fig. 5.43) and grapes are good examples of berries. Citrus fruits such as oranges and lemons are usually classified as berries but are sometimes given a special name due to the presence of the characteristic rind.

Bananas, squashes, melons, blueberries, gooseberries, and many other fleshy fruits similar to these are generally classified as berries due to the fleshiness of the ovary wall. However, in each of these

Fig. 5.43. Branch of a tomato plant with fruits (berries) in various stages of development.

other flower parts in addition to the pistil are involved in the fruit.

The fruits of apples and similar plants are called *pomes*. The inner cartilaginous core of the pome is formed from the inner zone of the ovary wall while the outer zones are fleshy and united almost indistinguishably with accessory structures that have become fleshy also (Fig. 5.44).

Fig. 5.44. Apple fruit (pome) entire and dissected.

The raspberry is an aggregate fruit composed of miniature drupes, each of which matures from one of the many pistils in the flower.

The strawberry is an aggregate fruit

in which the receptacle enlarges and becomes fleshy. The numerous pistils in the flower mature into tiny achenes that are imbedded in the fleshy receptacle (Fig. 5.45).

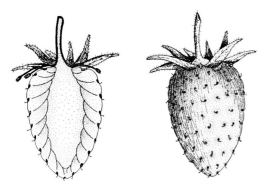

Fig. 5.45. Strawberry (an aggregate fruit with numerous small achenes embedded in the fleshy receptacle).

The pineapple is a multiple fruit in which the receptacles, pistils, and sepal bases of several flowers are united to form the fleshy tissues of the fruit (Fig. 5.46).

## SEED DISPERSAL

It may be assumed that the evolutionary development of fruits has been influenced by natural selection; i.e., the nature of the fruiting structure bears some positive relationship to survival of the species. Many factors might be considered in a full discussion of this topic but two of them are of major significance:

1. protection of the immature seeds
2. dispersal of the mature seeds

The fleshy portions of mature fruits are usually edible and form a part of the diet of numerous animal species. Before maturity such fruits are frequently inedible for reasons of taste or texture. This tends to prevent the destruction of the fruit before the seeds are mature.

When the seeds are fully mature it is advantageous for the species if they are transported away from the immediate vicinity of the parent plant. Animals may

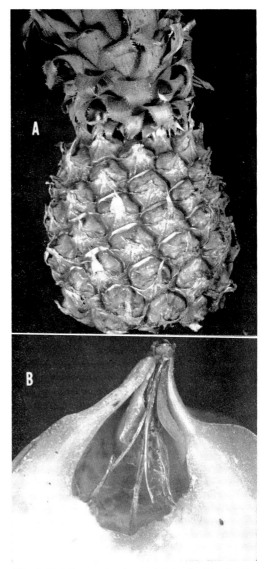

Fig. 5.46. Pineapple (a multiple fruit). A. Entire fruit. B. One of the segments, dissected and enlarged, showing the floral cavity with shriveled floral parts.

provide transportation by eating the flesh and throwing away the seeds.

In many cases the fruits are swallowed whole and the flesh is digested while the seeds pass through the digestive tract unharmed. Birds are particularly effective in spreading seeds this way. Such circumstances provide an explanation for the common occurrence of red cedars along fence rows.

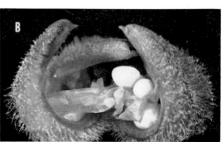

Fig. 5.47. Seed dispersal by shattering capsules. Capsule of a cultivated species of *Impatiens* before and after the violent dehiscence.

Among the fruits classified as being dry and dehiscent are many types in which tensions develop as the mature fruit dries. When the fruit opens these tensions are released violently. The fruit seems to shatter and the seeds are hurled away from the parent plant. The fruit of a common barnyard weed known as touch-me-not or jewel weed *(Impatiens biflora)* provides a classical example of such a mechanism. The illustrations in Fig. 5.47 are of the fruits of a cultivated species of *Impatiens*.

Many fruits develop hooks or spines of various kinds that become caught in animal fur and are carried away as the animal moves about. Some of these (beggar's ticks, cockleburs, stickseeds) are familiar to anyone who has tramped through a weed patch in the fall of the year (Fig. 5.48).

The fruits of dandelions and many similar plants are small and light in weight. As they mature each of the many fruits in the characteristic compact heads develops a whorl of hairlike appendages. These serve as parachutes that allow the fruits to float for long distances in air currents (Fig. 5.49).

In each of these cases a majority of the seeds are destroyed in various ways or transferred to places where successful growth of the next generation is impossible. However, seeds are produced in prodigious numbers and enough of them germinate in suitable environments to insure survival of the species.

## SEED GERMINATION AND ESTABLISHMENT OF THE NEW SPOROPHYTE

The germination period may be defined to include the time between planting of the seed and development of the seedling to a point when it is self-supporting. Many seeds begin to germinate shortly after being placed in a suitable environment while others are truly dormant and do not germinate for long periods unless specially treated in one of several ways. Some of the important environmental conditions that may influence the germination of seeds are temperature, soil moisture, oxygen supply, and light.

Temperature affects the rates of physical and chemical reactions necessary to the germination process. Absorption of water and digestion of reserve foods are examples of important reactions influenced by temperature. Seeds of different kinds of plants vary in their optimum temperature requirements for germination. Soil temperature has another vital influence on successful germination in that it affects the growth rates of disease-producing soil organisms which might destroy the germinating seeds.

The water content of mature seeds is usually low and the protoplasm of seeds has a low rate of metabolic activity. This dehydrated condition is important to the survival of seeds through periods of unfavorable environmental conditions. When

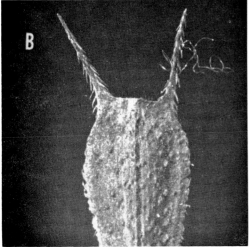

Fig. 5.48. Seed dispersal by hooks and spines. A. Cocklebur. B. One segment of the fruit of beggar's ticks.

Fig. 5.49. Seed dispersal by parachutes. Compact head of dandelion fruits, each of which is an achene with a whorl of radiating appendages.

Fig. 5.50. A comparison of dried peas (inner circle) with peas fully imbibed with water (outer circle).

water becomes available the cell walls and the protoplasm imbibe water and swell (Fig. 5.50). As more and more water enters the seed the speed of chemical reactions increases, particularly in digestive and respiratory processes. The formation of sugar from reserve foods lowers the relative concentration of water in the cell sap resulting in a further absorption of water by osmosis.

Oxygen is essential to the respiratory processes that release the energy necessary for growth and that also form many intermediate products of metabolism. Some seeds can germinate in low percentages of oxygen but most seeds need the normal concentration of this gas found in the atmosphere.

The light requirements of seeds vary. Some germinate only in the dark; some germinate in the light; others require alternating periods of light and dark. Many are not sensitive to light conditions and germinate in either light or dark.

Soil fertility does not have an appreciable effect on seed germination since most seeds contain enough inorganic and organic nutrients to insure good germination.

SEED DORMANCY

Dormant seeds frequently fail to germinate for considerable periods of time after being planted. Since many crop plants (clover, for instance) exhibit seed dormancy, this is an important matter and considerable research has been done with factors controlling dormancy.

In many legumes the seed coat remains impermeable to water for long periods of time. In some species of plants the seed coat may be permeable to water but resists swelling because of its mechanical strength. In nature various physical, chemical, and biological processes in the soil act on such seed coats, eventually softening them to the point where germination can occur. This process may take months or even years. However, germination can be hastened by a mechanical cracking of the seed coats or by one of several means of wearing them down *(scarification)*.

In many plants a mature ripened seed may contain a very small or otherwise immature embryo. For considerable periods of time after planting such embryos develop slowly within the seed. Perhaps months later they reach a size that permits them to emerge from the seed coat. Not much can be done in a practical way to hasten this slow process of embryo maturation.

Seeds of many plants have fully developed embryos as well as permeable seed coats yet still fail to germinate for long periods of time. During this period of after-ripening important biochemical changes occur in the seeds. One such change commonly cited is a gradual increase in acidity of the cell sap due to accumulation of carbon dioxide from respiration. Many other changes of similar nature have been investigated in various research laboratories. One practical method of decreasing the length of the after-ripening period is storage under moist conditions at low temperatures.

THE GERMINATION PROCESS

The first important phase in germination is the absorption of water by imbibition (Fig. 5.50). Dried seeds commonly have less than 20 percent water by weight. By way of contrast, young actively growing plants may consist of 95 percent or more water. The initial water that enters the seed becomes bound to the submicroscopic bundles of cellulose molecules in the cell walls and to the colloidal particles of protoplasm itself. Bound water molecules adhere tenaciously to the surfaces of such particles, increasing their volume and forcing them apart. Such imbibitional swellings, as noted previously, create tremendous pressures capable of moving or even cracking large stones.

A common laboratory demonstration of imbibition pressure involves placing a mixture of sand and dried peas in a jar. Enough water is added to thoroughly moisten the sand. Within a few hours the swelling peas exert enough outward pressure to break the jar. By filling the spaces between the peas the sand prevents the peas from swelling into these spaces.

Once the forces resulting in the imbibitional swelling are satisfied, free water accumulates in the cell wall and the protoplasm. This is free in the sense that molecules may escape from the wall into the cytoplasm and from the cytoplasm into the vacuole (or in the opposite direction). This condition contrasts to that of the bound water molecules which are not mobile. Under these circumstances it is possible for water to enter the cell and increase the volume of the cell sap. The soil solution usually has a higher concentration of water than does the cell sap and thus water tends to diffuse inward.

The hydration of the protoplasm also results in an increased enzymatic activity. Stored food in the endosperm or in the cotyledons is acted upon by digestive enzymes. The resulting simple soluble foods go into solution and this lowers the relative concentration of water in the cell. As a result more water enters the cells from the soil. The soluble foods are also available for normal metabolic processes including respiration.

As a result of the above changes the assimilation rate increases and more protoplasm is formed. This in turn makes it possible for nuclear and cell divisions to begin again and the number of cells in the embryo is thus increased.

The embryo increases in size to a point where it bursts out of the seed coat. The first part to emerge is the radicle or primary root, which develops from the growing point at the basal end of the hypocotyl. The primary root grows downward and becomes anchored in the soil (Fig. 5.51).

massive cotyledons and the other end is connected to the primary root the stretching hypocotyl is forced to bend upward, forming an arch (Fig. 5.52). This arch pushes upward through the soil. When it emerges into the light the arch is stimulated to straighten and pull the cotyledons and the epicotyl out of the soil.

Fig. 5.51. Seeds of bean before and after the start of germination. Note the emergence of the radicle.

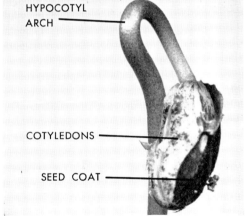

Fig. 5.52. Stages in the germination of bean seeds. A. Hypocotyl has elongated and bent upward to form the hypocotyl arch. B. Growth of the hypocotyl has pulled the cotyledons, epicotyl, and remains of seed coat above ground.

The root tip is continuously forced through the soil due to enlargement of cells in the region immediately behind the tip. This enlargement is due mainly to water absorption but a significant amount of new protoplasm is manufactured also.

In some plants, beans for example, the hypocotyl also begins to elongate. Since one end of the hypocotyl is attached to the

In other plants (peas for example) the hypocotyl does not elongate, instead the epicotyl begins to grow, forcing its way upward and out of the soil. In such plants the cotyledons and the unelongated hypocotyl remain underground (Fig. 5.53).

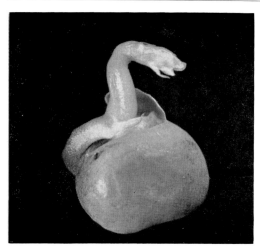

Fig. 5.53. Germination of a seed of the garden pea. Note that the epicotyl rather than the hypocotyl is emergent.

emerge from the soil without being mechanically injured (Fig. 5.54).

Perhaps the most important function of cotyledons is the absorption of food from the endosperm, either before the seed is mature or during the germination process. Sometimes the cotyledons shrivel up and drop off as soon as the young seedling has developed enough leaves to manufacture its own food (Fig. 5.55). However, in many plants the cotyledons may become green and function as leaves for long periods of time. The embryo and seedling development of castor beans provides a remarkable example. In the seed of this plant the cotyledons are thin, papery, and colorless, but in the seedling they become large and green.

The coleoptile of the corn embryo has a special function in germination. It has a pointed tip that penetrates upward as it grows. When the coleoptile emerges from the soil it stops growing and has no further function other than providing a convenient tube through which the epicotyl can

## GROWTH AND DIFFERENTIATION

In the life cycles of most plants there are stages when the future plant body consists of but a single cell. One of these stages is the zygote; the diploid plant body or sporophyte develops from this single cell. As has been emphasized, the major subdivisions of this growth process are cell

Fig. 5.54. Stages in the germination of a grain of corn. A. Emergence of radicle. B. Emergence of coleoptile. C. Upward growth of the coleoptile. D. Emergence of the epicotyl through the tube of the coleoptile.

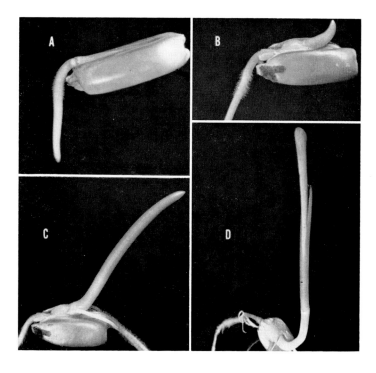

Fig. 5.55. Bean seedling with primary leaves fully expanded. Note the shriveled cotyledons which will fall off shortly.

division (including both mitosis and cytoplasmic division), cell enlargement, and cell differentiation.

Since all the cells in a plant are derived from the zygote they all have, theoretically, the same inherent potentialities; i.e., they all contain the same numbers and kinds of chromosomes and genes. Why then should one cell become a palisade parenchyma cell while another becomes a root hair? Why should one cell in a procambium strand become a xylem vessel while another just a few microns away becomes a sieve tube element? Why should a mature parenchyma cell in the cortex of a stem suddenly revert to the meristematic condition and form a cork cambium cell?

This general topic was discussed briefly when the nature of chromosomes and genes was considered and a conclusion was reached that the position of a particular cell in a mass of similar cells had a definite effect on its future development.

The shapes of cells can be cited as a case in point. Individual cells free of restraint tend to become spheres. Cells in masses, however, exert pressure against each other and become polyhedral rather than spherical. Cells in an apical meristem have an average of 14 faces per cell. Each face represents a contact with a neighboring cell. Immediately following a cell division each of the 2 daughter cells has fewer than 14 faces but during the succeeding interphase the number of faces gradually increases to more than 14 because of divisions in neighboring cells (Fig. 5.56).

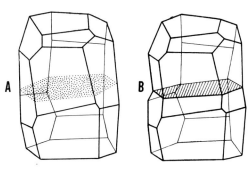

Fig. 5.56. Three-dimensional schematic representation of the division process in a cell of an apical meristem. A. Position of the cell plate as it is formed during cell division. B. Readjustment of newly formed cross wall and adjacent walls after completion of cell division.

The walls of meristematic cells are not rigid and are subject to surface tension adjustments. New cell walls often join existing walls at right angles. Such angles, however, are not stable in a system affected by surface tensions; they soon become readjusted to angles that are stable.

It is evident therefore that the shape of a particular cell in a meristem is subject to constant change due to division in neighboring cells and readjustments due to surface tensions as well as its own growth.

The position of the epidermal cells as the outermost layer of cells places them in a situation different from all other cells in the plant. Their direct exposure to the environment apparently fosters the chemical changes that result in the formation of the cuticle in leaves and stems.

The position of the cambium between a source of food in the phloem and a source of water and chemicals in the xylem undoubtedly facilitates its function as a meristem.

The position of an otherwise normal cell in the young ovule affects its nutrition so that it becomes a megaspore mother cell and ultimately undergoes meiosis.

The supplies of food, oxygen, and water available to the epidermal cells of roots are important factors in the characteristic elongation of root hairs.

These examples indicate that a cell in a mass of other cells is affected by its immediate microenvironment. The relative availability of organic foods, water, oxygen, inorganic salts, and so on all depends on position. Sometimes position also affects the ability of a cell to eliminate the products of its own metabolism. The accumulation of some of these products is known to have an important effect on the ultimate nature of the cell.

It is known further that chemical substances produced in one part of a plant may influence the growth of other parts. A classical example of such an effect is the inhibition of the lowermost buds on a twig by substances transported downward from the buds above. This phenomenon is termed *apical dominance* and it plays an important part in the normal growth habits of plants. An understanding of apical dominance makes possible a certain degree of control over plant growth habits. Certain varieties of *Chrysanthemum,* for example, are most desirable if they are bushy. To achieve this aim the growth of branches of the main shoot is stimulated by pinching out the terminal bud, thus releasing the lateral buds from their normal inhibition.

Plant hormones that affect growth in such fashion are often called *auxins*. One of them has been isolated and identified as the relatively simple organic chemical *indoleacetic acid*. The original discovery of this compound in plants stimulated a vast amount of research with both natural and synthetic growth-regulating substances. A very significant result of this research was the discovery that certain of the synthetic compounds when applied in excess would kill many plants. One of the most notable of these compounds, widely known as 2-4-D, selectively kills dandelions and many other broad-leaved weeds while having no permanent effect on lawn grasses if applied in proper concentration.

Growth substances apparently are controlling factors in the upward growth of stems and the downward growth of roots. The production of these substances seems to be associated with new cell formation in meristems; they are known to be transported away from such meristematic regions. Neither the actual pathway nor the mechanism of movement is known but it is evident that the movement is affected by both light and gravity.

If a stem is held in a horizontal position, growth substances accumulate along the lowermost side and stimulate the stem to grow more rapidly on that side than on the upper. This causes a stem to bend upward.

In roots the effect of gravity is similar but the reaction is reversed. The accumulation of excess growth substances on the lower side of a root held horizontally inhibits the elongation of cells on the lower side. Cells on the upper side continue to grow at a normal or possibly an accelerated rate and thus the root bends downward.

When light shines on a stem from only one side it is known that the growth substances on the shaded side accumulate in excess of those on the lighted side. The shaded side is stimulated to grow more rapidly than the lighted side and the stem bends toward the light.

Explanations such as these, which are based on scientific evidence, are more acceptable than are the popular fallacies that the stems are reaching for light while the roots are reaching for water.

However, it is not yet known how it is that light and gravity affect the distribution of the growth substances. Nor is it known exactly how they affect the growth of plant cells. In a given cell there is a balance between the tendency of the protoplast to swell under the influence of osmotic absorption of water and the resistance of the cell wall to expansion. The growth substances might increase the first or decrease the second since either event would shift the balance toward greater cell enlargement.

Growth responses such as those which have been described are *tropisms*. The response to gravity is called *geotropism*. (Stems are negatively geotropic while roots are positively geotropic.) The response to light is called *phototropism*.

In addition to the use of growth-regulating substances as weed killers there are many other interesting applications of these compounds. For example, it was discovered early that if cuttings or slips are soaked in dilute solutions of indoleacetic acid before being placed in propagating benches they develop roots more rapidly and in larger numbers. The related chemical indolebutyric acid is even more effective in this regard and is widely used by commercial horticulturalists.

The loss of fruit from orchard trees by premature fruit drop has long been a problem to orchardists. Fruit fall is usually due to the formation of an abscission layer across the base of the stalk. Following the discovery that certain growth-regulating substances such as napthaleneacetic acid prevent the formation of this layer it has become more or less standard practice to spray orchards with them.

Similarly such sprays are useful in preventing the sprouting of potatoes in storage. This is an interesting practical application of the results of studies into the nature of bud inhibition.

One of the applications of growth-regulating substances which has been of great popular interest is the stimulation of certain plants to produce fruits without seeds. In the normal development of fruits the growth of pollen tubes and meristematic activity in the developing ovules seem to stimulate the ovary to begin its enlargement into the fruit. Apparently the artificial application of appropriate growth regulators stimulates the ovary in a similar manner. If pollination has not occurred the resulting fruits are likely to be seedless.

The seedlessness is in itself of no great significance in commercial practice but the ability to accomplish fruit set in such plants as tomatoes when conditions for normal pollination are not satisfactory has been of great practical value.

The actions and interactions of plant growth-regulating substances are of intense interest to researchers and practitioners alike. Our knowledge of them is increasing so rapidly that they cannot be treated adequately in a brief course and interested students are strongly advised to take a subsequent course in plant physiology. Several current textbooks of plant physiology deal with them competently and students may use their indices to find discussions under such headings as *auxins, gibberellins, cytokinins, abscissic acid,* and *ethylene gas,* each of which functions in one or more ways as a growth-regulating substance.

The term *vitamin* applies to a group of growth-regulating substances common to both plants and animals. Information concerning these important compounds has been accumulated intensively since the early part of this century. They occur in such small quantities that they cannot be considered significant as basic foods or building materials. In most cases where their functions have been elucidated they facilitate many essential chemical reactions as parts of enzymes.

Two general groupings of vitamins are recognized. The water-soluble vitamins include the vitamin B complex and vitamin C (ascorbic acid), while the fat-soluble vitamins include vitamins A, D, K, and E.

In general animals obtain their vitamins directly or indirectly from plants. However, the experimental feeding of animals with vitamin-free diets is complicated by the fact that the bacterial flora in the intestinal tracts of animals often accomplish a synthesis of some of the deficient vitamins.

The recognition of the relation of vitamins to certain human diseases has been a dramatic part of medical history. Two horrifying diseases in particular, beri-beri and scurvy, were shown to be due to deficiencies of vitamin $B_1$ and vitamin C, respectively.

In the Orient, where polished rice is a major ingredient of the diet, a long educational process is necessary in order to effect a realization that the polishing process removes the main natural source of vitamin $B_1$.

In days gone by sailors on long voyages, where supplies of fresh fruits and vegetables were not practicable, often developed scurvy. It was learned by empirical means (trial and error) that citrus fruits could be used to alleviate the symptoms of this disease. It has since been established that citrus fruits are outstanding sources of vitamin C and thus the empirical cure for scurvy has been given a scientific basis.

Deficiencies of vitamin D cause rickets in children. The discovery that radiation with ultraviolet light results in the formation of this vitamin from its precursor has made possible an inexpensive method for adding vitamin D to such common foods as milk.

Vitamin A is related to the night vision and color perceptive abilities of certain animals including humans. The precursor to vitamin A, beta carotene, is available from plants, particularly the green or yellow plant parts. The molecular structure of this vitamin is essentially that of a beta carotene molecule cut in half.

The necessity of vitamins for plant growth has been more difficult to establish, particularly since the green plants are able for the most part to make their own supplies.

Some years ago investigators developed techniques for growing roots in culture without stems and leaves. In these experiments it was found necessary to add certain vitamins, particularly members of the vitamin $B_1$ complex, to culture solutions in order for the roots to grow normally.

A vast amount of biochemical research has been carried on with microorganisms such as bacteria and fungi. Many of these organisms synthesize certain vitamins in such quantities that they can be extracted commercially. Other forms have been shown to need certain vitamins or their precursors in order to grow normally. These interesting organisms have enabled research workers to gain an insight into the biochemical reactions in which specific vitamins are concerned.

Researchers have investigated the relation of growth-regulating substances to changes in apical meristems as they shift from leaf production to flower production.

In many plants a direct relationship exists between the length of days and flowering: the *photoperiod*. *Long day plants* are those that flower when days increase in length beyond a critical point that varies with species. *Short day plants* are those that flower after days have decreased beyond a critical point in the latter part of the year or before they have reached a critical length in the first part of the year. In some plants day length is not a factor in flowering; they are said to be *indeterminate*.

Present evidence strongly suggests that the length of the *dark period* also has an influence on flowering and thus short day plants might be termed *long night plants* and long day plants might be termed *short night plants*. One example of greenhouse management practice that is of considerable economic significance is the preparation of poinsettias for the Christmas market. Days are shortened (nights lengthened) artificially by blacking out greenhouses at a certain time each day. During the critical period of induction of the desired changes the whole process may be aborted by the brief turning on of a small incandescent light at the time of the blackout.

The induction of flowering has to do basically with changes in apical meristems. During vegetative growth stem tips give rise to leaves, nodes, internodes, and axillary buds. Flowers are modified branches and flower parts are modified leaves. Thus flowering involves changes in the nature of parts produced by apical meristem.

Apparently the actual induction begins in leaves and a substance is produced that is transported to the apical meristems, which are induced to begin the changes leading to flowering. The substance is as yet unknown. Determined efforts for more than half a century have failed to result in the clearcut identification of a flowering hormone.

One important scientific advance, however, may lead to an understanding of the timing mechanism. A faintly blue-green pigment called *phytochrome* is now known to occur in leaves in extremely small amounts. This pigment exists in two forms that are said to be light reversible.

1. The $P_{red}$ form absorbs red light (660 nm) and is transformed to
2. the $P_{far\ red}$ form, which absorbs far red light (730 nm) and is changed back to the $P_{red}$ form.

$P_{far\ red}$ form is considered to be the one that is significantly active in the flower induction process. Apparently it converts back to the $P_{red}$ form gradually during periods of darkness and the rate of this conversion may in some way be related to the night length effect on flowering.

Consider a sunset sky and recall that the earth's atmosphere has a refractive influence on sunlight. It is evident that portions of the spectrum with longer wavelengths (at the red end) disappear at sunset before those with shorter wavelengths (at the blue end). The longer wave length portions also appear later in the sunrise spectrum. It follows from this that in the natural environment far red light would disappear at sunset before red light and that red light would have the last effect of the day on phytochrome which should then enter the night in the $P_{far\ red}$ form.

As has been noted above, day length effects on plant activities are varied and numerous. The opening of buds in spring and the shedding of leaves in autumn are two examples of major plant activities affected by day length. Since change in the length of days is one of the few environmental parameters that occurs in an orderly, predictable, and reliable procession each year it seems reasonable that evolutionary modifications of plant activities have been geared to this change as part of the biological clock.

# SIX

ALGAE are a familiar sight to most people. The brownish green rockweeds of the seashore, the greenish "frog-spit" on the surface of a pond, the misnamed "moss" in a watering tank, the green discoloration of the water in an aquarium, the green stain on the damp and shaded side of a fence post all are examples of algae. Some are green as grass. Others are yellow, brown, red, purple, blue, blue-green, or other colors.

The term algae applies to a vast array of plants, many of which are only distantly related to one another. They range in size from single-celled structures barely visible under the microscope to ocean giants 30m or more long. Primarily they are restricted to moist or wet environments. Most of them contain chlorophyll and are photosynthetic although a few species of non-chlorophyllous algae are known. In some groups of algae there are accessory pigments that alter the normal green color caused by the complex of chlorophyll pigments. In this chapter we will be concerned with the *grass-green algae* since they are considered to be the progenitors of the higher green plants.

In the green algae two chlorophylls (chlorophyll *a* and chlorophyll *b*) usually occur in sufficient quantities to mask the presence of carotenes and xanthophylls, thus creating the characteristic green color. In this respect the green algae are very similar to all of the higher green land plants. None of the other groups of algae (except the euglenoids) contain chlorophyll *b* and none have quite the same greenness as the green algae.

## *CHLAMYDOMONAS*—A COMMON FRESHWATER ALGA

Much of the following discussion is based on the nature and reproductive mechanisms of *Chlamydomonas,* a single-celled, motile, green alga which is a common component of freshwater algal floras (Fig. 6.1). Individual *Chlamydomonas* cells are so minute (perhaps 10–20µm in diameter) that they appear merely as green dots when observed with low magnification. Yet when a single individual is introduced into a medium suitable for its growth it will multiply so rapidly that the solution appears green within a few days' time.

Although *Chlamydomonas* may seem to be a simple structure, it is in reality a very complicated organism. Figures 6.1 and 6.2 reveal some of the parts of this cell that can be identified when a living individual is observed through a microscope.

In the living condition *Chlamydomonas* cells move about rapidly but occasionally one will become stuck for a moment. With proper lighting the two *flagella* may then be seen thrashing about. It is this whiplike motion of the flagella that moves

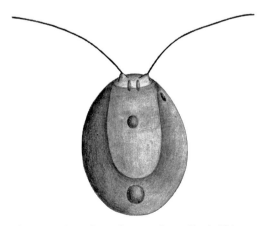

Fig. 6.1. Drawing of an entire cell of *Chlamydomonas* as seen in the living condition (about 1000X).

# THE GREEN ALGAE

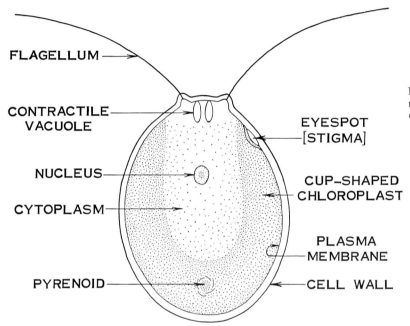

Fig. 6.2. Schematic optical section of a cell of *Chlamydomonas*.

the cell. Sometimes when a cell is oriented correctly a bright reddish spot may be observed against the green background of the cell. This is the *eyespot* or *stigma*. Experimental evidence indicates that this structure is light sensitive and in some way which is not understood controls the movements of the flagella so that the whole cell may move toward weak light or away from light that is too strong. Occasionally clouds of these green cells congregate in zones of optimum light concentration.

Characteristically the outer surface of plant cells consists of a definite cell wall in which cellulose is a major component. In many plants the wall is thick enough to be readily discernible but in *Chlamydomonas* it is thin and difficult to distinguish.

The *protoplasm* within the cell wall has several component parts but not all of them are clearly evident in living cells. The *nucleus* is a spherical body that occupies a more or less central position in the cell; it is often difficult to observe without special staining.

Surrounding the nucleus is the *cytoplasm*, a substance having a consistency similar to egg white. Very little evidence of the actual complexity of cytoplasmic structure is obtained by observations with the light microscope.

The green pigments are located in highly modified segments of the cytoplasm which are known as *chloroplasts*. In *Chlamydomonas* each cell has one chloroplast, a massive cup-shaped structure larger than any other structure in the cell. It appears at first glance that the chloroplast cup is filled with cytoplasm; in reality, the whole chloroplast is completely imbedded in the cytoplasm. However, that portion of the cytoplasm between the wall and the chloroplast is such a thin layer that it is rarely discernible with the light microscope.

*Chlamydomonas* does not possess a central vacuole but two small bubbles apparent in the cytoplasm near the base of the flagella are called *contractile vacuoles.* They alternately appear and disappear as the result of contractions which eject small drops of fluid from the cell.

Critical observations show that the paired flagella are associated with two small *basal granules. Mitochondria,* stacks of *Golgi membranes, ribosomes,* and the *endoplasmic reticulum* are clearly evident in electron micrographs of this organism.

Cells such as those of *Chlamydomonas* are liable to destruction by explosion if they absorb too much water by osmosis; starch formation plays an important role in the control of water absorption through the simple act of taking sugar out of solution. A dense area of variable size appears as a definite spot in the basal region of the chloroplast. This is the *pyrenoid* and tests with an iodine solution clearly demonstrate that it is a center of starch formation. In rapidly expanding populations having cells with high metabolic rates, starch does not accumulate rapidly and the pyrenoids remain small. On the other hand, when division rates slow and populations become static the pyrenoids tend to become large with stored starch.

THE LIFE CYCLE OF *Chlamydomonas*

*Chlamydomonas* plants normally occur in large numbers wherever they are found. One cell placed in a suitable environment can give rise to a vast number of progeny within a few weeks. This is accomplished by the division of 1 cell to form 2, 2 to form 4, 4 to form 8 and so on (Fig. 6.3). Cells may remain clumped together within the parent cell wall for 2 or 3 or even more divisions but eventually they separate and live as single individuals. The process of new cell formation involves a precise division of the nucleus *(mitosis)* and an equal division of the cytoplasm each time cell division occurs. In this plant the division of the cytoplasm is accomplished by an inward furrowing of the cytoplasm rather than by a cell plate. The daughter cells that are formed are identical in appearance with the parent cell as well as with each other. They are somewhat smaller than the parent but are capable of rapid growth. When first released these daughter cells are called *zoospores* but each soon becomes a normal-sized adult plant.

The term *mitospore* is sometimes applied to such spores because it emphasizes that mitotic nuclear divisions are associated with their formation and therefore that no genetic changes occur.

This type of reproduction is usually called *asexual reproduction.* As long as the environment remains suitable to the growth and division of *Chlamydomonas* it is probable that reproduction will occur in this manner for indefinite periods and the number of individuals becomes enormous. Careful estimates show that there may be more than 10,000 individuals per ml of the culture solution when it has a decided green color.

However, no environment, particularly a freshwater environment, is likely to remain unaltered indefinitely. A few of the more obvious changes which could occur are:

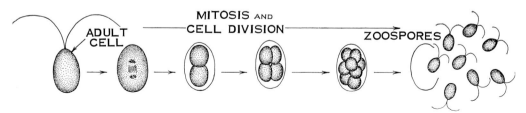

Fig. 6.3. Asexual reproduction in *Chlamydomonas.* (Mitosis and cytokinesis resulting in the production of zoospores.)

1. Natural ponds occasionally dry up.
2. Seasonal changes in temperature occur.
3. Many kinds of chemical changes may occur in the water.
4. Other plants compete for light, carbon dioxide, oxygen, mineral salts, and space.
5. When overcrowded in the environment individual cells of *Chlamydomonas* compete among themselves.

As a result of changing factors in the environment (perhaps the most important being a depletion of nitrogen) the cells of *Chlamydomonas* undergo physiological changes that cause them to unite in pairs instead of undergoing further growth and asexual divisions. Each member of a uniting pair is called a *gamete* and the fusion of gametes is an important step in the *sexual reproductive cycle* (Fig. 6.4).

In certain species of *Chlamydomonas* the gametes do not differ markedly in appearance from normal vegetative cells. However in some species of this genus, as well as in many other genera of algae, gamete formation involves the division of adult cells into large numbers of individuals that are much smaller than zoospores. These divisions are like the asexual divisions in that they involve mitosis and cytoplasmic division but they do not stop at 4, 8, or even 16 individuals. Divisions continue until 1 cell has been partitioned into a much larger number of cells, the number being on the order of 64, 128, or even more.

The many small individuals are gametes but except for their smaller size they are similar in appearance to zoospores. They differ physiologically, however, in that they do not have the ability to grow to the normal size for adult cells. Nor do they have the ability to divide again. Thus they have no future unless they pair with other gametes.

When 2 gametes which are capable of fusing come together the flagella become entangled and the cell walls are shed. Then the cytoplasmic membranes at the point of contact are dissolved and the cytoplasm of one blends with the cytoplasm of the other. Eventually the 2 nuclei fuse to form 1 nucleus that has twice as many chromosomes as does the nucleus of a single gamete (or a zoospore or an adult plant). The fusion process is called *syngamy* and is a natural event in the life cycles of all of the plants we will study. Syngamy actually involves two related processes: (1) *plasmogamy*, the fusion of the cytoplasms of the gametes, and (2) *karyogamy*, the fusion of

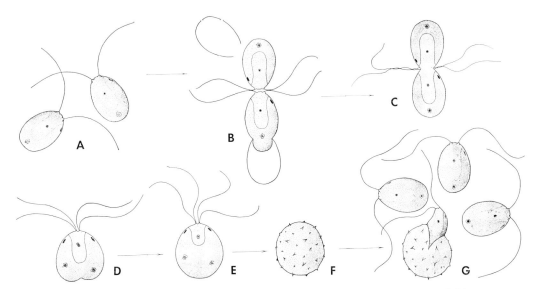

Fig. 6.4. Sexual union in *Chlamydomonas* resulting in the formation of a diploid zygote. Meiosis in the germinating zygote is followed by the release of 4 haploid zoospores.

their nuclei. In many plants karyogamy occurs imediately after plasmogamy but in some fungi long periods of time may elapse between the two steps.

The fusion cell itself is called a *zygote* and in *Chlamydomonas* the zygote soon loses its flagella, and becomes a thick-walled resting cell. In this condition it can survive desiccation and endue a greater range of temperatures than can vegetative cells. In a sense it is a survival mechanism that permits the inactive protoplasm within to remain alive whether frozen in icy mud or carried aloft in a dust storm. When conditions again become favorable germination of the zygote may occur. (A number of research workers are at present concerned with the actual conditions that induce zygote germination in various species of algae.) It is pertinent here to define two terms which may be applied to zygotes that become thick-walled resting cells. If the resting zygote was formed by the fusion of gametes which are similar in appearance *(isogametes)* then it may be termed a *zygospore*. By way of illustration, the resting zygotes of *Chlamydomonas,* described above, are zygospores.

If, however, the zygotes were formed by the fusion of gametes which are distinctly unlike in appearance, i.e., *eggs* and *sperms,* then the resting zygotes are termed *oospores*.

When a zygospore of *Chlamydomonas* germinates the wall cracks open and 4 zoospores emerge. These soon grow to normal adult size; each can initiate a new population. It is an important point that each of the 4 zoospores has a haploid nucleus. This is the result of *meiosis* which occurs in the zygote prior to its germination. Because they are the products of meiosis these particular zoospores also may be called *meiospores*. (Meiosis was described in the Introduction.)

If, for example, the diploid nucleus of the zygote has 16 chromosomes, this number is halved during meiosis and the nucleus of each one of the 4 zoospores has 8 chromosomes.

Meiosis plays a significant and necessary role in the sexual life cycle. If the chromosome number were not reduced at some point the number of chromosomes in a nucleus would be doubled by gamete fusion in one generation, redoubled in the next, and so on until an unwieldy number of chromosomes accumulated. Also meiosis accomplishes a reshuffling of the chromosomes and genes that affects the hereditary nature of succeeding generations.

It is now impossible for science to fix a time when the evolution of sexual reproduction began. It seems very likely that asexual methods of reproduction evolved first and were the only mechanisms for a long period of time. Possibly the union of two cells that we recognize as the initiation of the sexual cycle was a successful accident that resulted in a somewhat larger vegetative cell rather than a resting zygote. This may then have given rise to a whole population of similar diploid cells by asexual reproduction. It is even conceivable that the chromosome number may have doubled several times in this manner (as well as in other ways now well known to research workers). Furthermore, when the chromosome numbers became so large that the normal mechanisms of mitosis became inadequate, the stage would have been set for the evolution of meiosis.

## SEXUAL DIFFERENCES BETWEEN GAMETES IN THE ALGAE

In the most primitive condition all the gametes of a species are so alike that any two may fuse even though they come from the same parent cell. However, from this primitive condition a wide range of differences between gametes has evolved. These differences can be summarized by a consideration of three morphological terms: *isogamy, anisogamy,* and *oogamy,* and two physiological concepts: *homothallism* and *heterothallism*.

When all the gametes of one species appear alike, the condition is termed *isogamy*. Commonly *isogametes* are small, highly motile, and produced in very large numbers. However, in at least one group of green algae isogametes are formed that lack flagella entirely and consist of the undivided contents of previously vegetative cells.

*Anisogamy* is a condition in which the gametes are similar in apearance but differ in size and motility as well as in the numbers produced by the parent cell. Here we see evidences of the beginnings of morphological sexuality. Female gametes are like male gametes in general shape, number of flagella, and color. They are larger than the male, less numerous, and move about much more slowly.

*Oogamy* is a state in which female gametes have become highly modified while male gametes have retained such primitive features as being of small size, highly motile, and very numerous. Female gametes have become much larger and have completely lost their motility. Usually they accumulate large amounts of reserve food. Female gametes of this type are referred to as *eggs* while the smaller, motile, male gametes are called *sperms*.

Sometimes the eggs may be released from the parent before fertilization but in many algae they are retained within the cell wall in which they are formed. The enclosing wall is referred to as an *oogonium* and in most algae the zygotes are released from the oogonium before germination takes place, usually by a decay or breakage of the wall.

In summarizing the nature of morphological sexuality in the algae it is evident that male gametes tend to be smaller than females, are produced in larger numbers, have little reserve food, and usually have some means of locomotion that enables them to reach the female gametes. Many more male gametes are produced than females and therefore most of them perish. However, the active movements of large numbers of male gametes tend to insure that at least one will make contact with each female gamete.

Female gametes tend to be larger than males, are produced in fewer numbers, are less motile or completely nonmotile, and have greater food reserves. They may have the added protection of being retained within the parent cell wall. The chief advantage of the egg lies in the fact that it contains a large supply of stored nutrients. It is able to build a heavier wall around the zygote and when germination occurs the new haploid plants that emerge will be larger and more vigorous than those emerging from zygotes formed by isogametes. Possibly this advantage outweighs the concomitant disadvantage that eggs are produced in smaller numbers.

(It should be noted here that some authorities combine anisogamy and oogamy into a single term, *heterogamy,* which applies to the union of gametes that are different in appearance without defining the degree of difference.)

Functional differences in sexuality do not always parallel structural differences between gametes. For example, some isogametes will fuse only with isogametes from different plants of the same species while, on the other hand, some species produce eggs and sperms capable of a normal fertilization on the same plant body. Careful analysis of both physiological and morphological expressions of sexuality is necessary in order that the nature of a plant can be correctly interpreted. The basic terms used to describe physiological sexuality of the gamete-producing plants are *homothallism* and *heterothallism.* (The word thallus refers to the plant body in the lower plants.) When homothallism exists the gametes produced by the same thallus or in a clonal culture are capable of fusion to form zygotes. (A clonal culture is derived from a single individual by means of asexual reproduction.) In the heterothallic condition the gametes must come from different thalli or clones in order to fuse.

When the gametes involved are recognizably different as male and female gametes the thalli which produce them may be called male and female plants. However, when isogametes are involved there is no basis for determining which of the two strains is male or female. In such cases they are referred to as plus and minus strains or mating types.

Very often the separation of genetic factors that control maleness and femaleness or plusness and minusness occurs during meiosis and 2 of the 4 resultant nuclei carry the determiners of maleness while the other 2 carry determiners of femaleness. In such cases the term heterothallism has a clear-cut meaning and its use can lead to no confusion.

In some organisms that produce both male and female gametes on the same thallus these gametes are incapable of uniting with each other and must find appropriate gametes from other thalli in order to accomplish fertilization. Such plants have the appearance of homothallism but are functionally heterothallic.

The genetic factors controlling this type of sexuality are seemingly more complex than those controlling the simple heterothallism described above. Many organisms have been described as being homothallic on morphological evidence only and in such cases careful experiments involving clonal cultures may be necessary to determine the actual type of sexuality involved. In cases where such determinations have been made the species may be described as being either *bisexual and self-fertile* or *bisexual and self-sterile* in order to avoid misunderstandings.

The terms *monoecious* and *dioecious* are often used in place of homothallic and heterothallic. A rough expression of the meaning of monoecious is "one household" and it is used to indicate that both gametes are produced by the same plant. Dioecious means "two households" and implies that one kind of gamete develops in one plant and the other kind in another plant. However these terms are also subject to misunderstanding because they are used much more commonly in describing the *heterosporous* nature of the diploid generations of higher plants. This is an entirely different matter than sexuality in the haploid generation.

*Chlamydomonas* has been cited as an alga demonstrating both isogamy and homothallism. However, there are many species in this genus and some of them have highly advanced types of sexuality, ranging from isogamy to oogamy and from homothallism to heterothallism. A considerable amount of research with *Chlamydomonas* is being done in the hope of finding a biochemical basis for sexuality.

## STRUCTURAL VARIATIONS AMONG THE GREEN ALGAE

Many green algae differ greatly from *Chlamydomonas* in size and appearance. Since most of them can be observed critically only with a microscope, structural details become important as recognition features as well as in deriving systems of classification.

### Cell Size

Among the numerous species of *Chlamydomonas* cell sizes range from less than 5 μm to more than 25 μm in diameter, while individual cells of certain other genera may be measured in cm. As noted previously, central vacuoles do not occur in *Chlamydomonas* but they do occur in many other green algae (for example, *Spirogyra*) and greatly affect the cell size. It is important to recognize that cells with approximately equal masses of protoplasm may occupy vastly different amounts of space depending on whether or not central vacuoles are present.

### Cell Shape

One would expect a delicate-walled, single-celled, green alga to assume a spherical shape. The cells of a number of species are spherical and many others are slightly modified into smoothly rounded egg shapes. But there are hundreds of species of single-celled green algae and among them are forms with polyhedral shapes, disc shapes, needle shapes, and twisted shapes, as well as those with various types of projections, including spinelike outgrowths.

In species where individuals are made up of cell masses the shapes of individual cells are affected by contacts with other cells and tend to be polyhedral. However, the free surfaces may be rounded or be as variously modified as the single cells noted above.

The threadlike *filaments* of many

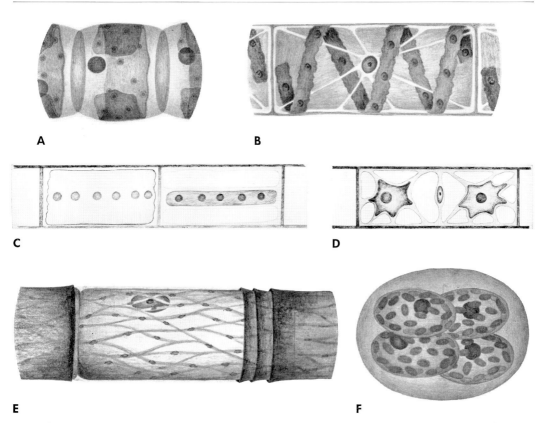

Fig. 6.5. Chloroplast types in the green algae. A. Bracelet-shaped chloroplast of *Ulothrix*. B. Spiral chloroplast of *Spirogyra*. C. Axial plate chloroplast of *Mougeotia*. D. Stellate chloroplast of *Zygnema*. E. Reticulate chloroplast of *Oedogonium*. F. Numerous small lens-shaped chloroplasts of an *Oocystis* species.

algae are made of cylindrical cells joined together end to end. When cell divisions occur in such algae the new cell walls form in planes perpendicular to the long axis of the filament; the filament becomes longer as the daughter cells grow in length.

CHLOROPLAST VARIATIONS

The massive cup-shaped chloroplast characteristic of *Chlamydomonas* may be thought of as a basic type, particularly in the green algae. Through evolutionary change this basic shape has been modified in various ways. Factors such as cell enlargement, which stretched the chloroplast, and development of the central vacuole, which displaced it, were probably involved in such modifications. Some of the more common types of chloroplasts are illustrated in Fig. 6.5. In considering these one should keep in mind three major points: (1) each of the cells illustrated has a central vacuole; (2) chloroplasts are always surrounded by a layer of cytoplasm and never lie free in the vacuole; and (3) each chloroplast may have one to several pyrenoids.

*Ulothrix* (Fig. 6.5A) has a single chloroplast, a broad band partially encircling the cell in the cytoplasm that is pressed against the wall by the fluids in the central vacuole. This type of chloroplast is often said to be *bracelet-shaped*. In *Spirogyra* (Fig. 6.5B) the band has been stretched to a long narrow ribbon twisted into a spiral pattern, hence the term *spiral chloroplast*. (Incidentally, *Spirogyra* cells often illustrate a rather common situation in the algae wherein the nucleus is more or less centrally located in a small mass of cytoplasm that is suspended in the vacuole by slender threads of cytoplasm connected to the cytoplasm lining the cell wall.) As shown in Fig. 6.6, *Spirogyra* cells may have two or more chloroplasts.

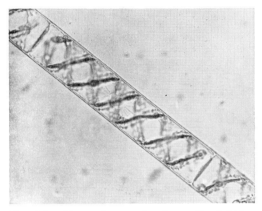

Fig. 6.6. Photomicrograph of a cell of *Spirogyra* with 2 chloroplasts twisting in opposite directions.

In *Mougeotia* (Fig. 6.5C) the flattened chloroplast is an *axial plate* completely imbedded in a thin layer of cytoplasm that extends lengthwise across the vacuole. This apparatus may rotate under the influence of light so that it may be seen in its entirety or only in edge view. In *Zygnema* (Fig. 6.5D) there are two massive chloroplasts, one in each cell half, with the nucleus suspended in a cytoplasmic bridge between them. Each of these chloroplasts is completely imbedded in cytoplasm and suspended in the vacuole by cytoplasmic threads. Since the chloroplasts often protrude somewhat into the cytoplasmic threads they tend to have a number of pointed projections, giving them a *stellate* appearance. The chloroplasts of *Oedogonium* (Fig. 6.5E) form a finely divided network and are said to be *reticulate*. In order to determine this one should examine young cells if possible because in older cells enlargement of the pyrenoids frequently destroys the network.

The cells of a few green algae contain many small chloroplasts, each shaped like a lens (more or less the condition found in the green cells of the higher plants). The choice of a green alga to illustrate this type of chloroplast proved surprisingly difficult. *Vaucheria,* a very common alga, was used for this purpose for many years but this genus is no longer included in the green algae. The species of *Oocystis* (Fig. 6.5F) finally chosen to illustrate the point is one of the common inhabitants of the community of microscopic, free-floating, living organisms known as *plankton*.

PLANT BODY TYPES

In many algae the entire body of a single individual may consist of a single cell. This cell may be motile or not, depending on whether it has flagella. In others individuals may consist of two to many cells joined in various ways. The list below gives the names of several types of plant bodies found in the green algae and the names of some common genera which are illustrated on following pages. The names applied to the body types are essentially self-explanatory.

1. Motile unicell—*Chlamydomonas* (Fig. 6.1)
2. Nonmotile unicell—*Chlorella* (Fig. 6.7)
   —most desmids (Figs. 6.8, 6.9)
3. Motile colony—*Pandorina* (Fig. 6.10)
   —*Volvox* (Fig. 6.11)
4. Nonmotile colony—*Scenedesmus* (Fig. 6.14A)
   —*Pediastrum* (Fig. 6.14B, C)
   —*Hydrodictyon* (Fig. 6.15)
5. Unbranched filament—*Ulothrix* (Fig. 6.16)
   —*Spirogyra* (Fig. 6.18)
   —*Oedogonium* (Figs. 6.19, 6.20, 6.21, 6.22, 6.23)
6. Branched filament—*Stigeoclonium* (Fig. 6.34)
   —*Cladophora* (Figs. 6.25, 6.26)
7. Membrane—*Ulva* (Fig. 6.27)
8. Tubular filament—Compare *Vaucheria* (Fig. 10.23A) (without cross walls)

Many complex forms of green algae of macroscopic size have been derived from simpler forms. The charophyte body (Fig. 6.28) is cited as an example.

# THE GREEN ALGAE

## SOME COMMON GENERA OF GREEN ALGAE

*Chlorella* is a single-celled, nonmotile, green alga only a few μm in diameter. The cell is basically spherical and contains 1 large chloroplast, which may or may not have a pyrenoid. Only asexual reproduction is known to occur in this genus; this process commonly involves 2 successive mitotic divisions after which the 4 nonmotile daughter cells are released (Fig. 6.7).

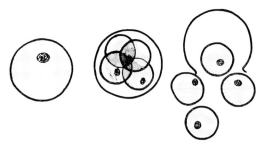

Fig. 6.7. A single-celled, nonmotile, green alga, *Chlorella,* showing process of asexual reproduction.

Because of its rapid growth rate and its lack of reproductive complexity, *Chlorella* has become a favorite research organism among plant physiologists, particularly in studies of photosynthesis. It is also an important food source for many microscopic animals. Some species of *Chlorella* have become adapted to living in the bodies of certain aquatic animals and are given the name of *Zoochlorellae*. Species of a closely related genus *(Trebouxia)* are the algal components of many lichens.

The *desmids* compose a large group of green algae in which individual cells exhibit a marked bilateral symmetry (Figs. 6.8, 6.9). For the most part they are nonmotile and unicellular but a few filamentous species exist and some forms exhibit an almost imperceptible jerky motion due to the spasmodic secretion of gelatinous materials through minute pores in the cell wall. In most desmids each cell has 2 symmetrical *half-cells (semicells)* with a single nucleus located in the *isthmus* that connects the half-cells. As mitosis occurs the isthmus stretches somewhat and a minute crosswall forms across the isthmus between the 2 daughter nuclei. Thus each of the 2 new cells consists of 1 of the original 2 half-cells plus a minute new half-cell which grows rapidly to full size. Commonly each of the original half-cells

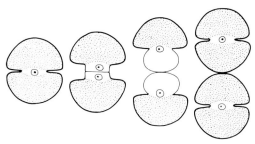

Fig. 6.8. Cell division in the desmid genus *Cosmarium*.

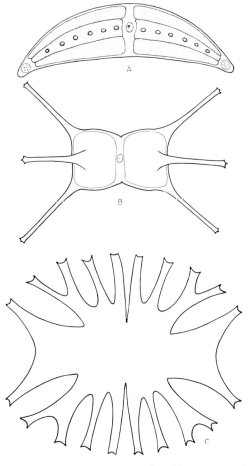

Fig. 6.9. Some common desmid genera. A. *Closterium*. B. *Staurastrum*. C. *Micrasterias*.

contains 1 large chloroplast that divides in 2, with 1 of the new chloroplasts migrating into the new half-cell. This type of division is shown in Fig. 6.8 for the genus *Cosmarium*.

*Pandorina* is a small motile colony usually consisting of 8 or 16 cells closely pressed against one another in a spherical array. Each cell is basically like that of *Chlamydomonas;* its 2 flagella protrude through the prominent gelatinous sheath (Fig. 6.10). When asexual reproduction occurs the colony loses its motility; the cells enlarge somewhat and separate slightly in the matrix. Then each cell divides 3 or 4 times by mitosis to create a new colony. The daughter colonies become motile within the matrix of the original colony and gradually move out of it.

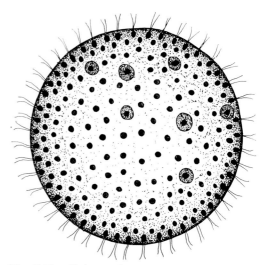

Fig. 6.11.  *Volvox* colony.

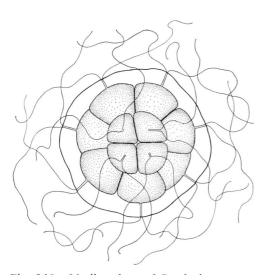

Fig. 6.10.   Motile colony of *Pandorina*.

*Volvox* is a common green alga in ponds and lakes which sometimes become so numerous that it causes a water bloom. It exists as a large motile colony of 500 to several thousand haploid cells, most of which are like individual cells of *Chlamydomonas*. They are held together in a spherical gelatinous matrix which is firm on the outside and fluid in the center. In some species delicate protoplasmic strands can be seen connecting each cell with its neighbors. A few cells of each colony are somewhat larger than the rest and are non-flagellated. They serve reproductive functions (Fig. 6.11).

During asexual reproduction these cells divide repeatedly in a definite pattern to form spherical *daughter colonies* (Fig. 6.12) which are liberated into the central portion of the parent colony before escaping from it when the old colony breaks open. The nuclear divisions are mitotic;

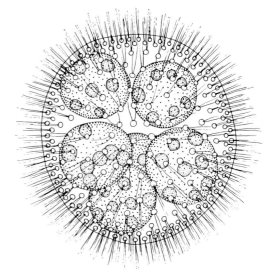

Fig. 6.12. Drawing of a *Volvox* colony with well-formed daughter colonies in which reproductive cells for a succeeding generation are already established.

no change from the haploid condition of the vegetative cell nuclei is involved.

During the oogamous sexual reproduction process the special cells function otherwise (Fig. 6.13). Some of them enlarge in place without division to become nonmotile female gametes *(eggs)*. Others divide repeatedly (by mitosis) to form disc-shaped groups of male gametes. These *sperm packets* escape abruptly and swim rapidly to the vicinity of the eggs, apparently attracted by a secreted chemical substance. (Some species are bisexual, producing both kinds of gametes in the same colony. In other species the gametes form in separate colonies.) The sperm packets break up and individual sperm cells penetrate the gelatinous material around each egg. Ultimately 1 sperm fuses with each egg to form a diploid zygote. Each of the several zygotes in a colony becomes a thick-walled resting cell called an *oospore*. After a long resting period the nucleus of the oospore undergoes meiosis as part of the germination process. It is thought that 3 of the 4 meiotic nuclei degenerate as only 1 zoospore usually emerges. This divides within itself to form a small daughter colony which is then able to begin a new population by forming normal-sized daughter colonies asexually.

*Scenedesmus* is a small nonmotile colony of (usually) 4 or 8 cells arranged in 1 (or 2) linear series (Fig. 6.14A). A new colony is formed within each cell of an existing colony and is released through a slitlike opening in the wall.

*Pediastrum* exists as a flattened disclike colony of 4, 8, 16, 32, or even 64 cells, so arranged that each cell is in contact only with other cells lying in the same plane (Fig. 6.14B, C). When asexual reproduc-

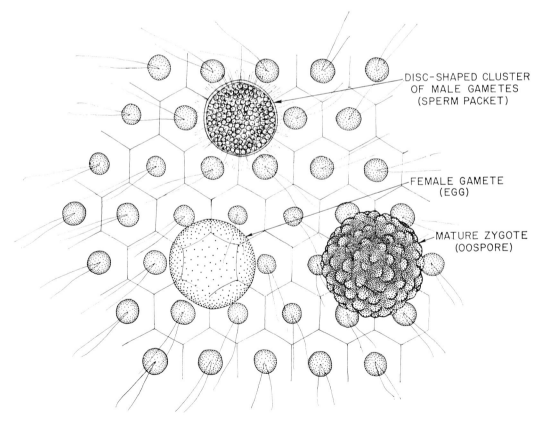

Fig. 6.13. Portion of a *Volvox* colony in which an unfertilized egg, a cluster of sperm cells, and a mature oospore are shown.

tion occurs each cell divides repeatedly within itself to form the characteristic number of cells for the species. A delicate membranous vesicle then protrudes through a crack in the cell wall and the daughter cells emerge into this as free-swimming zoospores. After swarming for a time these cells come together in a flat plate and lose their motility. They then become firmly jointed to each other to form the typical nonmotile colony.

*Hydrodictyon,* the "water net," is one of the few green algae that has an adequate common name (Fig. 6.15). The cells are

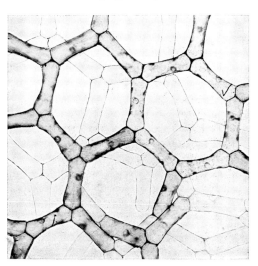

Fig. 6.15. Portion of a young net of *Hydrodictyon* as seen with the microscope.

joined together in such a way that they truly form a conspicuous net in the water. Each cell of an existing net may become 10 cm or more in length and is able to form a complete new net within itself. Thus a net with 1,000 cells can form 1,000 new nets which may rapidly fill a pond as they grow in size.

*Ulothrix* is a green alga with an unbranched, filamentous, plant body and bracelet-shaped chloroplasts. Each filament increases in length as a result of vegetative cell division with the new cell walls always forming at right angles to the long axis of the filament.

The number of filaments is increased by an asexual process involving the produc-

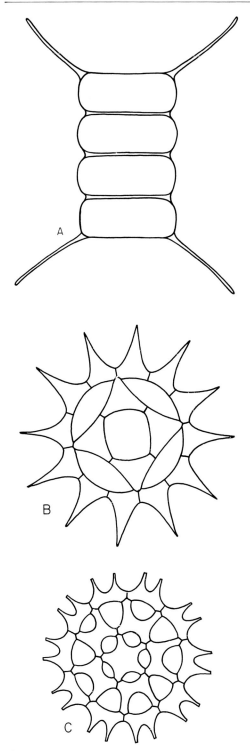

Fig. 6.14. Nonmotile colonies. A. *Scenedesmus.* B and C. Two common forms of *Pediastrum.*

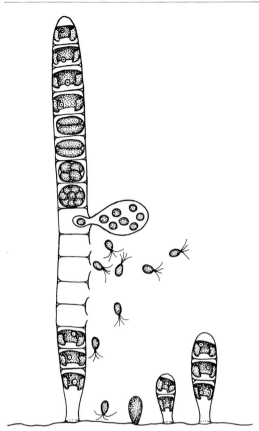

Fig. 6.16. Asexual reproduction in *Ulothrix*. Contents of cells of the filament are divided and released as zoospores. Each zoospore ultimately gives rise to a new filament.

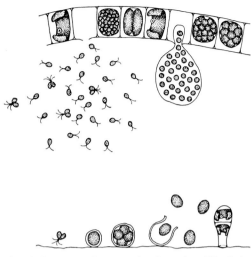

Fig. 6.17. Sexual reproduction in *Ulothrix*. Contents of the cells of the filament are divided into numerous small isogametes. Upon release they unite with gametes of opposite strains to form diploid zygotes. Meiosis occurs in the germinating zygote and each of the 4 resulting spores is capable of initiating a new filament.

tion of zoospores (Fig. 6.16). Any cell (except the basal attaching cell) can undergo a series of divisions by which its protoplasm is segmented into several cells. These escape through an opening in the wall and become zoospores. Except for the fact that they have 4 flagella, these zoospores look remarkedly like *Chlamydomonas* cells.

Each zoospore swims about rapidly for some time but eventually attaches itself to an object in the water. A new filament then develops from the attached zoospore as a result of repeated cell divisions with the new walls forming in parallel planes.

Sexual reproduction in *Ulothrix* (Fig. 6.17) begins in the same way as asexual reproduction. However, the number of divisions within a parent cell is greater than if zoospores were being formed; i.e., gametes are more numerous than zoospores.

Also the gametes are smaller than the zoospores and have only 2 flagella apiece instead of 4. In appearance they resemble the gametes of *Chlamydomonas*.

Most species of *Ulothrix* are heterothallic and therefore plus and minus strains must be mixed together in order for gametic unions to take place.

The zygote becomes a thick-walled resting cell (zygospore) that may have an extended period of dormancy. Meiosis occurs prior to germination and segregation of plus and minus strains occurs at that time. When the zygote does germinate 4 haploid spores emerge and each is capable of giving rise to a new filament. Since the nuclear divisions involved in filament growth are mitotic all of the cells of a given filament will be plus if the zoospore was plus or minus if the zoospore was minus.

This particular life cycle has been presented in some detail in order that its essential parallelism to the life cycle of *Chlamydomonas* be made clearly evident. It seems a logical speculation, based on

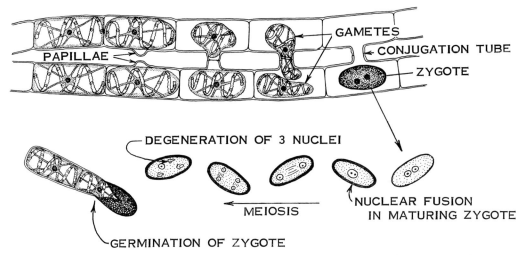

Fig. 6.18. Sexual reproduction in *Spirogyra*. Formation of conjugation tube; conjugation of nonflagellated gametes; maturation of the zygote; meiosis in the zygote; degeneration of 3 of the 4 haploid nuclei; emergence of new filament from the germinating zygote.

this parallelism, that *Ulothrix* evolved from a *Chlamydomonas*-like ancestor. And since *Ulothrix* is never classified as an animal, one can add this evolutionary relationship to other pertinent arguments favoring the classification of *Chlamydomonas* (and its relatives, *Pandorina, Eudorina, Volvox,* and so on) in the plant kingdom.

*Spirogyra* is a green alga with unbranched filaments and ribbonlike spirally arranged chloroplasts. Most species are heterothallic but it is impossible to recognize plus and minus strains in the vegetative condition. Normally they both grow tangled together in quiet waters.

When sexual reproduction is initiated (Fig. 6.18) small swellings become evident on each cell. The *papillae* from cells of opposite strains grow towards each other and eventually come into contact. The wall between them at the point of contact dissolves and an open *conjugation tube* is formed between each pair of cells. The gametes comprise entire protoplasts of individual cells. They are nonflagellated but commonly all of the gametes of one strain move through the conjugation tubes to join the gametes of the other strain. Zygotes are formed and develop very thick walls. The walls of the cells containing zygospores break open eventually and they fall to the bottom of a pond or stream where they undergo several months of rest.

When germination occurs each gives rise to a filament. Prior to germination the zygote nucleus undergoes meiosis to form 4 haploid nuclei of which 3 degenerate. The surviving nucleus contains either the gene for plusness or the gene for minusness and thus the emerging filament can be one or the other but not both. This is a matter of chance; since many zygotes germinate at the same time the filaments of the new population are a mixture of plus and minus strains.

*Oedogonium* is a filamentous green alga that is reognized by its reticulate

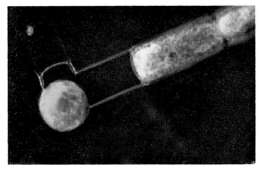

Fig. 6.19. Filament of *Oedogonium* showing the release of a zoospore (asexual reproduction).

# The Green Algae

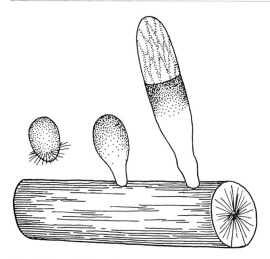

Fig. 6.20. Origin of a new filament of *Oedogonium* from a zoospore. (The zoospore could have been the product of either the gametic or the nongametic cycle.)

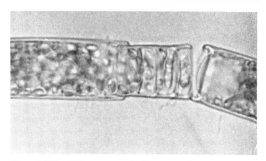

Fig. 6.22. Filament of *Oedogonium* with antheridia.

chloroplast and a peculiarity of some of its cell walls, a series of circular ridges *(apical caps)* at one end (Fig. 6.5E). Also in most species individual cells of the filaments are slightly broader at one end than the other.

During asexual reproduction any cell except the basal cell may break open to release its entire protoplast which functions as one zoospore (Fig. 6.19). (This is in contrast to *Ulothrix* where the contents of a cell are divided into many zoospores.) *Oedogonium* zoospores are unusual in that they possess a whorl of flagella instead of 2 or 4. The zoospores are powerful swimmers; when they come to rest each forms a new filament (Fig. 6.20).

All species of *Oedogonium* are ooga-mous and the egg is always retained in the oogonium which has a small entrance pore for the sperm (Figs. 6.21, 6.23). The sperms are small and produced in pairs in special cells called *antheridia* (Figs. 6.21, 6.22). When released they swim rapidly and are chemically attracted towards oogonia that contain unfertilized eggs.

The zygotes become resting spores (oospores) (Fig. 6.23) that are released by decay of the oogonium wall. Meiosis occurs prior to germination. After germination each of the 4 resultant zoospores can initiate a new filament.

Some species are heterothallic; others are bisexual and either self-sterile or self-fertile. Certain species are said to be *nannandrous* meaning that they produce dwarf male plants. This unusual situation is brought about by the production of modified zoospores which have enough physiological maleness to cause them to swim to female plants. There they attach themselves and germinate to produce dwarf filaments capable of producing true sperm.

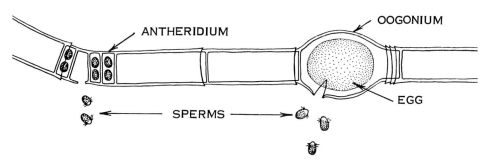

Fig. 6.21. Filament of *Oedogonium* with sperms released from antheridia swarming about the oogonium.

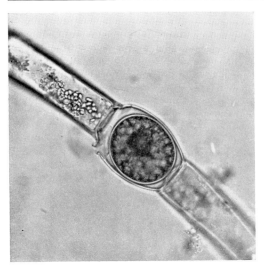

Fig. 6.23. Filament of *Oedogonium* with an oogonium containing a mature oospore.

*Stigeoclonium* has a system of erect branching filaments which are anchored to submerged rocks and twigs by prostrate branching filaments (Fig. 6.24). It has a single bracelet-shaped chloroplast much like that of *Ulothrix*. Since it somewhat tolerant of certain types of pollution it is frequently encountered in pollution studies. Usually *Stigeoclonium* secretes

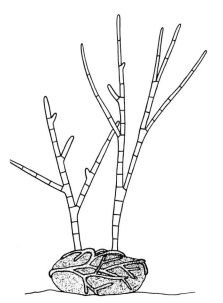

Fig. 6.24. *Stigeoclonium*, a branching, filamentous, green alga.

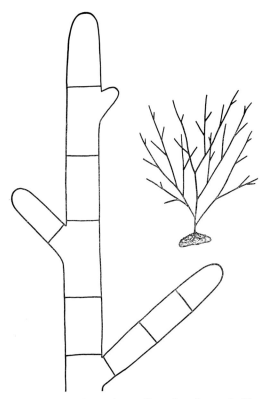

Fig. 6.25. Schematic outline drawings of *Cladophora*.

large amounts of a soft gelatinous material which gives it a slimy feel. The possible confusion with *Cladophora* is noted below.

*Cladophora* is another green alga (Figs. 6.25, 6.26) with branching filaments which is usually anchored to submerged rocks by a prostrate branching system. However, it is generally much coarser in appearance than *Stigeoclonium*. It has much larger cells, thicker cell walls, and reticulate chloroplasts; it is generally much more visible. Also it does not produce gelatinous secretions and its cell walls serve as attaching places for many microscopic organisms including vast numbers of diatoms.

*Cladophora* is closely related to *Basicladia*, the erroneously called "moss" on the backs of turtles. It is also related closely to a major nuisance alga called *Rhizoclonium* which forms dense mats in some lakes, ponds, and aquaria.

# The Green Algae

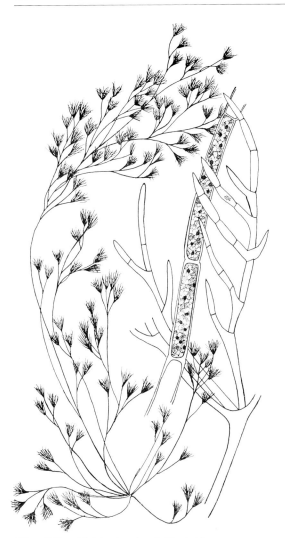

Fig. 6.26. Habit sketch of *Cladophora*.

Fig. 6.27. *Ulva*.

*Ulva* is a marine green alga, commonly known as "sea lettuce" (Fig. 6.27) from its quite obvious resemblance to lettuce leaves. It grows attached to rocks in the tidal zone and is often washed up on beaches after storms.

The *charophytes* compose a major group of aquatic plants classified usually as green algae even though their relationships to other members of that division are somewhat remote. Two common genera are *Chara* and *Nitella*. Both have a jointed appearance due to the presence of plates of cells at the *nodes* alternating with long, undivided, *internodal cells* (Figs. 6.28, 6.29) which are among the largest cells known to science. The whorls of lateral appendages, sometimes called *branchlets,* diminish in size toward the apex, giving members of this group the appearance of underwater specimens of the common horsetail (Fig. 6.28). A single cell at the tip of each stem functions as a meristem so that these plants demonstrate apical growth.

In many species of *Chara* each internodal cell is surrounded by a single layer of corticating cells (Figs. 6.29, 6.30). All species of *Nitella* lack such cells. *Chara* species are more frequent than *Nitella* species in hard waters and frequently are encased in a brittle layer of lime, which ac-

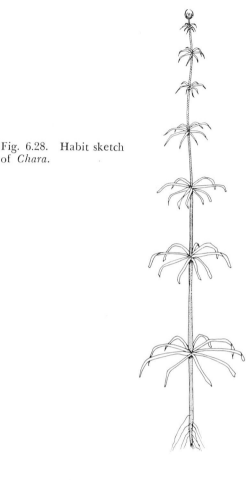

Fig. 6.28. Habit sketch of *Chara*.

counts for their common names of "stoneworts" and "brittleworts." This is due to the rapid withdrawal of bicarbonate ions from the water for use in photosynthesis, causing a shift in the ionic equilibrium that results in the deposition of the calcium carbonate crust. Charophytes often grow in extensive "meadows" on the bottoms of ponds and lakes; these growths are an important food source for many forms of aquatic life. The flaking off of calcium carbonate particles is a major source of a white deposit of *marl* on many lake bottoms. Internodal cells, particularly those of *Nitella* species, are of interest to physiologists for several reasons, including a remarkable ability to accumulate potassium ions to concentrations far exceeding the natural concentrations in pond waters. Also the inner layer of cytoplasm exhibits

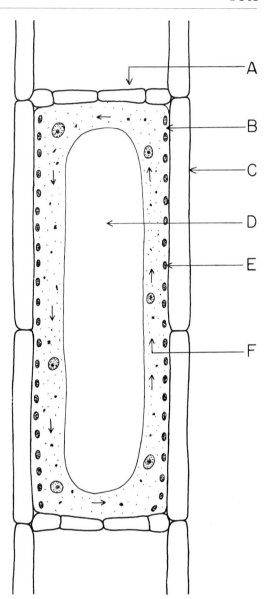

Fig. 6.29. Portion of a lateral appendage of *Chara* that is more simply constructed than the main axis. A. Nodal plate. B. Outer nonmotile layer of cytoplasm. C. Corticating cell. D. Central vacuole. E. Chloroplast. F. Inner layer of cytoplasm with direction of cytoplasmic streaming indicated by arrows.

a massive flow that is a classical demonstration of cytoplasmic streaming *(cyclosis)*. The chloroplasts are restricted to the outer layer of cytoplasm which does not move.

The structures that contain the male

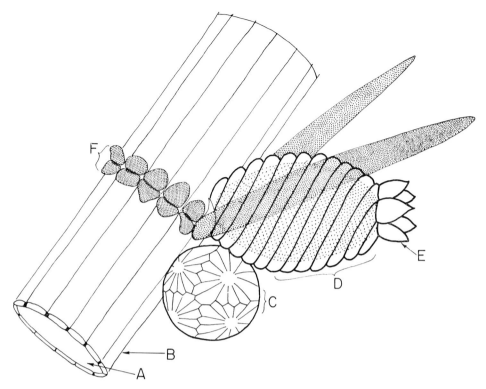

Fig. 6.30. Portion of lateral appendage of *Chara* showing attachment of gametangial structures. A. Internodal cell. B. Corticating cell. C. Globule (the male structure). D. Nucule (the female structure). E. Corona. F. Node.

and female gametes are unique. The male structure, classically termed a *globule,* is spherical in appearance, as the name implies. Its outer surface consists of a series of large, polygonal, plate cells, each of which is attached to one of the radiating cells that arise from a central stalk cell. Within the globule are many threadlike *antheridial filaments*. Each cell of such a filament is an *antheridium* containing one sperm. At maturity vast numbers of sperm are released from the globule. In some respects the globule is analogous to the bryophyte type of antheridium.

The female structure is classically termed a *nucule*. It consists of an oogonium containing one very large egg cell and a single layer of outer cells spirally twisted around it. The tips of these cells are cut off by cross walls and form a *corona* at the outer end of the nucule.

After fertilization the zygote becomes a thick-walled resting *oospore,* characteristically marked by a series of spiral ridges derived from walls of the spiral cells of the nucule. In the long evolutionary history of this group there were many more forms than presently exist. The oospores have persisted in certain types of geological formations and are of some use to geologists in interpreting the history of the earth's surface.

# SEVEN

FROM the preceding discussions it is clear that many significant features of higher plant life cycles originated in the algae. Many of them probably originated more than once. It seems likely, for instance, that oogamy evolved from isogamy in several lines; this makes unnecessary the awkward assumption that *Volvox* and *Oedogonium* are closely related because both of them are oogamous. Practically all of the land plants demonstrate a clear-cut oogamy and it is reasonable that the green algal ancestors of the land plants had evolved this characteristic before the invasion of the land began.

It is generally assumed that primitive cells had haploid nuclei. This means that each nucleus had one set of chromosomes. The number in a set varied from species to species but was constant for the individuals of a particular species. The symbol $n$ is used to designate this fact; i.e., a haploid nucleus has the $n$ chromosome number.

It seems probable that life forms existed in the haploid condition for long periods of time before the origin of sexual cycles. When gametes evolved they may have been at first little more than vegetative cells which fed upon each other in times of environmental stress.

In the green algae discussed above cells of the vegetative plant body are haploid and nuclear divisions which precede gamete formation are mitotic. Thus the gamete nuclei are haploid and have the same chromosome number as nuclei of vegetative body cells.

*With a few notable exceptions,* the gametes of plants are produced by plant bodies having haploid chromosomes numbers; this gamete-producing generation is called a *gametophyte*. (This point should be considered carefully by students who are familiar with the life cycle of animals because in practically all animals the nuclear divisions which precede gamete formation are meiotic instead of mitotic.)

## ZYGOTIC MEIOSIS

When 2 gametes fuse and their nuclei unite the resulting nucleus of the zygote has 2 sets of identical chromosomes; this is the *diploid* condition and the symbol $2n$ is used to designate it. In sexual life cycles the haploid ($n$) condition is restored ultimately by the process of meiosis.

The occurrence of *zygotic meiosis* (meiosis in the germinating zygote) is characteristic of many green algae but does not occur among land plants. This feature is especially common in those green algae where the zygote becomes a thick-walled resting cell (zygospore or oospore).

Algal life cycles with zygotic meiosis may be summarized with diagrams such as Fig. 7.1. In such cycles the zygote is the only diploid cell.

In life cycles such as those summarized

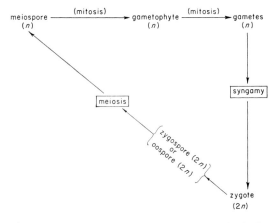

Fig. 7.1. Diagram of the sexual cycle of algal species with zygotic meiosis.

# Evolutionary Directions: Green Plants, an Overview

in Fig. 7.1 the products of meiosis may be strictly termed *meiospores* but often are referred to simply as spores.

## PREGAMETIC MEIOSIS

When zygotes germinate promptly after their formation without a resting phase, the nuclear divisions are likely to be mitotic rather than meiotic. In such cases the resultant plants or populations have diploid ($2n$) nuclei. Eventually some of the cells in such plants may undergo meiosis to form haploid ($n$) cells which in certain species function directly as gametes. Species behaving in this way are said to have *pregametic meiosis*. Or, because cells giving rise to gametes are often called *gametocytes*, the process may be called *gametocytic meiosis*. Pregametic meiosis is the usual mechanism among a very large group of algae, the *diatoms,* and among many species of brown algae. (See the discussions in Chapter 10.) It does not occur among the green land plants but does occur here and there among the green algae; life cycles in the green algae with this feature are not commonly cited in beginning discussions. Figure 7.2 shows the general pattern of such a life cycle. Note particularly that there can be no meiospores in cycles with pregametic meiosis.

## LIFE CYCLES WITH AN ALTERNATION OF GENERATIONS AND PRESPORIC MEIOSIS

In all of the green land plants and many algae zygotes divide by mitosis to form diploid plants (or populations) as in the preceding situation. However, when meiosis occurs in specialized cells the products of meiosis function as spores *(meiospores)* and give rise to haploid plants which function as gametophytes. The generation which bears the meiospores perhaps should be called the *meiosporophyte* but it is much more commonly called the *sporophyte.*

As has been emphasized, each haploid spore gives rise to a new gametophyte generation and in this process all nuclear divisions are mitotic.

From this discussion it may be seen that many algae have an alternation of a gamete-producing generation, the gametophyte, with a spore-producing generation, the sporophyte. Figure 7.3 illustrates the basic outline of this *alternation of generations.*

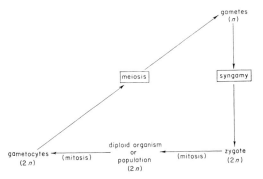

Fig. 7.2. Diagram of the sexual cycle of species with pregametic (gametocytic) meiosis.

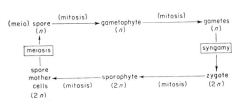

Fig. 7.3. Diagram of a basic life cycle involving presporic meiosis and alternation of generations.

In any discussion of plant life cycles with alternation of generations it is difficult to avoid creating an impression that the diploid chromosome number determines that a plant will be a sporophyte while plants with haploid chromosome numbers are automatically gametophytes. Meiosis and fertilization are depicted as being critical points in the endless alternation between sporophyte and gametophyte generations. Enough exceptions exist, however, to warrant considerable caution in assuming that chromosome number alone determines the nature of a given generation.

For example, sporophytes of certain mosses can be stimulated by injury to give rise to diploid protonema and diploid gametophytes with diploid gametes. When these unite tetraploid ($4n$) zygotes and tetraploid sporophytes result. When meiosis occurs the spores are diploid.

Another well-known situation exists in certain ferns in which diploid gametophytes develop directly from unreduced cells in the sporophyte leaves. In some such cases certain cells of the diploid gametophytes develop without fertilization into diploid sporophytes.

It is also well documented that in a number of species of flowering plants unfertilized eggs may develop into mature sporophytes which therefore have the haploid chromosome number.

Despite such exceptions, however, the doubling of the chromosome number at fertilization and the halving of the chromosome number during meiosis are significant events in the normal sexual cycle. Furthermore they play critical roles in the mechanisms of inheritance and evolution.

If one looks carefully at the details of the life cycles one will see that the cell which becomes the embryo (usually a zygote) is retained within tissues of the gametophyte. On the other hand, the cell that becomes a gametophyte (usually a meiospore) is thrown out of the sporophyte with some kind of a violent explosive reaction and is thus totally removed from any protection. The resultant gametophyte must survive in a hostile environment. Thus it appears that the conditions of life under which each generation begins are totally different; this fact may be of greater significance in its development than whether the chromosome number is haploid or diploid.

## SPORE DISPERSAL IN AIR

The term *spore* appears in so many contexts (meiospore, mitospore, microspore, megaspore, zoospore) that precise definition fails even though the significance of spores as reproductive units is generally accepted. One great consequence of the invasion of the atmosphere by emergent green plants was a change in the nature of various spore types that allowed them to become adapted to dispersal in air instead of in water.

Such adaptations included loss of the flagella, development of a thick and often cutinized wall, and partially dehydrated protoplasm. Spores modified in this way may be blown about in air currents for long distances without having to expend energy of their own. (By way of contrast, the energy requirements of flagellar motion limit the distances to which algal zoospores can travel in water.) Furthermore, spores modified in the described manner can survive as independent units for long periods in the unnatural new environment of dry air.

(An intriguing concept should be interpolated here. The transport of spores by insects also requires no expenditure of energy by the spore and insect transport may have become significant much earlier in evolution than previously considered likely.)

Spore containers are variously named also: *sporangia, zoosporangia, moss capsules, microsporangia, megasporangia*. They may be involved in either sexual or asexual cycles or both. The sporangium wall may be simply a modified cell wall, as in most green algae, or it may be a multicellular structure with an outer layer of sterile cells, as in ferns and mosses.

Sporangia of the hypothetical primitive emergent plants were probably terminal in position. In a number of fossilized primitive vascular plants terminal sporangia contain spores in groups of four (tetrads). This indicates that the spores were meiospores and therefore that the plants producing them were diploid sporophytes.

Although we do not know the exact nature of the gametophytes that alternated with the first aerial sporophytes, we can proceed with a reasonable assumption that selective factors in the evolution of the gametophyte differed from those affecting the sporophyte in at least one major way: sperms were dispersed in water while spores were dispersed in air. Thus aerial dissemination of reproductive cells would not have been a primary selective factor in the evolution of aerial branches of the gametophyte.

On the other hand, the atmosphere does represent an enormous reservoir of carbon dioxide for photosynthesis, and the surface of the land contains many of the essential inorganic chemicals for plant growth. The available concentrations of carbon dioxide and inorganic chemicals are often limiting factors in the growth of algae. Also the absorption of light by dense growths of algae near the surface often means that light becomes a limiting factor in deeper waters. Thus emergence of the gametophyte from the water would have certain beneficial aspects.

Another important factor influenced the gametophytic invasion of the land. Wind-blown spores would be just as likely to come to rest on the land as in the water. If the land were moist enough to support growth of the germinating spores then there would be nothing to prevent the beginnings of such growth on land. The formation of the algalike protonemal stage in the life cycle of mosses may be cited as an illustration of such an event.

## THEORETICAL ORIGINS OF ALTERNATION OF GENERATIONS IN LAND PLANTS

There are two major hypotheses concerning the origins of alternation of generations in land plants. The proponents of one of these maintain that the primitive algal ancestors of land plants had an alternation of two independent generations that were alike or *homologous*. This means that the diploid generation was quite similar to the haploid generation and that the existing differences between them were acquired gradually. The proponents of the other hypothesis do not deny that an alternation of homologous generations does occur in some algae. They maintain, however, that the sporophyte of land plants has always been basically different from (or *antithetic* to) the gametophyte generation; i.e., it is a new structure interpolated between two successive haploid generations.

In mosses and their relatives the liverworts the alternation of haploid gametophyte generations with diploid sporophyte generations is well established. The gametophyte is wholly self-nourishing and independent but the sporophyte is epiphytic and partially dependent on the gametophyte. Moreover the gametophytes of many species of mosses and liverworts have evolved the most varied and elaborate modifications for independent life on land of any known gametophytes. These, plus the fact that neither generation has true xylem and phloem tissues, sharply separate the group from all the higher land plants.

In the past it has been considered possible that the vascular plants may have evolved from the bryophytes by a progressive increase in the complexity of the sporophyte and a corresponding decrease in the complexity of the gametophyte. This hypothesis was generally abandoned, however, when investigations of the fossil psilophytes indicated a very distinct possibility that the vascular plants and the bryophytes arose independently from the green algae.

In many green algae the zygote is formed by fusion of two gametes outside the parent body. In others fertilization occurs inside the oogonium but the oogonium wall eventually breaks open, allowing the zygote to separate from the parent plant. In either circumstance the germination of the zygote occurs outside the parent gametophyte tissue, often long after the gametophyte has disappeared.

In the bryophytes as well as in all of the lower vascular plants the archegonium not only retains the egg, so that fertilization occurs within its tissues, but also retains the zygote. While an archegonium as such is not clearly defined in the embryo sac (female gametophyte) of flowering plants, it is still essentially correct to state that both the egg and the resultant zygote are retained in place where formed.

The *embryo* stage of the sporophyte is the result of the retention of the zygote within the structure in which it is formed. It does not become a resting spore but begins to divide soon after its formation. The divisions involve mitosis instead of meiosis and as a result a mass of diploid cells is formed inside gametophyte tissue. This is the *embryo* and its appearance marks an important highlight in plant evolution.

It has been suggested herein that the conquest of the land began with the invasion of the atmosphere by an erect branch of an algal system and that this erect axis developed a terminal sporangium as well as a cutinized epidermis with stomates before it developed vascular tissue.

If such structures ever existed, they may have been modified on the one hand into the primitive stems of vascular plants; on the other hand they may have been modified without vascularization into the sporophytes of the bryophytes.

The bryophyte sporophyte has only two growing points, one of which becomes wholly modified into the capsule while the other becomes completely converted into the foot. There is no leftover meristem from which innovative tissue systems might be derived to aid in severing the dependent relationship with the gametophyte. In this way the sporophytes of bryophytes may have lost the evolutionary initiative and become subordinate to the gametophytes as they are today.

If such an interpretation is correct, then sporophytes such as those of mosses would be close to the ancestral type, while the simpler sporophytes of liverworts could well have resulted from degenerative evolution.

In the case of the primitive vascular plants the lack of fossil evidence concerning their gametophytes is disappointing and we may never know what they were like. The fact that in most living groups of vascular plants sporophytes become independent from gametophytes at an early stage makes reasonable a hypothesis that primitive vascular plants behaved in a similar manner. The structure of these unknown gametophytes must have permitted the separation readily.

As a conclusion to this discussion the hypothesis is presented that the divergence between the bryophytes and the vascular plants began before the aerial sporophyte axis became vascularized and that the divergence was influenced by basic but unknown differences between the gametophytes of these two evolutionary lines of development.

## ORIGIN AND SIGNIFICANCE OF THE SEED HABIT AS IT AFFECTS BOTH ANGIOSPERMS AND GYMNOSPERMS

In mature seeds the embryo is packaged in a protective tissue, the seed coat, and is provided with an adequate supply of food for the resumption of growth during germination. In the dormant state the embryo can withstand severe environmental conditions and yet is stimulated to rapid growth when conditions become suitable. Seeds cast on periodically dry surface soil germinate during periods of favorable surface moisture. The roots then penetrate deeply into the soil where they come in contact with more constant sources of water. When the surface soil dries out again the new plant is able to survive because it can transport water from the underground source to all of its parts that are exposed to the atmosphere. The fossil record suggests the seed habit evolved several times in heterosporous plants. In each case, a profound change in the nutritional role of the female gametophyte was involved. The food stored in the megaspore (as in *Selaginella*) or immediately available from the sporophyte (as in the modern seed plants) made possible the essentially parasitic growth of the female gametophyte. One advantage of such an arrangement is that when the female gametophyte is parasitic on the preceding sporophyte generation it is able to supply the embryo with far greater amounts of food than it could produce itself as a small independent entity.

The retention of the megaspore in the megasporangium was an important step in this direction. But this development posed a serious problem, for it tended to interfere with the normal mechanism by which the sperm reached the egg. In the primitive condition the enlarging female gametophyte may have exerted so much pressure on the megasporangium that it was cracked open enough to expose the archegonium, allowing fertilization of the egg by swimming sperms. It is very likely that the evolution of the pollen tube was an essential corollary of the complete retention of the female gametophyte within sporophyte tissue.

The pollen tube may not have been an entirely new device since it bears considerable resemblance to a rhizoid. If the male gametophyte as it developed within the microspore produced a rhizoid capable of penetrating the megasporangium, that rhizoid would have served as a very convenient passageway for the movement of the male gametes inward to the female gametophyte.

This is, of course, a highly speculative interpretation. Some authorities feel that the pollen tube had its origin as a type of haustorium, i.e., a device for penetrating and obtaining food from the nucellus tissue of the ovule.

No matter what its origin, the pollen tube has eliminated the dependence of the plants that possess it on the presence of surface films of moisture in which sperms must swim to reach the egg.

(In this connection the cycads illustrate an interesting transitional stage. In the ovules of these plants there is a small space between the enclosed female gametophyte and the megasporangium wall. When the pollen tube penetrates this cavity it discharges two sperms in a drop of fluid near the archegonia. These sperms do not swim to the egg but it has been observed that they are motile in the pollen tube just before the discharge.)

The megasporangium alone is not truly an ovule. In the seeds of modern plants it is equivalent to the tissue that is commonly called the nucellus. Each ovule has one or more additional layers (integuments) that overgrow the megasporangium leaving a small hole (the *micropyle*) at the tip. The concentration of stored food and other organic matter in primitive seeds would have been attractive to other organisms such as insects, birds, primitive rodents, fungi, and the like. Thus the development of the integument as an additional protective layer over the megasporangium would have had a survival value.

The integument presented an additional barrier to the eventual union of gametes. In gymnosperms this barrier is passed by the device of secreting a droplet of sticky fluid through the micropyle. Pollen grains become adherent to this fluid and are drawn through the micropyle when the fluid is reabsorbed.

In flowering plants the pollen tube passes the integument barrier by growing through the micropyle. A secretion of chemical substances from the micropyle guides the direction of growth of the pollen tube.

The female gametophyte of pines was discussed in some detail as part of the gymnosperm life cycle. This gametophyte exhibits considerable growth and produces recognizable archegonia. It nourishes the embryo even though it must obtain the necessary food from the sporophyte. At maturity it is the largest structure in the pine seed.

In the flowering plants the embryo sac is interpreted as a much reduced female gametophyte. The primary function of the embryo sac is the production of the egg. The nourishment of the embryo is accomplished by a special new tissue, the *endosperm,* which develops from the triple fusion nucleus formed as a result of the fertilization of two polar nuclei by the second sperm from the pollen tube. The endosperm has a tremendous capacity for rapid growth at first and later it accumulates large quantities of reserve food which are utilized by the embryo in its development.

It is possible to interpret the antipodal cells in the embryo sac as vestigial body cells of the female gametophyte and (by stretching the imagination considerably) to interpret the two synergids as vestigial neck cells of an archegonium.

Vascular plants had a long evolutionary history before the advent of the seed habit and it is possible that sporophytes were able to grow in more diverse habitats than they actually occupied. The independent gametophytes of the preseed plants may have been unable to survive in areas other than those with moist surface soils. Moreover, there was an added requirement that free surface water be available in order that fertilization by swimming sperms could occur. Since the sporophyte has to begin its existence as an embryo in the gametophyte, the distribution of the more vigorous sporophyte generation was limited to the habitats where the gametophyte could exist.

The retention of the female gametophyte in the megasporangium eliminated the first of these limitations and the concurrent evolution of the pollen tube eliminated the second. Thus the seed habit meant a profound increase in the ability of plants to continue the conquest of the land. When they left the swamps the plants left behind the conditions that favored fossilization of plant remains and we have very scanty and fragmentary fossil records of the early dwellers of the uplands.

Perhaps the seeds of these plants were rather heavy and not well equipped for dispersal. They fell near the parent plant and the spread of species was very slow at first. Later, devices evolved that served to increase the dispersal rate. One simple and very effective device is that of the pine where a layer of epidermis from the seed-bearing scale remains attached to the seed. This wing causes the pine seed to whirl like a single helicopter blade as it falls from the tree. The rate of fall is slowed and wind movements carry pine seeds as much as a quarter of a mile away from the parent tree. Everyone is familiar with the parachutes of dandelion achenes and the efficiency of this particular device in spreading the dandelion.

## EVOLUTION OF SPOROPHYLLS

Primitive vascular plants bore their sporangia at the tips of dichotomous branch systems. In plants with fern-type leaves (macrophylls) the sporangia are borne on the leaf and the spore-bearing leaf is called a sporophyll. At various times and in various ways a division of labor has occurred between vegetative and fertile parts of plants. Even in some of the psilophytes portions of a branch system are known to have been smaller and more branched than others. In such cases the sporangia were restricted to the tips of the smaller branches while the larger ones were presumably vegetative.

The nutrition of the developing sporogenous tissue in such situations must have depended in part on the ability of the rest of the plant to produce and transport food to the sporangia.

In many ferns portions of a single frond are fertile and nonphotosynthetic, while other parts of the same frond are sterile and vegetative. In the interrupted fern the middle pinnae of a frond are fertile, while in the royal fern it is the upper pinnae that bear the sporangia. In some other ferns whole leaves are fertile and entirely different in appearance from the much larger vegetative leaves. This is true of the cinnamon fern, the sensitive fern, and the ostrich fern.

In the cycads, the oldest living group of seed plants, the vegetative leaves are large and much like the fronds of the an-

Fig. 7.4. Cycad. A species of *Cycas*. Note the large fernlike leaves.

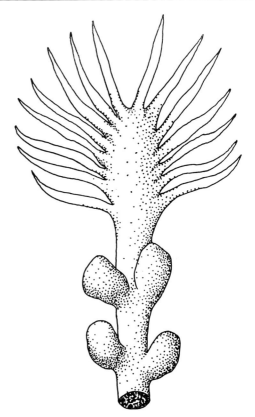

Fig. 7.5. Megasporophyll (with ovules) of *Cycas revoluta* in which the resemblance to a vegetative leaf is evident.

cient ferns (Fig. 7.4). The sporophylls, however, are much reduced in size. The megasporophylls in many species are scalelike and bear two seeds apiece. But in at least one species, *Cycas revoluta*, the sporophyll (Fig. 7.5) has a vestigial resemblance to a vegetative leaf and bears several seeds.

The microsporophylls of the cycads are scalelike and bear a great many microsporangia which are scattered over the leaf surface (Fig. 7.6). This is in contrast to the microsporophylls of pine which bear only two microsporangia apiece.

There seems little question that the sporophylls of the cycads are modified leaves. Nor is there much question as to the basically leaflike nature of the microsporophylls of pine.

The nature of the ovule-bearing scale of pines and related types is another mat-

Fig. 7.6. Microsporangiate strobilus (pollen cone) of a cycad cut crosswise to show the numerous microsporangia on each of the microsporophylls.

ter. The presence of the subtending bract is scarcely evident in pines but in larches and some other conifers it is a highly developed structure which is larger than the ovule-bearing scale. Since the scale is formed in the axil of the bract, a well-substantiated hypothesis has been advanced that the scale is a reduced branch system.

In recent years there has been some discussion of the possibility that the stamen of the flowering plants might have been derived directly from a branch system without passing through a leaf stage. However, most interpreters of the evidence consider the stamen as being a true microsporophyll in which the leaf tissue has been reduced.

The development of the simple pistil was a basic step in the evolution of flowering plants. This structure is considered to be a modification of the megasporophyll. Possibly the first step was a failure of the megasporophyll to flatten out from the folded condition as an immature leaf in the stem tip. This folded condition resulted in the enclosure of the ovules so that they were provided with added protection against predators and dessication.

Because of the added protection this device had considerable survival value. There followed a gradual sealing of the suture and progressive migration of the area receptive to pollen to the position that we now recognize as the stigma.

The postulated series of changes that led to the modern pistil must have been accompanied by evolutionary changes in the abilities of the pollen tube since it must often grow for long distances through the stigma, style, and ovary cavity before reaching the ovules.

## EVOLUTIONARY TRENDS IN FLOWERING PLANTS

Although we have no fossil evidence of the exact nature of the first flowering plants nor of the intermediate forms between them and the fernlike plants from which they arose, it is possible to arrive at a concept of the probable nature of the primitive flower on the basis of comparative studies of living plants.

The primitive flower (Fig. 7.7) was probably a bisporangiate strobilus having both stamens (microsporophylls) and simple pistils (megasporophylls) on the same axis. In addition the primitive flower had modified sterile leaves attached to the basal portions of the axis. These have become further modified into the petals and sepals

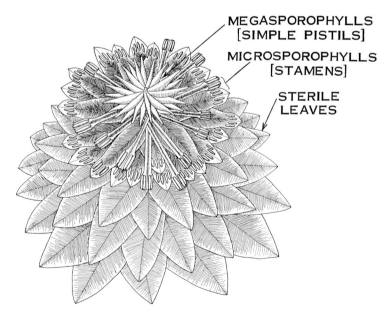

Fig. 7.7 Appearance of a hypothetical primitive flower with numerous parts spirally arranged and separately attached.

of the modern flower. The receptacle of the flower is interpreted as having been modified from the axis of a strobilus.

Some of the characteristics of primitive flowers were these:

1. The axis was more elongated than the present day receptacle.
2. The different parts of the flower were all numerous.
3. They were spirally arranged on the axis.
4. They were separately attached.
5. The megasporophylls were clustered near the tip of the axis.
6. The megasporophylls resembled present-day simple pistils.
7. The microsporophylls were below the megasporophylls.
8. The microsporophylls were more leaflike than present-day stamens but possibly had only four microsporangia.
9. The basal leaves were sterile but not differentiated into sepals and petals.

The plants that produced the first flowers probably were treelike and had woody stems with secondary growth. Among living flowering plants, the magnolia is an example of a plant that has retained many primitive features. The buttercups also have flowers with primitive features but most of the species are herbaceous rather than woody. Members of the rose family have also retained many of the characteristics of primitive flowers.

The evolution of the vast array of modern flowering plants can be appreciated best as being a matter of progressive change along any or all of the following lines:

1. Reduction in the numbers of parts.
2. Change from the spiral arrangement to a whorled or cyclic arrangement.
3. Fusion of the parts in a whorl.
4. Fusion of one whorl to another.
5. Change from radial symmetry to bilateral symmetry.
6. Change from hypogyny (a condition in which the flower parts seem to arise below the ovary) to epigyny (a condition in which the flower parts seem to arise above the ovary).
7. Elimination of one or more whorls.
8. Reduction in the number of ovules in each carpel.

The legume family, which includes the familiar beans and peas, provides an example of the reduction of the numbers of simple pistils. The flowers of members of this family have but one simple pistil and this bears many resemblances to the hypothetical primitive pistil (Fig. 7.8).

Fig. 7.8. Flower of the perennial sweet pea (legume family) with sepals and petals removed to show the single, simple pistil and fusion of the filaments of 9 to the 10 stamens.

The change from the spiral arrangement to the whorled arrangement is related to the apparent shortening of the receptacle (not so much a shortening as a failure to elongate). The unexpanded spiral pattern of flower parts has become broken up into separate whorls of sepals, petals, stamens, and pistils. Sometimes there are 2 or more whorls of the same kind of part.

The true geraniums, for instance, have 5 whorls of 5 members each: 5 sepals, 5 petals, 2 whorls of 5 stamens each, and 5 carpels (Fig. 7.9).

Flowers of the mustard family have 1 whorl of 4 sepals, 1 whorl of 4 petals, 1 whorl of 2 short stamens, a second whorl of 4 longer stamens, and a single compound pistil in the center (Fig. 7.10).

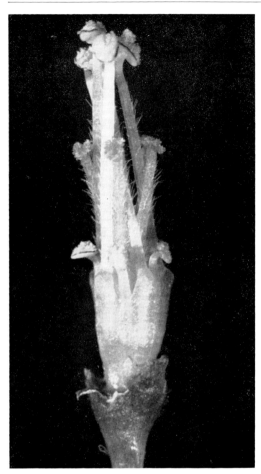

Fig. 7.9. Flower of *Oxalis* (geranium family) with sepals and petals removed. In order from top to bottom note 5 long stamens, 5 stigmas, and 5 short stamens.

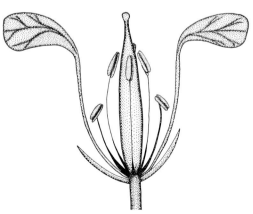

Fig. 7.10. Partially dissected flower of a mustard. Note the 1 whorl of 4 long stamens and the second whorl of 2 short stamens.

Fig. 7.11. Flower of a member of the lily family *(Ornithogalum)* with 5 whorls of 3 members each.

Lily flowers have 3 sepals, 3 petals, 2 whorls of 3 stamens each, and a compound pistil in the middle (Fig. 7.11).

Tomato flowers have only 4 whorls: 5 sepals, 5 petals, 5 stamens, and a pistil in the center (Fig. 7.12).

There is an interesting comparison between the number of parts per whorl in monocot flowers and in dicot flowers. The

Fig. 7.12. Flower of the tomato plant with 5 sepals, 5 petals, 5 stamens, and a compound pistil in the center.

basic number of parts per whorl in monocots is 3 while dicots have 4 or 5 parts per whorl. (This comparison holds best for those members of both groups that do not exhibit extreme reduction.)

The fusion of parts in a whorl has occurred to varying degrees in many plant families. The flowers of the common petunia serve to illustrate the fusion of petals (Fig. 7.13).

Fig. 7.13. Flower of *Petunia* illustrating fusion of petals.

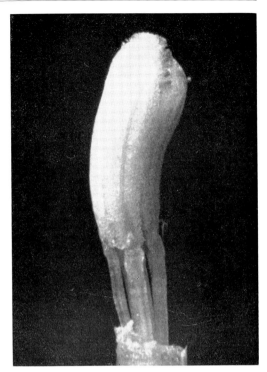

Fig. 7.14. Dissected flower from the compact flower head of a marigold (composite family). The portion illustrated shows the characteristic fusion of the anthers.

In the flowers of the sweet pea and many other legumes 9 of the 10 stamens have their filaments fused to form a tube around the ovary (Fig. 7.8).

In dandelions, daisies, and other members of the composite family the anthers of the 5 stamens are fused to form a cylinder through which the stigma of the pistil emerges like a piston when the style elongates (Fig 7.14).

In the grass family the floral branches *(inflorescences)* are reduced to *spikes* composed of spikelets, each consisting of a short axis with 1 to several flowers enclosed in sterile bracts. Each flower is quite simple, having 1 pistil with 2 feathery styles and 3 stamens. The perianth is missing although the scalelike *lodicules* at the base may be reduced perianth parts (Fig. 7.15). The whole flower is enclosed by 2 bracts, the *palea* and the *lemma*.

One of the most interesting of all evolutionary studies concerns the relation of

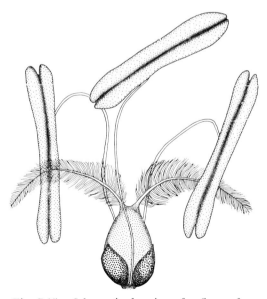

Fig. 7.15. Schematic drawing of a flower from the spikelet of a grass with 3 stamens, 2 feathery stigmas, and a pair of lodicules at base of ovary.

insects to flowering plants. As is well known, insects are the natural pollinating agents for many kinds of flowers. In numerous cases there is abundant evidence that evolutionary modifications in flower structure have been matched by evolutionary modifications of the insects which pollinate them.

One especially remarkable example is a species of moth that accomplishes pollination in a species of *Yucca* by carefully packing pollen into the stigma. Having insured the normal development of the *Yucca* fruit in this way she proceeds to lay a few of her own eggs in the ovary where the larval stages develop.

# EIGHT

THE primary advantage of chlorophyll to plants is in the manufacture of simple foods during photosynthesis. In these simple foods the energy of sunlight is stored so that it can be released in controlled amounts within the energy-demanding biochemical systems of living protoplasm.

It follows then that living organisms can survive without chlorophyll providing they have mechanisms for incorporating available and usable organic foods into their own protoplasm. This is true for the vast majority of animals and for a few higher plants such as the parasitic weed dodder *(Cuscuta)* (Fig. 8.1). This plant obtains its basic food supply through absorbing organs *(haustoria)* which penetrate the tissues of a host plant. The dodder plant is normally yellow-orange in color but its seedlings have been observed to be green. The Indian pipe *(Monotropa)* (Fig. 8.2)

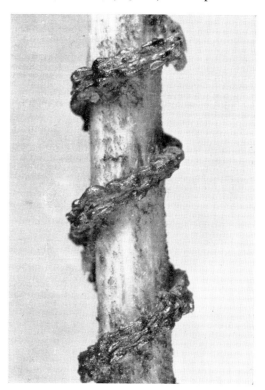

Fig. 8.1. Stem of a dodder plant twisted about the stem of a host plant.

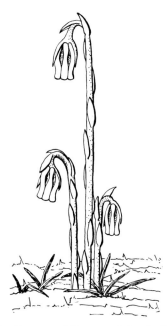

Fig. 8.2. Stems and flowers of the Indian pipe.

is completely without pigmentation and is dependent for its basic nutrition on an association with a fungus that penetrates its roots.

While such plants are exceedingly interesting, they are not particularly important. The major emphasis of this chapter will be on a study of the fungi.

A fungus can be defined in a very general way as a simple plant without chlorophyll. Of course no living organisms are truly simple. When one attempts to draw

# Life Without Chlorophyll: The Fungi

firm lines among the plant kingdom, the animal kingdom, and the prokaryotes, one finds that many of the simple plants without chlorophyll have apparent relatives on both sides of the lines.

Such a broad definition of fungi includes bacteria, actinomycetes, and slime molds but the "true fungi" include only the algalike fungi, the sac fungi, and the club fungi. Some of the differences among these groups are indicated in a brief annotated outline of the plant kingdom included in Chapter 11.

In the following discussions examples from the various groups of fungi are used to illustrate the salient features of their growth, reproduction, and biological significance.

## NATURE AND GROWTH OF THE FUNGUS PLANT BODY

The plant bodies of various fungi parallel in many ways the types of plant bodies found in the algae. They range from single cells that may be motile or nonmotile to complex densely intertwined masses of filaments in which the mass has a characteristic shape. Many of the filaments of fungi are septate; i.e., they have cross walls that divide the filament into a cellular series. Others generally lack septa and are thus without division into cellular units. None of the fungi, even those with highly complex plant bodies, have evolved vascular tissue.

The similarity in growth habits between fungi and algae as well as certain similarities in mechanisms of reproduction have led many persons to suggest that some fungi may have evolved from algae following loss of the ability to manufacture chlorophyll. However, authorities are generally averse to this hypothesis.

Some of them hold the opinion that fungi had diverse origins and that any similarities between present-day algae and fungi have resulted from parallel evolution.

Most forms of fungi have stages in which they exist as single cells for a period of time. The cells referred to are reproductive structures (spores and gametes) and are not considered here as plant bodies.

Much emphasis is placed on the significance of single-celled, motile, plant bodies in the evolution of the algae and it is logical to speculate that certain groups of fungi had a similar ancestry. The type of ancestral cell envisaged would have been without chorophyll but would have had cytoplasm, a true nucleus, and one or more flagella. It may or may not have had a cellulose wall. Many fungi produce reproductive structures (zoospores and gametes) that fit this description. Many protozoa (primitive animals) could be described in a similar manner. In addition, some unicellular motile algae can be induced to develop without chlorophyll and grow in an otherwise normal manner provided they are able to obtain simple soluble foods. It is not clearly evident, however, that there are any living organisms classified as true fungi that exist only in the single-celled motile condition. Possibly they did exist at one time but have been supplanted by more complex forms.

The single-celled, nonmotile, plant body is characteristic of many of the cultivated yeasts (Fig. 8.3) and is the result of centuries of cultural selections favoring this habit of growth.

However, it is much more common for the wild-type (noncultivated) yeasts to occur as branched filaments. New cells arise

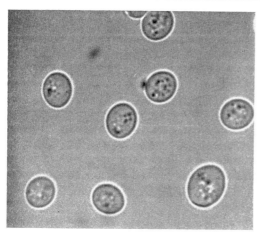

Fig. 8.3. Cells of a cultivated yeast.

from existing cells as small swellings that gradually increase in size. This process is usually called budding (Fig. 8.4).

The *slime molds* make up a small group of organisms with such a mixture of plant and animal characteristics that they have been classified in both the plant and animal kingdoms. These interesting organisms have been the subjects of considerable research. Modern studies have clearly shown that there are at least three groups in this general category; they probably bear little or no evolutionary relationship to one another.

In the largest of these groups, which might be referred to as the *true slime molds*, the vegetative body consists of a multinucleate noncellular mass of protoplasm called a *plasmodium*. Protoplasmic synthesis and nuclear divisions occur in the plasmodium, which grows rapidly when suitable conditions exist. It exhibits an amoeboid type of movement in which portions of the organism are pushed forward or retracted in relation to a massive flowing motion of the cytoplasm.

It continues to move in this way throughout its vegetative existence, flowing over as well as permeating its available food supply. Although food particles may be imbedded in the plasmodium, they actually are separated from it by membranes through which enzymes are secreted that accomplish an *extracellular digestion*. (See a discussion of this topic later in this chapter.) A plasmodium of the slime mold *Physarum polycephalum* is illustrated in Fig. 8.5.

Fig. 8.5. Plasmodium of the slime mold *Physarum polycephalum* growing on agar.

Fig. 8.4. Branching growth habit and new cell formation (budding) in a yeast.

Commonly this vegetative phase is passed in protected places such as partially rotted woody stems and leaf mold. Prior to the reproductive phase, however, the plasmodium flows into a position that is more exposed to light and desiccation. It then produces fruiting structures from which numerous spores are released at maturity (Fig. 8.6).

Fig. 8.6. Sporangia of a species of the slime mold genus *Stemonitis*.

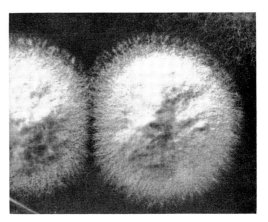

Fig. 8.7. Mycelium of a species of *Aspergillus* growing on agar.

In the *cellular slime molds* many individual vegetative cells become massed together to form a *pseudoplasmodium* in which they do not lose their separate identities. When the reproductive phase is initiated individual cells move into seemingly preordained positions and contribute to the formation of an intricate fruiting body by encysting in place without further division. Elaborate discussions of this group, particularly of the genus *Dictyostelium*, are often included in general biology courses.

A third group, the *endoparasitic slime molds*, is characterized by producing small plasmodia that infect plant tissues but do not form sporangia. For the most part members of this group have become so highly specialized that many clues to their possible relationships have been lost.

In filamentous fungi, whether septate or nonseptate, the individual thread is called a *hypha* and a mass of intertwined hyphae is referred to as a *mycelium* (Fig. 8.7). Fungi with unbranched hyphae do occur; more commonly they are profusely branched. The growth of a mycelium on a suitable medium is often extremely rapid. The mycelium of the common black bread mold, for example, can cover many square inches of a suitable nutrient medium in a few hours' time (Fig. 8.8).

Hyphae may be modified in various ways to form specialized vegetative structures. The black bread mold mentioned above is organized into growth centers. At each center some of the hyphae form rhizoidal branches that penetrate the food substance and anchor the mycelium. From each center as it matures numerous other hyphae spread in a radiating pattern. When each of these *stolons* comes in contact with the surface of the food supply it establishes a new growth center (Fig. 8.9). This development is analogous to the method by which strawberry plants are spread. When each center reaches a certain stage in maturity it gives rise to reproductive structures which will be discussed presently.

Some filamentous fungi that are dependent on living organisms for their food supply develop specially modified hyphae called *haustoria* that penetrate the living cells of the host. In the formation of a haustorium an ordinary hypha comes in contact with a host cell. At this point a small hole is formed in the host cell wall as a result of the digestive activity of enzymes secreted by the fungus. Through this small hole a minute hyphal branch presses inward where it establishes an intimate contact with the host cell proto-

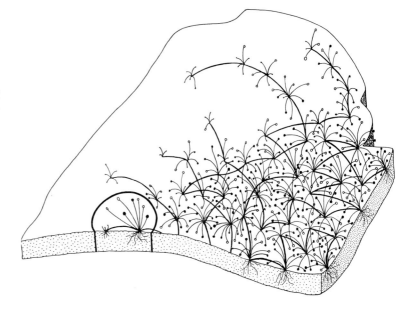

Fig. 8.8. Mycelium of the black bread mold spreading across a grain of rolled oats.

plasm. Sometimes the haustorium enlarges into a small peglike protrusion, as in the case of *Albugo* sp. infecting the tissues of the weed shepherd's purse (Fig. 8.10). In other cases it may enlarge into a rhizoidal system. This is true of the haustoria of *Erysiphe* sp. (illustrated in Fig. 8.11 in the epidermal cells of a grass host).

In some species of a primitive group of fungi called chytrids the haustoria become more extensively modified into *rhizoidal mycelia*. In species of the genus *Rhizophidium* a zoospore swims to the wall of a host cell where a small pore is digested in the wall. Some of the protoplasm of the spore then passes through this opening into

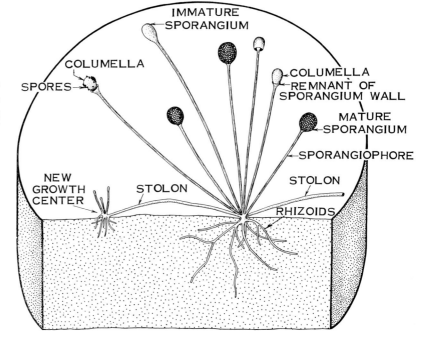

Fig. 8.9. Detail of stolons, growth centers, rhizoidal branches, and sporangia of black bread mold.

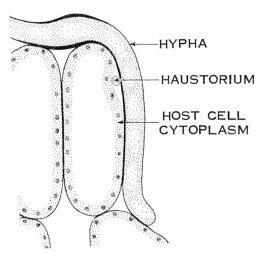

Fig. 8.10. Haustorium from a hypha of *Albugo* penetrating a host cell.

Fig. 8.12. Ropelike strands of mycelia of a fungus growing under the bark of a rotten log.

the host cell protoplasm where it proliferates into a finely branched system of nonseptate hyphae. This rhizoidal system presents a large total surface area in contact with the host protoplasm through which food is absorbed rapidly (see Fig. 8.15).

Some mycelia have a diffuse organization with the branching hyphae being loosely intertwined. Others may form felt-like layers of tightly woven hyphae.

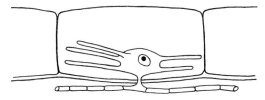

Fig. 8.11. Haustorium of *Erysiphe* in the epidermal cell of a grass host.

Growth conditions, particularly humidity, may cause the same species of fungus to react differently under different circumstances. In some fungi there may be such a dense compacting of the mycelium that it takes on the characteristic appearance of a pseudoparenchyma. Frequently the interwoven hyphae form ropelike strands. Such strands are of common occurrence under the bark of rotting logs (Fig. 8.12). The complex fruiting bodies of many fungi are formed by highly modified mycelia. These are illustrated elsewhere along with a more detailed discussion of fungus fruiting bodies.

One very highly specialized group of plants is known as the *lichens* (Fig. 8.13). Their plant bodies consist of an algal component and a fungal component (Fig. 8.14). It has been demonstrated that each component can be grown separately but when

Fig. 8.13. Lichens growing on bark.

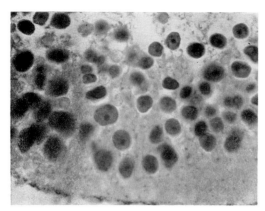

Fig. 8.14. Photomicrograph of a section of a lichen showing algal cells among the fungal hyphae.

this is done neither the alga nor the fungus resembles the lichen from which it came. The alga may be a green alga or a blue-green alga while the fungus is commonly a member of the sac fungi. Lichens grow in many habitats from swamp forests to exposed rocky surfaces. In the latter case they hasten the weathering of rock, which leads eventually to soil formation. The mycelium of the fungal component is highly modified to resist desiccation and provides an internal environment that permits the algal component to carry on photosynthesis. Apparently the fungus obtains its basic food supply from the alga.

## MODES OF NUTRITION AMONG THE FUNGI

Since plants without chlorophyll are unable to manufacture simple carbohydrates they must obtain them elsewhere and in the fungi evolutionary selection has fostered many different mechanisms for obtaining access to basic foods.

Many fungi lack only the photosynthetic mechanism and are able to synthesize the myriad other organic compounds that are essential to their existence from simple carbohydrates. However, modern research has shown clearly that numerous species of fungi lack mechanisms for one or more other biochemical syntheses. Some of them are unable to synthesize essential vitamins; others fail to manufacture one or more of the essential amino acids or lack some other process.

The fungus *Neurospora,* which gives rise to a red mold of bread, has been used extensively for experimental purposes in such research. Wild strains of *Neurospora* are generally able to accomplish most of their own biochemical syntheses and will grow on simple media containing glucose. Research workers have selected several strains with deficiencies, i.e., inabilities to synthesize specific compounds essential to their metabolism. Other deficient strains have been developed as a result of mutations induced by such treatments as exposure to injurious radiation.

Frequently the loss of a specific gene has been correlated with the loss of the ability to form a specific enzyme. Such experiments have been largely responsible for the development of the modern concept that the probable function of a gene is related to the synthesis of a specific enzyme.

It is now widely understood that complex metabolic processes such as respiration involve enzymatically controlled reactions in which the end products of one reaction become the raw materials of the next. Thus if a cell is unable to synthesize one of the enzymes that control a given reaction, the whole process is interrupted. If the end product of the missing reaction becomes available from another source the series of reactions can continue.

It is also possible that some compound that limits the synthesis of the missing enzyme by its absence can be obtained from a source outside the cell. This too would permit the cycle to continue.

In many natural environments deficient strains disappear because of the difficulties involved in obtaining the compounds they are unable to synthesize. However, in biotic communities such as those which exist in the soil there is a delicately balanced relationship among many organisms; each provides something that another needs. Also, many fungi have become adapted to growing as parasites on organisms that are able to manufacture the compounds that the particular parasite lacks.

Although many fungi require only simple energy-supplying foods, these foods often occur in forms that cannot be absorbed directly. For instance much of the food manufactured by primary producers is converted to storage products such as starch and oil or to building materials such as cellulose and protein. These compounds have large and complex molecules that are relatively insoluble and unable to pass through living cell membranes.

Many fungi have acquired mechanisms for secreting enzymes into such complex food materials. These enzymes accomplish an *extracellular digestion* that breaks down the complex molecules into simpler and more soluble molecules that can be absorbed through the living cell membranes of the fungus.

This process gradually alters the texture and other properties of the substance being digested. Woody tissues invaded by fungi with cellulose- or lignin-digesting enzymes gradually lose their mechanical strength and may be reduced eventually to a pulpy mass.

Much of the digested food is absorbed into the protoplasm of the fungus where it enters into diverse metabolic reactions. Some of it is used in respiration. The resulting carbon dioxide passes off into the air as a gas while the water formed during respiration may evaporate or accumulate. In the latter case a substance that was dry originally may become wet and soggy even though no water from other sources has been added.

The conversion of organic compounds to carbon dioxide and water results in a loss in dry weight of the total mass of both fungus and substrate. A rotted log is much lighter than a sound one of the same dimensions.

The soluble foods formed during the extracellular digestion process may benefit other organisms than the fungi that secrete the enzymes. The beneficiaries may be other fungi, plants of other groups, or members of various categories of animal life. Most animals lack the enzymes necessary to digest cellulose and the bacterial flora in the rumens of many herbivorous animals accomplishes the digestion of cellulose in the plants that have been ingested. This is an important function since herbivorous animals provide us with most of our meat and dairy products.

The destruction of wood as well as numerous other organic substances by fungi represents a serious economic loss but is, on the other hand, a vital part of various biological cycles. As has been pointed out, the carbon dioxide content of the atmosphere is constantly diminished by the process of photosynthesis. The respiration of fungi thus assumes significance as the major mechanism by which carbon dioxide is returned to the atmosphere. Furthermore the decay process gradually reduces the bulk of dead material on the surface of the earth and makes room for the growth of new organisms.

UTILIZATION OF THE METABOLIC PRODUCTS OF FUNGI

While carbon dioxide and water are the most abundant substances released as end products of the respiration of fungi, other substances are released in significant amounts. In an earlier discussion of respiration it was pointed out that under aerobic conditions the carbon source is oxidized completely to $CO_2$ and $H_2O$ while under anaerobic (without air) conditions the carbon source is incompletely oxidized and the end products are in part organic compounds with residual energy values. Such compounds are frequently utilized in the diet of other organisms.

A number of compounds produced during the respiration of fungi have important industrial uses. Among them are alcohol, citric acid, acetic acid, and butyric acid. Techniques for growing fungi under controlled conditions with lowered oxygen tensions are important in several biochemical industries and much research has been done to improve the quality and production rate of such processes. This type of research involves experiments with growth conditions as well as genetical selections of strains with the ability to produce higher yields.

The discovery of the antibiotic chemical penicillin was the direct result of an observation that a substance secreted by a species of the common mold *Penicillium* caused an inhibition of the growth of certain bacteria in the same culture. This type of reaction had been known for a long time but Sir Alexander Fleming proceeded to extract the chemical in crude form. He was then able to demonstrate that it could be administered to human beings and that it would inhibit the growth of many disease-causing bacteria.

The search for other types of antibiotics has been carried on actively and has resulted in an arsenal of new wonder drugs for use in the fight against disease. Many of the organisms from which antibiotics have been extracted are soil dwelling actinomycetes. Possibly the antibiotics inhibit competitors for survival in the microenvironment of the soil. Organisms possessing the ability to produce antibiotics would be favored in the process of natural selection.

SAPROPHYTES AND PARASITES

From the preceding discussions it is evident that the fungi are extremely diverse in their growth habits and requirements. Their modes of existence may be grouped into two large categories. Those that exist only on dead organic matter are called *saprophytes* while those that exist only in a dependent association with a living host are called *parasites*. The borderline between the two categories is not sharp. Some parasites can function as saprophytes; certain saprophytes can function as parasites. The common black bread mold has been known to be a weak parasite of strawberries and sweet potatoes and has been identified as the causal agent of one very painful type of ear infection in human beings.

Many of the fungi which exist only on or in living organisms actually kill the host cells in the path of the advancing mycelium, apparently through the secretion of toxic substances. As the permeability of the host cell membranes is altered in this way, substances within the host cells become more readily available by simple diffusion processes. Furthermore, living host tissues are often able to resist penetration by the infecting mycelium and death of the host cells becomes a necessary prelude to growth of the fungus.

The invasion of host cell protoplasm by haustoria creates a very delicate relationship since the fungus is able to absorb nutrients from the host without killing it. The general vigor and growth rate of the host are reduced by such infections but the fungus is often able to grow for much longer periods of time than if the host tissues were killed immediately.

One very common example of such a relationship is powdery mildew of lilac. The mycelium of the fungus that causes this disease covers the surface of the leaves while haustoria penetrate into many of the epidermal cells. However, lilac leaves are rarely killed by the fungus which is universal on this plant. Apparently such fungi have a great many nutritional requirements as they are the most difficult types to establish in cultures on artificial media.

The *mycorrhizal* relationships between fungi and the roots of many plants were mentioned briefly elsewhere. This type of relationship should be considered as something other than strict parasitism. The fungus grows both within the root and externally in the soil; apparently this relationship increases the ability of the root to absorb necessary chemicals from the soil.

## ASEXUAL REPRODUCTION IN THE FUNGI

Since fungi are dependent organisms it follows that they can be successful when they are able to come in contact with and penetrate potential food supplies. The spread of a mycelium may permit coverage of an entire unit of the food supply such as a loaf of bread, a dead insect in the water, or a leaf on a tree; but the spread to another loaf, another insect, or another

# The Fungi

leaf frequently presents a barrier that the mycelium cannot bridge.

The rapid spread of many fungi is actually accomplished by spores or sporelike structures produced by various reproductive processes. Some of these swim in water, some are dispersed in air, others are carried on or in the bodies of animals. Special mechanisms of spore dispersal are characteristic of a particular type of fungus. The numbers of spores produced are large; this insures that any new and usable food sources in the vicinity are subject to infestation.

## Production of Zoospores

The production of zoospores is limited to the algalike fungi and does not occur in all members of this class. The chytrids comprise one order of this class; they commonly occur in aquatic environments. Earlier in this chapter it was noted that when a zoospore came to rest on the wall of a host cell infection was accomplished by digestion of a small hole in the host cell wall through which the parasite entered the host protoplasm.

In many chytrids the orginal spore remains outside the host cell and gradually enlarges as food is absorbed from the host. This enlarged structure is densely protoplasmic and at maturity becomes a sporangium (Fig. 8.15). Its contents become divided into a large number of small cells that are released as uniflagellate zoospores. Each one is able to swim to a new host cell and repeat the infection.

The aquatic mold *Saprolegnia* frequently develops in the bodies of dead insects and other dead organisms in the water. After the infestation becomes established many long hyphae grow outward in a radiate manner from the infested tissue (Fig. 8.16). These hyphae are without cross walls except at the tip where the terminal portion of each one becomes modified into an elongated zoosporangium. The hyphae that bear such sporangia are called *sporangiophores*.

The protoplasm of each of the zoosporangia becomes subdivided into many small cells that are discharged through an opening at the tip (Fig. 8.16B). Each of these becomes a biflagellated zoospore that after a brief period of activity enters a temporary, nonmotile, resting stage. When this

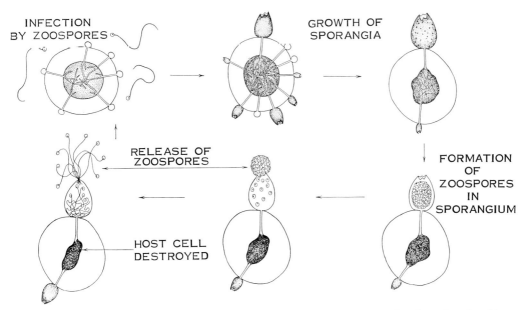

Fig. 8.15. Series of drawings showing the growth and asexual reproduction of a chytrid and the simultaneous destruction of the blue-green algal host cell.

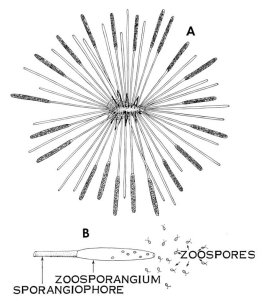

Fig. 8.16. Water molds. A. Dead fly infected with *Saprolegnia*. B. Discharge of zoospores from the zoosporangium of *Saprolegnia*.

structure becomes active again it gives rise to another zoospore that is able to initiate a new infestation.

AERIAL DISPERSAL OF FUNGUS SPORES

The evolution of windspread spores has been as important a factor in natural selection among the fungi as it has among the higher green plants. Probably the ability to achieve aerial distribution of spores or sporelike bodies evolved independently in various groups of fungi.

SPORES OF THE BLACK BREAD MOLD

A very common example of wind dissemination is provided by the spores of the black bread mold. Earlier it was noted that the growth centers give rise to erect hyphae with swollen tips (Figs. 8.9, 8.17, 8.18). These densely protoplasmic tips become modified into sporangia when delicate cross walls form to separate the tips from the rest of the erect hyphae that support them. Then cleavage begins in the protoplasm and numerous spores are formed within each sporangium. These spores become pigmented and are so densely packed that light is not transmitted through the mature sporangium. As a result it appears black. When the spores are mature the outer wall of the sporangium breaks and the spores float away in the air. A *columella* is left behind as a central

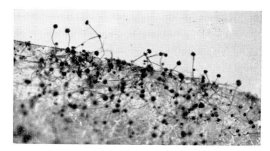

Fig. 8.17. Mycelium and sporangia of the black bread mold. Compare with Fig. 8.8.

dome with a little collarlike rim at the base representing the broken sporangium wall (Figs. 8.9, 8.18).

Bread mold spores are produced in enormous numbers and are almost universally present in the atmosphere. Thus any suitable medium that is exposed to air may become infested provided moisture and temperature conditions are satisfactory for

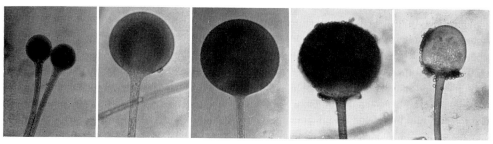

Fig. 8.18. Series of photomicrographs showing maturation of the sporangia of black bread mold. Compare text description and Fig. 8.8.

# THE FUNGI

the germination of spores and growth of the mycelium.

## THE SPORANGIUM OF PILOBOLUS

The fungus *Pilobolus* is similar to the black bread mold in that it forms a sporangium at the tip of an erect sporangiophore (Fig. 8.19). However, the entire sporangium is distributed as a unit rather than as single spores. This sporangium is discharged in a most interesting manner which is related to the growth habits of the fungus.

Fig. 8.19. Sporangium and sporangiophore of *Pilobolus*. Note inflation of the sporangiophore.

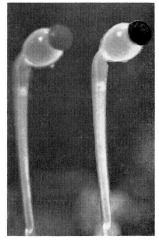

Fig. 8.20 Photograph of *Pilobolus* sporangiophores showing bending toward a light source.

The sporangiophore below the sporangium becomes highly turgid and inflated. This inflated portion is sensitive to light and bends in such a way that the sporangium is aimed toward a light source (Fig. 8.20). When the turgidity reaches a maximum the sporangiophore explodes violently and hurls the sporangium toward the light.

*Pilobolus* is a dung fungus; in grazing areas the droppings of herbivorous animals are commonly infested with it. The aiming mechanism insures that the discharged sporangium will pass between the leaves of overhanging plants and come to rest on the surfaces of leaves at some distance away. These leaves are more apt to be ingested by another animal and in this way the continuous presence of this fungus in dung is assured.

## CONIDIA

The *conidium* is a type of spore which is one of the most efficient means of asexual reproduction in the fungi. In a broad sense a conidium is a fragment of a hypha that becomes adapted to survival during aerial dissemination before it breaks away from the parent hypha (Fig. 8.21). Two of the evident modifications are a thickening of the wall and a dehydration of the protoplasm. In some species the conidium is unicellular but conidia with two, three, or several cells are common.

Conidia may be formed in a linear series as the direct result of fragmentation of a hypha or they may be formed at the tips of special branches called *conidiophores*. The tip of a conidiophore may become pinched inwards so that the conidium is formed by an abstriction process. In other cases the conidium is formed by a protrusion process in which a small droplet of protoplasm is forced out of the open end of the conidiophore. Both processes are frequently repeated so that conidia form in series at the tips of the conidiophores.

From the above discussion it is evident that there are at least three basically different processes by which conidia are formed. Special names have been proposed for the spores resulting from each process but as yet they are not in common use.

The word conidium is derived from a Greek word meaning dust and the connotation is apt. For example, an orange infested with *Penicillium* will appear white as the mycelium grows and then blue as the conidia mature at the surface of the mycelium (Fig. 8.22). A slight puff of wind will displace vast numbers of these spores and they appear as a cloud of dust drifting away from the orange.

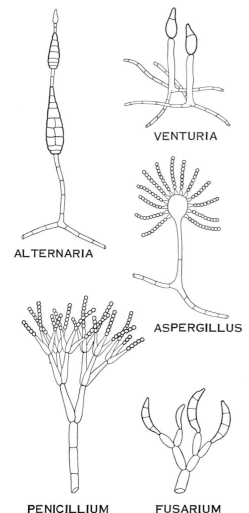

Fig. 8.21. Examples of conidiophores and conidia of several common fungi.

Fig. 8.22. Orange infested with *Penicillium*.

Just as the spores of bread mold are widely distributed, the conidia of *Penicillium* and the related genus *Aspergillus* are everywhere present. Thus infestation can occur wherever usable organic material is available and suitable growth conditions prevail.

Many diseases of plants become serious because of the efficiency of the method for spreading the causal agents by means of conidia. The sexual process, on the other hand, is often no more than a survival mechanism that results in the localized initiation of infections when the environment becomes favorable to the growth of the fungus. Once the disease is established the spread of the fungus from plant to plant is most commonly effected by spores such as conidia. Diseases like apple scab and the brown rot of stone fruits would be of little significance were it not for the mechanism of spreading the infection by means of conidia.

UREDOSPORES OF STEM RUST

In the life cycle of the fungus that causes the black stem rust of cereals several types of spores are produced (as will be discussed in more detail below). One of these spore types is the *uredospore* (see Fig. 8.42). It resembles a conidium in being a spore produced by an asexual process at the tip of a hypha and is the only means by which the fungus can be spread from one plant to another in a field of susceptible cereal crop plants (wheat, barley, oats, rice, etc.).

THE MODIFIED SPORANGIUM

One structure which might be considered as an intermediate step in the evolution of a type of windspread spore is the modified sporangium characteristic of an important group of phycomycetes. Usually it is formed by the abstriction of a multinucleated terminal portion of a special hypha called a *sporangiophore* (Fig. 8.23). In several species the modified sporangia are formed in a series at the tips of these hyphae. They develop thickened walls and their protoplasm becomes dehydrated. As a result they are able to survive

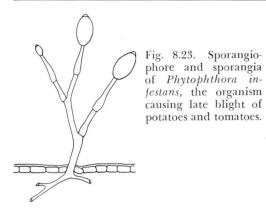

Fig. 8.23. Sporangiophore and sporangia of *Phytophthora infestans,* the organism causing late blight of potatoes and tomatoes.

in dry air as they are transported from one host to another.

This type of sporangium may germinate in two ways. At warmer temperatures (above 20 C) it may germinate directly to produce a single hypha that causes a single infection in the host plant.

At cooler temperatures (below 20 C) the contents of the sporangium divide several times to produce zoospores. If the leaf to be infected has a surface film of moisture the zoospores are released and begin to swim about on the leaf. Each one is capable of initiating a new infection and thus a much higher rate of infection is achieved. The zoospore may enter an open stomate or may come to rest directly above an open stomate. It is also possible that it can form a penetration hypha that causes a direct infection by perforating the cuticle.

Weather conditions that induce the occurrence of water films on leaf surfaces (cool nights after humid days, for instance) will favor the rapid spread of organisms that produce this type of sporangium. Diseases such as the late blight of potatoes and tomatoes, the blue mold of tobacco, and the downy mildew of cucurbits become serious under such conditions, so serious that the U.S. Department of Agriculture maintains a warning service that predicts danger periods. These predictions are based on weather forecasts and records of local infection. The information serves as a guide for the timely use of expensive fungicidal sprays.

## SEXUAL REPRODUCTION IN THE FUNGI

In the fungi, as in the algae, many variations of the basic cycle of sexual reproduction may be found. The examples that are discussed below serve to illustrate a few of these variations.

As mentioned in Chapters 6 and 7, reproduction involving motile isogametes is thought to be primitive. The chytrids are considered to be a primitive group of fungi and the production of motile isogametes has been demonstrated in some of them. Occasional observations have been made of the fusion of such gametes in pairs followed by the formation of motile zygotes that penetrate a new food supply before entering a resting phase. Presumably meiosis occurs during germination of such zygotes. Details of the gametic reproduction of a great many chytrids are unknown and the above description is not meant as a general one for the group.

It may be assumed that advanced forms of gametic reproduction such as oogamy evolved in many separate series in the fungi. Quite possibly the sequence of events in some of them followed the pattern of increasing size, loss of motility, decrease in numbers, and increased protection for the female gamete, while the male gamete retained many of the features of the primitive isogametes.

OOGAMOUS REPRODUCTION IN *Saprolegnia*

The water mold *Saprolegnia,* which belongs in an order of the algalike fungi, provides a classical example of oogamous reproduction. The egg case (oogonium) consists of the swollen tip of a hypha with a cell wall across the base. Within the oogonium one to several eggs (oospheres) are formed (Fig. 8.24). Antheridia develop from the less prominently inflated tips of nearby hyphae. These may or may not be branches of the same hyphal system as that bearing the oogonia since *Saprolegnia* species may be bisexual or monosexual.

When an antheridium comes in contact with an oogonium one or more very slender penetration tubes pass into the oogonium and come in contact with the

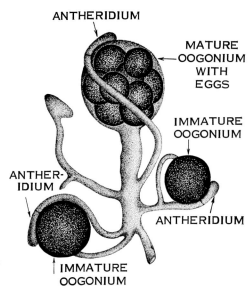

Fig. 8.24. Oogonia and antheridia of the water mold *Saprolegnia*.

eggs. These function as fertilization tubes through which small amounts of cytoplasm and a single nucleus pass into each egg. Nuclear fusions occur in each of the fertilized eggs and the diploid zygote becomes a thick-walled resting cell *(oospore)*. It has been assumed that meiosis occurs in the germinating zygote.

ZYGOSPORE FORMATION IN THE
BLACK BREAD MOLDS

Sexual reproduction in the black bread molds resembles conjugation of nonflagellated gametes in the green alga *Spirogyra*. *Rhizopus stolonifer* is heterothallic; sexual reproduction will not occur unless mycelia of plus and minus strains grow close to each other. When this happens slightly modified hyphae from one strain are stimulated to grow directly toward similar hyphae from the other. These hyphae are somewhat inflated and may be considered as either *progametes* or *progametangia*. When they come into contact each one is stimulated to form a cross wall which delimits a multinucleate cell at the tip (Fig. 8.25).

The contents of these cells do not become further subdivided. For this reason they may be interpreted either as gametangia or as multinucleate gametes. The walls at the point of contact then dissolve and plasmogamy occurs. This is followed by the development of a very thick black wall and the zygote becomes a resting spore *(zygospore)*.

When the zygospore of *Rhizopus* germinates it gives rise to a sporangium similar to the sporangia on the vegetative mycelia. In this species all of the spores produced in such a sporangium give rise either to plus strains or to minus strains. Details of when and where genetic segregation occurs are somewhat obscure but good evidence exists that nuclear fusions and meiosis are followed by degeneration of all but one of the haploid nuclei prior to germination of the zygote.

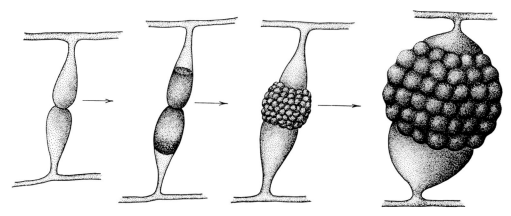

Fig. 8.25. Series of drawings illustrating sexual reproduction in the black bread mold *Rhizopus stolonifer*.

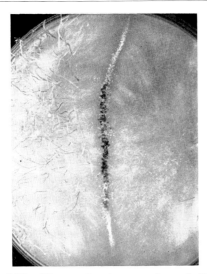

Fig. 8.26. Plus and minus strains of *Phycomyces* growing on an agar plate with a line of zygospores forming at the juncture of the two strains.

*Phycomyces* is a genus somewhat closely related to *Rhizopus*; it has a similar sexual cycle. When plus and minus strains are grown on the same culture plate a line of zygospores forms more or less across the middle of the plate where the mycelia meet (Fig. 8.26). A series of developmental stages of these spectacular zygospores usually can be found by examining the line. (The antlerlike spines that surround the mature zygospore arise from supporting hyphae rather than from the zygospore itself [Fig. 8.27]).

ASCI AND ASCOSPORES IN THE SAC FUNGI

Sexual reproduction in the *sac fungi* is an elaborate process that terminates in the production of a type of spore called an *ascospore*. In most cases the ascospores occur in groups of eight within saclike structures called *asci* (sing., *ascus*) (Fig. 8.28).

The female structure involved in the production of asci is called an *ascogonium*. While similar in basic function to an oogonium it does not contain egg cells as such. Instead, it contains a mass of noncellular protoplasm with several nuclei. The *antheridium* is similarly a multinucleate structure. In some of the sac fungi the antheridia and ascogonia are scarcely distinct from normal vegetative cells. In others the ascogonium may be considerably larger and more elaborate than the antheridium (Fig. 8.28). In many species a slender tube passes to the antheridium from the ascogonium. It is called a *trichogyne* and the protoplasm of the antheridium flows through it into the ascogonium.

With respect to time, the three major steps in the gametic cycle (plasmogamy, karyogamy, and meiosis) are spaced differently in this group than in any other previously described. *Plasmogamy* occurs when the cytoplasms mix but nuclear fusions do not occur immediately. Instead, the nuclei come together in pairs, each one retaining its own identity throughout several succeeding mitoses. Each ascogonium may have several pairs of such nuclei. Eventually pairs of nuclei migrate into special hyphae that arise from the ascogonium. These are called *ascogenous* (ascus-bearing) *hyphae* since each one gives rise ultimately to one or more asci (Fig. 8.28).

Each young ascus receives a pair of nuclei. Early in the further development of the ascus these nuclei fuse (*karyogamy*) and then go through the normal meiotic process immediately to form four haploid nuclei. In those species with sexually different strains two of these haploid nuclei contain genes for one type and two contain the other.

Normally one further mitotic division occurs in each of these nuclei so that eight haploid nuclei result. Each one of these becomes incorporated into an ascospore when the cytoplasm of the ascus is partitioned. In this manner eight ascospores are formed within each ascus. Figure 8.29 shows several mature asci of the soil fungus *Neocosmospora vasinfectans*.

It has been mentioned that the fungus *Neurospora* is used extensively in studies of biochemical genetics. This organism forms ascospores in a similar manner to the one just described. Part of the reason for its value as a research organism is that the individual ascospores can be isolated one at a time in the order of their occurrence in the ascus. This enables research

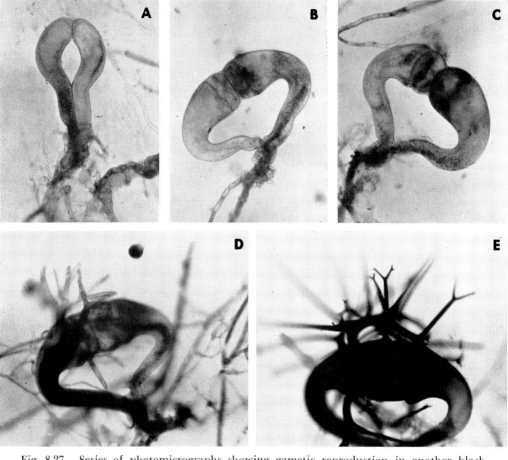

Fig. 8.27. Series of photomicrographs showing gametic reproduction in another black bread mold, *Phycomyces blakesleeanus*.

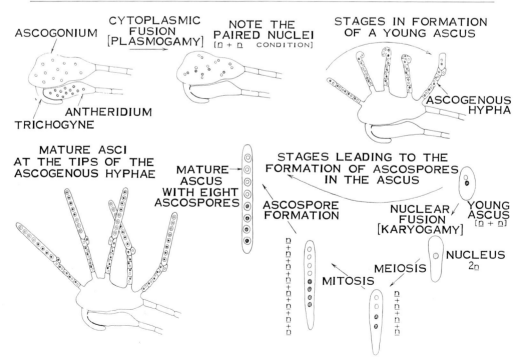

Fig. 8.28. Schematic illustration of the gametic process in an ascomycete.

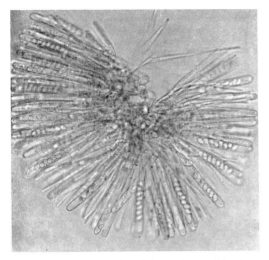

Fig. 8.29. Cluster of asci from the perithecium of *Neocosmospora vasinfectans*.

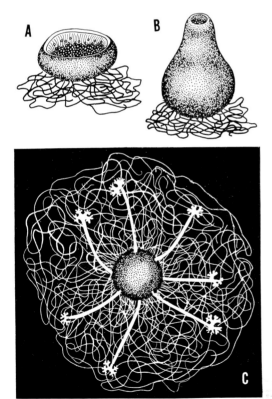

Fig. 8.30. Some common shapes of ascocarps. A. Disc-shaped. B. Flask-shaped. C. Spherical. (The elaborately branched appendages of the ascocarp shown in C are characteristic of the fungus which causes powdery mildew of lilac.)

workers to relate experimental results obtained from subsequent cultures with the genetic segregations that occurred during meiosis.

In many of the sac fungi numerous asci occur together within the bodies of complex fruiting structures called *ascocarps* that develop from intertwined and compacted masses of sterile hyphae. Ascocarps are of various shapes: spheres, cups, discs, flasks, as indicated in Figs. 8.30, 8.31, 8.32.

Ascocarps show considerable range in size. Many are small and inconspicuous, approximating the size of the head of a pin. Others are much larger and showy like the scarlet cup fungus (*Sarcoscypha* sp.) and the morel *(Morchella esculenta)*. The latter fungus is highly prized as an edible mushroom (Fig. 8.32).

Among the numerous sac fungi that do not develop ascocarps of any type are the yeasts and species in the genus *Taphrina* that cause such diseases as peach leaf curl and plum pocket.

## IMPORTANCE OF UNDERSTANDING LIFE CYCLES IN CONTROLLING PLANT DISEASES

The nature and control of plant diseases are matters of large concern to humankind and require the constant attention of numerous specialists who must be highly trained in several related areas of science and practice. Most of the diseases of plants are caused by fungi and an understanding of the life cycles of the fungi involved is of paramount importance. Otherwise, considerable time, effort, and money may be expended on inefficient procedures. In the following discussion the life cycle of the fungus causing brown rot of stone fruits is cited as an example.

### BROWN ROT OF STONE FRUITS

This disease is caused by one of the sac fungi, *Monilinia fructicola* (formerly *Sclerotinia fructicola*), and is responsible for millions of dollars in economic losses each year to growers of peaches, cherries, plums, and other fruits. The name of the disease is derived from the brown spots

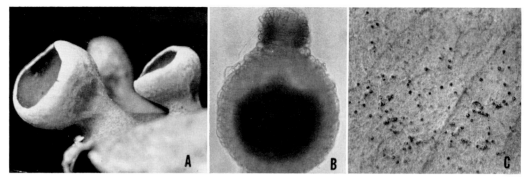

Fig. 8.31. A. Cup-shaped ascocarp formed by the fungal component of a lichen. B. Photomicrograph of a flask-shaped ascocarp formed by the soil fungus *Neocosmospora vasinfectans*. C. Minute spherical ascocarps formed by a fungus causing powdery mildew on a leaf.

that appear on infected fruits and destroy their economic usefulness.

A convenient point of entry into a discussion of the life cycle of the fungus causing this disease is with the ascospores which may be found floating in air currents in an orchard when fruits are beginning to form in spring. If an ascospore falls on an ovary and germinates the hypha will penetrate the host tissues, forming a mycelium that shortly destroys the ovary. However, trees of this type generally produce so many fruits that the loss of a few in this manner would not be particularly significant to the final harvest.

The disease becomes serious because the mycelium next begins to produce large numbers of conidia asexually and they are distributed throughout the orchard by air currents. Each conidium can give rise to an infective mycelium that is genetically the same as the primary mycelium and equally destructive of host tissue.

It should be clearly understood that while the ascospores may cause a relatively few initial infections early in the season the rapid spread of the fungus until the disease reaches epidemic proportions is due to the conidia.

As the season progresses the disease becomes more and more prevalent. By the time the fruits begin to ripen most of them show some signs of infection, particularly in seasons when weather conditions have favored the fungus. Chocolate brown patches (the brown spots) develop on the fruits, which become inedible and worthless. Healthy fruits normally develop an abscission layer across the fruit stalk and fall from the tree. This does not occur in infected fruits, which dry and shrivel but remain hanging from the branches. In this

Fig. 8.32. The large and edible ascocarps of a morel.

condition they are referred to commonly as *mummies*.

Even though the fruits are dead the mycelia in them remain alive and eventually give rise to cup-shaped ascocarps on slender stalks about 2.5 cm long (Fig. 8.33).

Fig. 8.33. Cup-shaped ascocarps arising from mycelia of the fungus causing brown rot, growing within the mummified tissues of an infected peach.

The mummies fall from the trees some time during the winter and the ascocarps emerge from the fallen mummies in spring, at the flowering time of the host trees. When the asci that line the inner surface of the cup are mature they explode and hurl the ascospores upward into the air. (As was noted in the general discussion of ascospore formation, they are the product of a sexual cycle.)

With respect to possible control measures, removal of infected fruits during the summer or the hanging mummies in winter would be excessively expensive since it would require much tedious hand labor. After the mummies fall off and before blossom time they may be buried or gathered and burned at much less expense. Once infections have been initiated locally the best protections against spread are through the use of sprays that leave a residual substance on the fruit surface. The residue kills the hyphae arising from germinating conidia. (Public health considerations require that this substance be non-injurious to humans or that it be completely removed before marketing.)

REPRODUCTION IN THE CLUB FUNGI

The *club fungi* are so named because a club-shaped structure, the *basidium*, bearing four externally maturing *basidiospores* is formed at one point in the sexual life cycle. The cells of the mycelium that give rise to basidia are all *dikaryotic* (binucleate). This is sometimes referred to as an $n + n$ condition in order to emphasize that the two nuclei are separate and haploid. As will be seen subsequently this condition arises in various ways from monokaryotic (uninucleate) mycelia. If the monokaryotic mycelia happen to exist as plus and minus strains the dikaryotic cells will have one plus nucleus and one minus.

BASIDIA AND BASIDIOSPORES OF THE CLUB FUNGI

The basidium is a cell of the $n + n$ mycelium that becomes somewhat inflated and club-shaped. In the young basidium the two nuclei fuse and then undergo meiosis to form four haploid nuclei (Fig. 8.34).

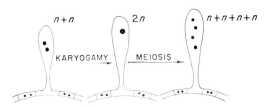

Fig. 8.34. Karyogamy and meiosis in a young basidium resulting in 4 haploid nuclei.

In those cases where the fusing nuclei were plus and minus two of the four nuclei will be plus and two minus. (Nuclear abortions or failures in the meiotic process may result in fewer than the normal four haploid nuclei in the young basidium.)

Following meiosis the basidium may become septate or remain without cross walls. In either event delicate peglike structures grow out from the wall of the basidium. These are called *sterigmata* and through each one a drop of cytoplasm containing one of the haploid nuclei is ex-

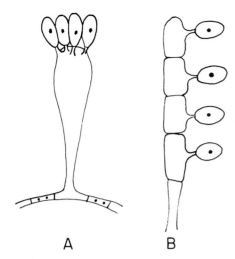

Fig. 8.35. Formation of basidiospores externally on sterigmata arising from basidia. A. Nonseptate basidium. B. Septate basidium.

truded. This droplet then develops a wall and functions as a spore. Appropriately it is called a *basidiospore* (Fig. 8.35).

Basidiospores commonly give rise to monokaryotic mycelia with rather limited growth potential, particularly when compared with the much more vigorous dikaryotic mycelia of the same species. Establishment of the dikaryotic condition occurs in various ways including: (1) fusion of hyphae of appropriately matched monokaryotic mycelia, (2) fusion between a spore and a hypha, (3) fusion of two spores, and (4) nuclear transfers within hyphae.

SPOROPHORES OF THE CLUB FUNGI

Many of the club fungi live in humus and are especially significant in biological cycles such as the carbon cycle. A number of them grow in the woody tissues of trees and are thus responsible for serious economic losses in forest-related industries.

The mycelium of such organisms may grow for many years, accumulating food reserves that permit the formation of large and complex fruiting structures or *sporophores* (Fig. 8.36). The sporophore of a mushroom, for example, consists of two major parts, the *stalk* and the *cap*. On the underside of the cap there are a series of radiating plates called *gills*. Each gill is a compacted mass of mycelium and its whole surface becomes lined with nonseptate basidia standing side by side. The details of the formation of the basidiospores are essentially as described above.

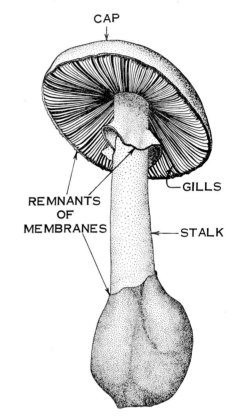

Fig. 8.36. Sporophore of a mushroom.

Mycelia that produce mushroom sporophores are generally unnoticed, particularly when growing in soil humus. With uniform soil and growth conditions this mycelium tends to grow in a radial fashion and to have a circular margin. The sporophores develop periodically near the perimeter of the mycelium. When first formed they are tiny, compacted, buttonlike masses just under the surface of the soil. Warm and humid weather conditions stimulate a very rapid growth of these buttons into mature sporophores. The growth may be completed in one night and this has resulted in many legends about mushroom growth. The discovery of a perfect ring of mushrooms appearing overnight in a pasture is a somewhat startling event and it is small wonder that folklore describes such *fairy rings* as "evidence" of the conclaves of diminutive magical beings.

Although the edibility of mushroom sporophores has always fascinated people, some of them contain deadly poisons and there is no safe field test by which the layman can predict the edibility of a strange mushroom. It is evident also that small doses of certain mushroom toxins result in severe psychological disturbances. Some of the visionary dreams induced by these compounds have a dramatic illusion of reality. In some instances individuals go berserk and become physically dangerous. In fact it has been suggested that the leaders of certain peoples with a reputation for fearlessness and invincibility in battle fed their troops with mushroom toxins to induce this type of madness.

Many of the sporophores of the club fungi have unusual and interesting shapes. Among these are the puffballs, the earth stars, the bird's nest fungi, the coral fungi, and many types of brightly colored mushrooms (Fig. 8.37).

## The Rusts and Smuts

The rusts and smuts include most of the truly parasitic species of club fungi. They are of particular interest because many plants of economic importance such as wheat, corn, oats, and barley are susceptible to diseases caused by members of these two groups.

### CORN SMUT

In corn that has been infected with the smut fungus *Ustilago maydis* cells are stimulated to grow abnormally and form distorted masses of nonspecialized tissues. This growth utilizes food that otheriwse would contribute to the nourishment of the ear. Furthermore if the grains themselves are infected they become enlarged and are eventually destroyed by the fungus, which utilizes such abnormal tissues as a food supply (Fig. 8.38).

Cells of the mycelium eventually become converted into masses of dark-colored teliospores which are released when the distorted grains disintegrate. They are long-lived and generally present in all areas where corn is grown.

When a teliospore germinates it gives rise to a basidium and basidiospores. The basidiospores are capable of initiating small localized infections when transported to living corn plants. These infections are heterothallic and each of their cells contains a single haploid nucleus. Mycelia with binucleate $n + n$ cells are initiated by cellular fusions when compatible mycelia come in contact with one another. It is the binucleate mycelium that induces the severe symptoms in the host.

### BLACK STEM RUST

One of the most critical plant diseases known to humanity is the black stem rust of cereal grains. It has been known ever since people began to cultivate these crop plants and its ravages have often affected the course of history. In addition to its economic importance it is interesting that

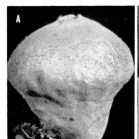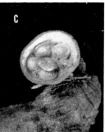

Fig. 8.37. Sporophores of some common club fungi. A. Puffball. B. Earth star. C. Bird's nest fungus. D. Coral fungus.

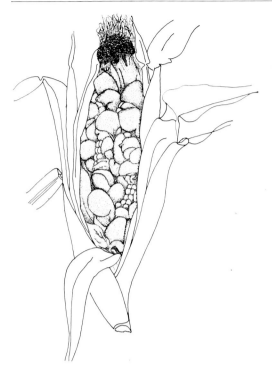

Fig. 8.38. Ear of corn showing the characteristically enlarged and distorted grains resulting from corn smut.

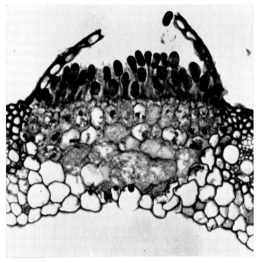

Fig. 8.39. Transverse section of a stem showing the production of uredospores by the infecting fungus.

the causal fungus *Puccinia graminis* must spend part of its existence in each of two entirely different host plants in order to complete its sexual cycle.

During the major part of the growing season *Puccinia graminis* lives in the stem tissues of various grass-type hosts including all of the cultivated cereal grains. When a particular mycelium has been growing for a week or ten days in the tissues of such a stem it begins to form a large number of *uredospores*. Each one of these is single-celled but binucleate ($n + n$) as are all the cells of the mycelium. They are produced by an asexual process and, as noted previously, are similar to conidia.

The uredospores develop just under the epidermis and their formation forces the epidermis to be raised somewhat forming a longitudinal blister or *pustule* on the surface. When the pustule cracks open large numbers of the reddish-colored uredospores are exposed for wind dispersal. This is the *red-rust stage* which is familiar to grain farmers the world over as a sign of approaching disaster (Figs. 8.39, 8.40).

Uredospores are single-celled and dikaryotic (Fig. 8.41). They are windspread among plants of the field and each one is capable of initiating a new infection provided environmental conditions are suitable for germination and growth of the newly formed hypha until it can penetrate the host. The resultant mycelia

Fig. 8.40. Red rust stage of the black stem rust of cereal grains. Pustules in stem are releasing uredospores.

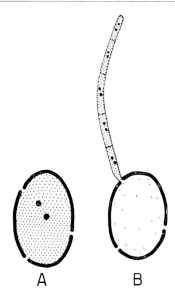

Fig. 8.41. A. Single uredospore. B. Germinated uredospore showing infection hypha. Note dikaryotic condition of cells.

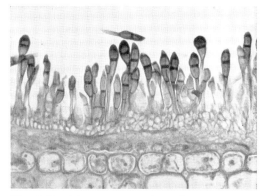

Fig. 8.42. Transverse section of a stem showing the production of teliospores by the infecting fungus.

are also dikaryotic. Within another week to ten days the mycelia in the new infections mature and begin to produce more uredospores. It takes but a few generations for a field of wheat to become so completely infected that the potential harvest of grain is seriously reduced in volume. Since the uredospore is the only mechanism by which this disease can be spread from plant to plant in the grain fields, this type of spore is responsible for most of the damage.

When the infected plants begin to mature existing mycelia and any new ones begin to produce a somewhat different type of spore, the *teliospore* (Fig. 8.42), which plays an important role in the sexual life cycle. The teliospore of *Puccinia graminis* is a two-celled spore rather than a single-celled spore as is the uredospore. Each cell, however, has two nuclei. The walls of the teliospore become very thick and the protoplasm becomes partially dehydrated. In this way the teliospore is able to withstand the desiccating effects of winter weather. In all localities where freezing is common during winter seasons the teliospore is the only stage of *Puccinia graminis* that can survive. Both the mycelia and the uredospores are eliminated by freezing weather.

The teliospore does not initiate any new infections directly. When it germinates in the following spring (Fig. 8.43) each of the two cells gives rise to a *basidium*. The plus and minus nuclei in each of the cells fuse and then migrate into the basidia. Thus each of the young basidia has a single diploid nucleus. This nucleus undergoes meiosis and of the 4 resultant haploid nuclei 2 are plus and 2 minus. Cross walls then develop between the nuclei to form 4 cells. A sterigma develops from each cell and basidiospores are extruded in the manner previously described.

These basidiospores cannot infect any of the grass-type host plants; the only plant they can infect is the common barberry, *Berberis vulgaris*. (This is not the familiar hedge plant, which is the Japanese barberry.) Cereal grain plants and barberry are *alternate hosts* for the fugus *Puccinia graminis*.

When a basidiospore lands on a barberry leaf it gives rise to a localized infection in which the *mycelium has monokaryotic cells*. After a brief period of growth this mycelium forms a flask-shaped structure called a *spermagonium* just under the epidermis (Fig. 8.44). It produces numerous, minute, sporelike bodies that are regarded as male gametes and are called *spermatia*. Also several specialized *receptive hyphae* grow through the sperma-

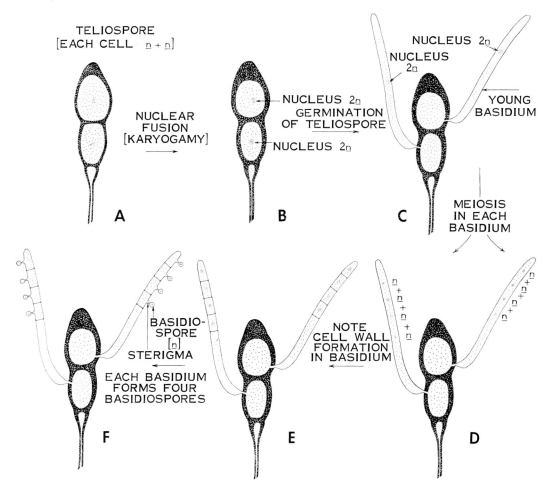

Fig. 8.43. Series depicting the germination of a teliospore, the formation of basidia, and the production of basidiospores.

gonia and protrude from their open necks. Each of these hyphae arises from a single monokaryotic cell called an *aecial initial*, which lies deep within the host leaf. The aecial initial may be considered as a female gametangium and the receptive hypha as a trichogyne (Fig. 8.45).

A drop of sugary solution that is secreted over each spermagonium contains many free-floating spermatia (Fig. 8.44). Small insects are attracted by the nectar as a food and crawl about the leaf in search of it. In so doing they inadvertently transfer spermatia from one spermagonium to another (Fig. 8.45).

The monokaryotic mycelia that produce the spermatia are heterothallic; spermatia will not fuse with receptive hyphae of the same strain from which they are derived. However, the insects sooner or later bring plus spermatia in contact with minus hyphae and vice versa. When this happens protoplasm of a spermatium passes into a receptive hypha and its nucleus migrates downward to the aecial initial at the base of the hypha. Within the aecial initial the plus and minus nuclei pair but do not fuse and in this way the dikaryotic $(n + n)$ condition is established (Fig. 8.45).

Meanwhile still another kind of reproductive structure is being formed just

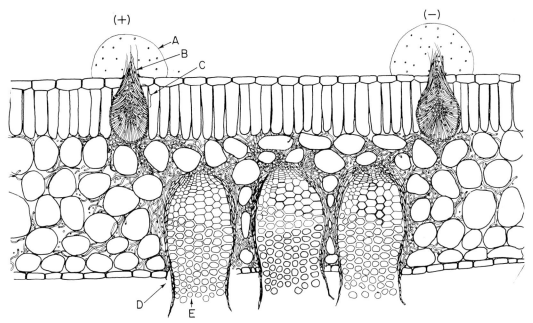

Fig. 8.44. Cross section of the leaf of common barberry infected with *Puccinia graminis*. A. Nectar drop containing floating spermatia. B. Tips of several receptive hyphae. C. Spermagonium in which spermatia are produced. D. Cluster cup (aecium). E. Aeciospores.

under the opposite epidermis from that in which the spermagonia are formed. These occur in clusters and since each of them will be cup-shaped at maturity they are commonly called cluster cups (Fig. 8.46).

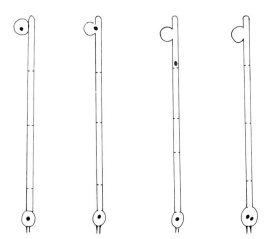

Fig. 8.45. Schematic and highly enlarged drawings showing fusion of a single spermatium with a receptive hypha (plasmogamy) and subsequent migration of the "male" nucleus downward to join the nucleus in the aecial initial, thus establishing the dikaryotic condition.

A more technical term for them is *aecia* (sing. *aecium*). Dikaryotic spores called *aeciospores* are formed in long chains within each cup (Fig. 8.44). These spores are unable to reinfect the barberry but are able to infect an appropriate grass-type host such as one of the cereal grains. *This is one important way in which the disease is initiated in grain fields in spring.*

Following the discovery of the basic facts of this alternation of hosts by *Puccinia graminis* it became evident that eradication of the barberry would break the cycle and possibly bring about some control of this most serious of all plant diseases. Much effort has gone into this task and the barberry has been eradicated in many agricultural areas. Despite the distinct advantages of the eradication of barberry in the control of stem rust, the disease has not been eliminated completely. The reasons for this are based on geography as well as the biology of the fungus concerned.

In areas where freezing weather does not occur, such as many localities in the

Fig. 8.46. Cluster cups (aecia) on the under surface of a barberry leaf.

southern part of the United States and Central America, the fungus remains alive the year round in cultivated crops and wild grass hosts. Thus the production of uredospores is never eliminated on the continent. In the spring as the frost season passes new crops are planted in succession northward. The uredospores spread the infection from one field to another and in this way the infection progresses slowly northward each spring and summer. In some years this progression is so slow that many crops are well along in their development before infections begin. In such years losses due to stem rust are light.

Uredospores are spread more rapidly when large scale cyclonic disturbances cause air masses to travel hundreds of miles northward in a few days' time. Uredospores float in these air masses and retain their viability. Eventually they drift down on fields throughout the agricultural regions of the continent and any susceptible variety of crop plant is liable to infection early in its growth.

Many years ago plant pathologists working with agronomists discovered that there were many different varieties of *Puccinia graminis*. Microscopic examinations of these varieties showed no detectable differences among them but cultural experiments demonstrated that they would infect different species of cereal crop plants. Furthermore within each variety there were shown to be many *races* of the rust; some varieties of a specific kind of crop plant were found to be more resistant to one race of the rust than to another.

This information has been used widely in crop breeding programs and new varieties of crop plants with inbred resistance to many races of rust have been developed. Widespread use of resistant varieties is an effective means of control of the disease; unfortunately this procedure seems to favor both the spread of previously insignificant races and the evolution of new ones.

Theoretically the evolution of new races of rust would be slowed down by elimination of the barberry since the rate of recombination of genetic factors in the fungus could not be significant if the life cycle were broken. However, the complete elimination of the barberry on the continent is a practical impossibility; thus the evolution of new races continues.

Moreover, recent evidence indicates the distinct probability that an exchange of nuclei can take place between cells in the $n + n$ mycelium. Such a bypass of the complete sexual cycle would abet evolutionary progress by the fungus.

Research workers in this area of science are taking a frank look at a rather grim future in which races of stem rust may develop for which there will be no genetic sources of resistance to use in further plant breeding programs. This event could mean a tragic curtailment of the food supply in many areas of the world.

However, breeding for resistance to disease is not the only answer to the problem. Chemical control of plant disease has long been practiced but it is an expensive matter when applied to field crops. Such controls could be used, however, in cases of dire necessity. Research work with this aspect of disease control is being carried out at an increasing rate and it is almost certain that such research will result in less expensive chemical controls before the current techniques of breeding for resistance become obsolescent due to successful evolution in the fungi.

# NINE

IN some modern classification systems bacteria and blue-green algae are termed *procaryotes* because they are alike in not having well-organized nuclei with distinct nuclear membranes typical of higher plants and animals. They also lack structures recognizable as plastids, mitochondria, Golgi membranes, and endoplasmic reticula. Several other similarities are recognized, including the occurrence of diaminopimelic acid in the walls of most genera. The size ranges of some of the smaller blue-green algae overlap those of the bacteria but in general algae are larger. Most blue-green algae are photosynthetic while most bacteria are heterotrophic. The relatively few bacteria which are photosynthetic use compounds other than $H_2O$ as hydrogen donors in the reduction of $CO_2$. Bacterial chlorophyll is slightly different chemically from chlorophyll *a* and has a light absorption peak at a different wave length. The characteristic phycobilin pigments of the blue-green algae are absent from photosynthetic bacteria.

## THE BACTERIAL CELL

The minute size of bacteria is difficult to comprehend. Many of them are no more than 1 μm ($1 \times 10^{-6}$ m) in diameter and 10,000 of them (plus or minus a few hundred) could form a line only 1 cm long. A cube, 1 cm along each edge, solidly packed with bacterial cells, might contain 1,000,000,000,000 individuals. On the other hand, a few thousand bacterial cells distributed at random in 1 ml of fluid would be relatively as far apart as are the moon and the earth. Bacterial cells are barely visible with the light microscope, yet since some of them may divide every 20 min or so one such cell could give rise to enough progeny in a day's time to completely fill the 1 cm cube mentioned above.

Most bacterial cells are composed primarily of a thin but definite cell wall and the enclosed protoplast; in a few genera the cell wall is absent. Most species have an outer slime layer that is analogous to the slime sheath of blue-green algal cells. This slime layer is frequently difficult to demonstrate; in cases where it is distinctly obvious in light microscope preparations it is often called a *capsule*. The cell walls of some bacteria are composed mostly of *mucopeptides* in which *diaminopimelic acid* is a frequent constituent. Cell walls of other bacteria are relatively low in mucopeptide content and rich in *lipoproteins* and *lipopolysaccharides*. The organization of the protoplast is more comparable to a blue-green alga than to a higher plant cell. The somewhat dense mass in its center is not limited by a membrane and thus cannot be considered as a true nucleus even though staining reactions show that it contains DNA. It is often referred to as a *chromatin body*. Modern techniques have shown that the DNA is organized as an extremely long, unbroken, looped and twisted thread which is packed densely (like a heap of wet spaghetti) to make the chromatin body. This thread is often called the *bacterial chromosome* although it differs in a number of respects from the chromosomes of eucaryotic cells.

As noted above, such membrane systems as mitochondria, plastids, Golgi apparatus, and endoplasmic reticula are absent from bacterial cells. However, ribosomes are numerous and masses of small membranous bubbles called *mesosomes* seem to be associated with the inward growth of new cell walls, by which process bacterial cells are divided passively. This

# Life Without Chlorophyll: The Bacteria

process is usually called *binary fission* and is another similarity between bacteria and blue-green algae.

One of the few microscopically visible characteristics of bacteria is cell shape. A *coccus* is spherical before division takes place but is temporarily hemi-spherical immediately after cell division; a *bacillus* is rod-shaped; as the name implies, a *spirillum* has a spirally twisted appearance. (A minor variant of the spirillum is the *vibrio*, which has the shape of a comma or lazy S.)

Unbranched filaments may arise when cells remain united after division provided each plane of division is parallel to the preceding one. *Streptococci* are unbranched filaments of spherical cells; *streptobacilli* are unbranched filaments of rods. (The genus *Diplococcus* is characterized by having coccus cells united in pairs.) Species in which cell divisions proceed alternately in two sets of planes to form flat sheets of cells are *staphylococci*; those dividing in three sets of planes to form more or less regular cubical packets are *sarcinae*.

## Motility

Motility or the lack of it is a useful feature in identification but is not always trustworthy because the same organism may exhibit both motile and nonmotile phases. The organelles responsible for motility are called *flagella,* but they have a different structure from the analogous organelles of eucaryotic cells and are essentially invisible without special preparation under the light microscope. In some cases a jerky type of motility is achieved by the alternate flexing and relaxation of an entire cell. Motility by means of flagella occurs in almost all spirilla and in some bacilli. However, it is generally lacking in cocci.

A type of gliding motion occurs in the *myxobacteria*. Individual cells in this group are rod-shaped and may exist independently or be joined to form slender flexuous filaments, as in species of *Cytophaga* that are common in aquatic habitats, growing superficially on algae. They are notable decomposers of cellulose and chitin. The gliding movements are thought to be associated with the excretion of gelatinous materials, as in the motile forms of blue-green algae. Some members of this group are known as the *slime bacteria* because of the conspicuous trails of slime left behind as they move about. Under certain environmental conditions, notably the depletion of usable food materials, vast numbers of these bacteria assemble in masses to form brightly colored *fruiting bodies* that are occasionally large enough to be seen without a microscope. Individual cells in some of these fruiting bodies become modified into thick-walled resistant units called *microcysts,* which function as survival and dispersal mechanisms.

Still another type of motility occurs in the *spirochetes,* a relatively small group that includes some major disease-producing organisms. The cell of a spirochete is basically cylindrical. However, a long axial filament (composed of one or more fibrils) is associated with the cell. This structure has been likened to a rubber band under tension. The effect is to contract the cell into a series of helical twists; since the tension appears to vary constantly the cell is continuously involved in a frenzied motility.

The *stalked bacteria* are nonmotile forms characterized by being attached to slender threads that are frequently branched so the cells are arranged in a rosette. The *budding bacteria* comprise another small group in which cell increase occurs by a budding process, vaguely similar to that of yeast cells, instead of by the more common process of binary fission.

The *actinomycetes* are sometimes called the *mycelial bacteria* because their long slender filaments frequently branch to form tangles of threads similar to fungus mycelia. Their affinities are still somewhat uncertain but they are of large significance in many habitats, particularly soils. They are sometimes the source of "musty" odors in public water supplies; more importantly, they are the source of several antibiotics including streptomycin.

## FEATURES FOR IDENTIFICATION OF BACTERIA

Simple recognition features alone are not sufficient for taxonomic determinations; many additional features such as growth characteristics, growth requirements, staining responses, and effects on the substrate must be determined before satisfactory identifications of unknown bacteria can be made. The following examples are but a few of the many which could be cited.

1. There is an extensive lexicon of terms that apply to the appearances of bacterial colonies on culture media. They cover shape of the colony, appearance of the margin, coloration, relative luster, surface topography, and other features.
2. The Gram staining technique is an especially valuable tool since it separates bacteria into two large groups, the members of which have a number of additional characteristics in common. Crystal violet and iodine preparations are used first to stain the bacteria densely. If the bacteria are then treated with a destaining solution such as alcohol and lose the stain, they are said to be Gram-negative. If they retain the stain they are Gram-positive. This procedure is one of the most important and widely used techniques in bacteriology.
3. The nutritional requirements of many bacteria are well known and clues to the identity of an unknown form can be obtained by attempting to grow it on a wide variety of culture media, some of which are deficient in particular compounds. If the unknown form does not grow in the absence of certain compounds, such as one or more of the essential amino acids, this evidence is useful in determining its growth requirements and ultimately its degree of relationship to known forms.
4. The ability to produce enzymes capable of digesting particular substances is an important characteristic of bacterial species. If, for example, the presence of a gelatin-digesting enzyme complex is suspected, then the organism could be grown on a gelatin medium. Liquefication of the medium would be considered evidence of the production of the suspected enzyme system.

    Starch digestion by bacterial enzymes can be detected readily by allowing the bacteria to grow for a time on a starch-containing medium and then pouring a solution of iodine in potassium iodide ($I_2.KI$) over the culture. The medium outside the influence of the bacteria will stain blue-black while the portion where the starch has been digested will not react.
5. Gas production by bacterial cultures provides much useful information as to the nature of the organism, particularly if the gas is readily identifiable. (Hydrogen sulfide and ammonia may be identified by smell and by simple chemical tests. Hydrogen gas can be induced to explode when mixed with oxygen and ignited. Carbon dioxide can be absorbed into sodium hydroxide solutions.)
6. Acid production can be determined by incorporating a suitable indicator dye into the culture medium. Bromthymol Blue, for instance, changes from blue to yellow if the pH is lowered from 7.6 through neutrality toward 6.0.
7. The degree of dependency on oxygen in the environment is an important characteristic of each bacterial species. Those that require molecular oxygen

from the environment are said to be *aerobic* while those able to live without oxygen are *anaerobic*. A *facultative aerobe* is one which normally uses oxygen but can survive without it while an *obligate aerobe* requires oxygen at all times. A *facultative anaerobe* is an anaerobic form which can survive in the presence of oxygen while an *obligate anaerobe* is one which cannot survive in the presence of molecular oxygen.

8. The production of *endospores* is a feature possessed by a number of bacilli. These spores have a very dense protoplasm and are resistant to extremes of environmental conditions that prove fatal to the unmodified cells of the species (and to most other forms of life as well). For example, some bacterial spores can survive long exposure to boiling water; this accounts for the need to use steam under pressure to insure complete sterilization of organic matter, as in the preparation of canned foods. The production of spores is probably a survival mechanism for the species rather than a method of asexual reproduction because one vegetative cell normally produces only one spore. Experimentally spores have been demonstrated to be viable after more than 70 years of storage in dry soil on a laboratory shelf.

## BACTERIAL NUTRITION

Although their nutritional requirements are extremely varied, bacteria may be classified as *autotrophic* forms if they are able to utilize a nonorganic source of energy to reduce $CO_2$ to carbohydrate or *heterotrophic* forms if they must rely on some existing organic compound as their basic source of energy for metabolism.

### Autotrophic Bacteria

*Autotrophic bacteria* may be *photosynthetic* if light is the energy source or *chemosynthetic* if energy is derived from oxidative changes of nonorganic substances.

Some of the more common *autotrophic photosynthetic bacteria* are listed below:

1. *The green sulfur bacteria* contain bacteriochlorophyll and carry on a type of photosynthesis in which $H_2S$ is used in place of $H_2O$ as the source of electrons in the reduction of $CO_2$ to the carbohydrate level. In some cases molecular sulfur is deposited as granules externally while in others it may be oxidized further to sulfates. These strictly anaerobic forms sometimes grow in dense concentrations in the deeper waters of lakes.

2. *The purple sulfur bacteria* contain an additional carotenoid pigment, *bacteriopurpurin,* that gives them the characteristic coloration. Hydrogen sulfide is commonly used in a similar manner to that of the green sulfur bacteria but the sulfur granules are deposited within the cells. In some sewage lagoons with an overload of dissolved organic matter the growth of purple sulfur bacteria may be so extensive as to give the water a deep red color.

3. *The purple nonsulfur bacteria* do not deposit sulfur in the protoplasm but are otherwise quite similar to the purple sulfur bacteria. They can function heterotrophically if oxygen is present by obtaining energy from the oxidation of $H_2S$ and other sulfur-containing compounds.

The photosynthetic pigments of these three groups are associated with membranes, some of which resemble thylakoids. Others are smaller than thylakoids and have an inflated appearance rather than being flattened. (These are commonly called *chromatophores* although this term has a prior accepted usage designating the nongreen chloroplasts of several algal groups.)

Some of the numerous examples of *autotrophic chemosynthetic bacteria* are listed below:

1. *The colorless sulfur bacteria* are mostly aerobic and obtain energy for the reduction of $CO_2$ to carbohydrate from the oxidation of sulfur compounds such as $H_2S$. Sulfur granules are formed internally or the oxidation may proceed

to the formation of sulfuric acid. A very common filamentous form, *Beggiatoa,* bears a strong resemblance to the filamentous blue-green alga *Oscillatoria.*

(Note: The ability of several bacteria to obtain energy from the oxidation of sulfur compounds to the level of sulfuric acid is of considerable interest ecologically, since the action of acid on carbonate rocks has a weathering effect on the rock particles and adds significant quantities of $CO_2$ to atmospheric cycles. The release of acid to natural bodies of water from sulfur-containing rock wastes of mining operations exposed to bacterial action under aerobic conditions is a pollution problem of some significance.)

2. *Iron bacteria* use energy from the oxidation of iron compounds to reduce $CO_2$ to the carbohydrate level. In the ferrous state iron is soluble in water, particularly if the water is also rich in dissolved $CO_2$. Iron is oxidized to the ferric state, which is insoluble and is frequently incorporated into the bacterial sheath. It is a common phenomenon where iron-rich waters emerge from rock formations as seeps or springs to find dense clouds of orange-red or ochre-colored iron bacteria. The color is due to ferrous hydroxide, which is imbedded in the thick gelatinous sheaths surrounding the bacterial filaments. This is an important geochemical process and may have resulted in large deposits of bog-iron ore.

3. *The nitrifying bacteria* are responsible for the oxidation of ammonia first to nitrite and then to nitrate. (The energy released is used to reduce $CO_2$ to the carbohydrate level.) This process is referred to as *nitrification;* it is a natural process of considerable significance. It also is part of a modern pollution problem. The breakdown of protein materials by bacterial action in efficiently operated sewage disposal plants results in considerable ammonia being present in the final effluent solution. This is often discharged into rivers and streams and oxidation of ammonia by nitrifying bacteria and the respiration of aerobic heterotrophs which use ammonia in their metabolism contribute to a temporary depletion of dissolved oxygen in the stream water. This so-called oxygen-sag extends downstream for varying distances and can adversely affect animal life in the stream. This is a matter for public concern and must be included in considerations of practical regulations of waste disposal.

4. *The organic autotrophs* provide an interesting contradiction in terms since they are able to oxidize CO and $CH_4$, and oxidizable carbon compounds are generally considered as organic compounds. However, in these instances the energy released is used to reduce $CO_2$ as in most autotrophs; the $CO_2$ which results from the oxidation may actually be cycled into the reduction process as a raw material.

5. *The methane-producing bacteria* comprise a group of several unrelated species that are able to reduce $CO_2$ to $CH_4$. The energy for the reduction process comes from the oxidation of such varying compounds as hydrogen gas, carbon monoxide, acetic acid, other organic acids, and various alcohols. The process is strictly anaerobic, occurring in such habitats as the muck layers of swamps, the rumens of many herbivorous animals, and some sewage treatment plants. (In the latter case the methane is often collected and burned as a source of energy to help operate the plant, a not commonly recognized financial relief to the taxpayer.) Note: The inclusion of this group with the autotrophic bacteria is open to some question but is based on the utilization of energy from oxidation to reduce $CO_2$.

HETEROTROPHIC BACTERIA

*Heterotrophic bacteria* must be structurally organized to permit the entrance of at least simple organic molecules for energy sources as well as building materials. Provided that essential chemical elements are also available from the environment (as inorganic compounds) many bacteria can use simple sugars such as glucose in the respiration cycle to release energy and in various other metabolic pathways to produce the basic units of proteins, lipids, and structural carbohydrates. The requirements of other bacteria for organic substances from the environment are more complex, however, and the growth of a particular form in a particular environment may be dependent on the availability of a particular amino acid, a particular vitamin, and so on. These special requirements are often related to an inability to carry out one particular step in a long chain of essential chemical reactions. If the environment can provide the compound normally produced during this step then the whole process continues and the organism survives.

A number of significant environmental processes may be summarized under the general heading of *heterotrophic nutrition*.

1. *Denitrification* processes more or less reverse the nitrification processes. They are energy demanding *(reducing)* rather than energy releasing *(oxidizing)*. The general direction of such reactions is from nitrate to nitrite to ammonia to elemental nitrogen. Denitrification generally occurs under anaerobic conditions. If the nitrites and ammonia are transferred to an aerobic environment they may be recycled, but when molecular nitrogen escapes into the atmosphere it is essentially lost to the biological environment.
2. *Nitrogen fixation* is the major process by which molecular nitrogen is brought back into biological systems; it is a process of equal significance to the photosynthetic fixation of carbon dioxide. Lightning discharges convert small quantities of nitrogen to ammonia, which becomes washed into soils and bodies of water in rainfall; manufacturing processes exist by which electrical energy is used to create nitrates and ammonia from nitrogen gas for use in agriculture and industry. But the vast bulk of the continuing need for available nitrogen by living organisms is provided by the nitrogen-fixing activities of a few bacteria and blue-green algae. In this process metabolic energy is used by these organisms to reduce molecular nitrogen to the level of ammonium ions that are utilized in the construction of amino acids. The ultimate death and decomposition of bacteria are integral parts of the total nitrogen cycle by which nitrogen compounds are made available for use by other organisms.

    The *free-living forms* of nitrogen-fixing bacteria are common inhabitants of the soil. *Azotobacter* and *Clostridium* are significant genera of this type.

    The *root-nodule bacteria* are common inhabitants of the roots of members of the legume family of flowering plants (soybeans, garden beans, peas, clover). These bacteria, which belong in the genus *Rhizobium,* invade the roots and stimulate root cells to divide abnormally and form large masses of parenchymalike tissue in which the bacteria proliferate rapidly. The masses of root cells are the root nodules; the food provided by them to the bacteria is used as the source of necessary energy for the reduction of molecular nitrogen (Fig. 9.1). As part of modern agricultural techniques genetic strains of *Rhizobium* adapted to each of the major legume crop species have been isolated, cultivated in laboratory situations, and inoculated into the soils in which the crops are grown.

    It appears probable also that the historical development of the rich prairie soils that provide so much of our basic food supplies was in large part dependent on the inclusion of legume plants with their associated root nodule bacteria in the native prairie flora.
3. *Ammonification* is a summary term applied to the several step process by which proteins are broken down to ammonia and various other compounds; the ammonia is then released to the environment. Many different kinds of bacteria are involved and in general this process occurs most efficiently under aerobic conditions. It is the major path-

Fig. 9.1. Root nodules on the roots of a soybean plant.

way by which organic nitrogen is made available for reuse and thus is essential to the continuation of life on earth.

This process has an additional special use in modern sewage treatment plants. Organic wastes in water are sprayed on beds of gravel in which they trickle downward and are constantly exposed to air. This encourages the ammonification process by aerobic bacteria and as a result the final effluent is usually rich in dissolved ammonia. Some nitrification may also occur during the passage of sewage through trickling filters.

4. *Putrefaction* occurs when proteins are broken down under *anaerobic* conditions. This process is not particularly beneficial and often results in products with obnoxious odors. One common example of putrefaction is provided by the sewage fungus, *Sphaerotilus,* which grows in dirty, greyish-white, felty layers on the bottoms of streams that are heavily loaded with organic wastes.

5. *Aerobic respiration* is carried on by most bacteria as well as the majority of plants and animals. As is generally well known, such simple organic compounds as glucose enter in the presence of oxygen into a series of reactions by which they are oxidized completely to carbon dioxide and water with the accompanying release of maximal amounts of energy.

6. *Fermentation* is an anaerobic respiration process in which organic compounds in the absence of molecular oxygen serve as the terminal electron (or hydrogen) acceptors. The products are organic acids, ethanol, $CO_2$, and others.

7. *Sulfur reduction* is a reversal of the processes noted earlier in which energy derived from the oxidation of sulfur compounds is used to reduce carbon dioxide to carbohydrate. In the present process energy derived from the oxidation of carbohydrates is used to reduce (return energy to) sulfur compounds. For example, some bacteria when grown in waters with high sulfate concentrations may produce large quantities of $H_2S$ through sulfate reduction.

8. *Phosphate release* by bacteria is a major topic barely touched on in many discussions of natural recycling processes. Most of the phosphorus in living organisms exists as units of phosphate which are incorporated in many important organic compounds (nucleic acids, ATP, TPN, and others). The removal of phosphate groups from such compounds by enzymatic degradation appears to be a relatively simple operation which is nonetheless of vast significance.

9. *Micronutrient cycling* is similarly a topic of rather large significance, which has come under intensive scrutiny so recently that summary generalities are not yet possible. One sidelight on this area of knowledge concerns the availability of iron. During the biological degradation of organic compounds in soil a

class of complex compounds called *humic acids* results. These compounds are natural chelates in the sense that they affect the relative availability of micronutrients to other organisms. Iron, which may seem to be sufficient in natural waters, may actually be unavailable because of its chemical state. Humic acids washed into bodies of water from soils may have the effect of suddenly making the iron available and thus stimulating the growth of algae that had been previously inhibited by its unavailability.

## BACTERIA AND THE ENVIRONMENT

The environmental relationships of bacteria may be summarized under three very general headings, *parasitism, saprophytism,* and *symbiosis.*

*Parasitic bacteria* are those which rely upon an intimate relationship with a living host. Of the 1500–2000 known species of bacteria a relatively small percent are *obligate parasites,* and of these a still smaller number are responsible for human diseases. It is indeed fortunate that most of these cannot exist outside the human body since this means, among other things, that isolation of infected individuals from the rest of the population may help control epidemic diseases.

*Saprophytic bacteria* are those heterotrophic forms that are able to survive without a living host provided an organic substance is available from the environment that may be used both as a source of energy and a source of building materials for the growth of new protoplasm. By far the largest number of bacteria fall in this category. They are of immense significance, along with the fungi and protozoa, in the natural recycling of chemical elements.

*Symbiotic relationships* of bacteria with other organisms were thought originally to be simply a living together with mutual benefit. Yet as more and more was learned about the complexity of such relationships it began to appear that one term was insufficient to cover all of them and new terms such as *commensalism, mutualism,* and even *helotism* (in which one member of the association is slave to the other) came into usage. One example of the complexity of this type of relationship is presented briefly.

The cow's rumen is a large compartment in the digestive tract. It receives a more or less constant influx of plant material that has been presoaked with saliva which contains starch-digesting enzymes. The animal cannot create cellulose-digesting enzymes, however; these are provided by the protozoans and bacteria that inhabit the rumen. The simple organic compounds released by the digestion process are then fermented and changed further by bacterial action. Some of these products enter the bloodstream of the animal and are metabolized. The rest are used by the microflora and microfauna in the rumen. Large quantities of gas including $CO_2$ and $CH_4$ are released. A large bulk of microorganisms passes from the rumen into the rest of the digestive tract where they are killed by rapid changes in acidity. The proteins of the dead organisms are then digested by enzymes produced by the animal and the resultant amino acids are incorporated into the animal body as a major source of food.

## *ESCHERICHIA COLI*

Finally, no introductory discussion of bacteria should avoid mention of the most famous bacterium, *Escherichia coli.* This is a widely studied organism with many genetic strains adapted to growing in the digestive tracts of various warm-blooded animals. (It is also highly adaptable to laboratory existence and responsive to various types of experimental procedures.) One of its more interesting practical uses provides a measure of the degree of pollution of natural waters by fecal wastes from animals, including humans. Samples of water to be tested are added to culture media suitable to the growth of *E. coli* and the cultures are incubated at a suitable temperature. In a few days individual bacteria, which are not visible on the culture plate, multiply to enormous numbers and form visible colonies. Counting colonies is a tedious but simple procedure and application of a suitable formula to the count gives a reliable estimate of the number of coliform bacilli in the polluted water.

Scientific studies of *E. coli* are numerous and exciting. One of them led to an understanding of a special type of sexual reproduction that occurs in some bacteria and ultimately to the award of a Nobel prize.

# TEN

THE several groups that make up the *Algae* are remote from one another with respect to their origins; their possible relationships extend far back in unknown evolutionary history. The title chosen for this chapter reflects the need for organization of the subject matter but also notes the fact that the green algae have had more attention due to their putative ancestral relationships to the land plants. All of the groups to be discussed contain chlorophyll but for various reasons they are not as green as grass. Some of them are essentially insignificant in the present ecological scheme of things while others rival the higher plants in biological significance and at least one group, the diatoms, probably exceeds the land plants in terms of total primary production of organic matter and oxygen.

Each group will be referred to by its common name in this chapter; a later chapter relates common names to the scientific terminology of modern classification schemes.

One cannot study all algae in a one-term course in general botany; the choice of those to be considered is frequently influenced by geography as well as the length of the term. The author's backyard, for instance, has not been a seashore since Cretaceous times and discussions of seaweeds do not seem as relevant as discussions of the blue-green algae, green algae, and diatoms that turn some of our waters into soup each summer. However, anyone living within sniffing distance of the ocean quite naturally would have different priorities.

## BLUE-GREEN ALGAE

The *blue-green algae* have more features in common with bacteria than with any other group of living organisms. This may seem difficult to comprehend at first since they are considerably larger than most bacteria and the vast majority of them are self-nourishing in that they are able to use a photosynthetic mechanism to reduce $CO_2$ to the carbohydrate level. However, modern studies of their structure made with the aid of the electron microscope have revealed details that strongly support the above contention.

A basic genus in this group is one in which individual cells are spherical at maturity. Species of this genus, *Chroococcus*, commonly exist in the single-celled condition but may form few-celled three-dimensional clusters that periodically revert to the single-celled state by breaking apart.

### Basic Cell Structure

Blue-green algal cells appear to have a simple structure when viewed with the light microscope (Fig. 10.1). Frequently

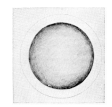

Fig. 10.1. A single-celled, spherical, blue-green alga showing gelatinous sheath, thin inner cell wall, and smooth, apparently unstructured protoplasm. Appearance of protoplasm may be variously modified as discussed in text.

only three parts may be distinguished: (1) an outer gelatinous sheath, (2) an inner cell wall, and (3) a uniform mayonnaise-like mass of protoplasm without any evident structure. In some forms the protoplasm may be resolved into an outer pigmented layer, the *periplasm,* and a denser, colorless, central area, the *centroplasm.* Particles of various sizes frequently give the protoplasm a granular appearance.

# The Nongreen Algae

The outer sheath has a gelatinous texture and varies in thickness among genera and species and with environmental conditions. It is a hydrophilic gel, mainly a derivative of pectin, and its water-holding capacity is of particular significance to terrestrial forms. Quite possibly it may also serve to adsorb ionic particles.

Photosynthetic pigments are located in thylakoids distributed throughout the periplasm instead of being aggregated within choroplasts, as in other algal groups and the higher green plants. (The existence of thylakoids was not recognized until they were revealed in electron micrographs.)

The term *central body* is often used for the centroplasm but perhaps suggests a more discrete organelle than actually exists. Chemical staining tests for nucleic acids reveal that DNA is present in the centroplasm and electron micrographs have shown the existence of DNA fibrils in this region. This means that the major chemical materials of a nucleus are present but not organized as a nucleus *per se*. There is no nuclear membrane and no evidence of chromosomes during division. Thus neither mitosis nor meiosis is possible and sexual reproduction in this group is unknown.

Cell division is accomplished by the inward growth of a new wall, which begins as a circular ridge on the inner surface of the existing wall and gradually extends across the middle of the cell, much as a closing diaphragm crosses the aperture of a camera lens (Fig. 10.2). The wall pushes the outer cytoplasmic membrane ahead of it as it grows and the genetic material of the central body is divided approximately in half by this procedure. No chromosomes are formed and this passive division bears no resemblance to mitosis.

## Coloration of Blue-Green Algae

Blue-green algae exhibit a wide range of colors extending across the whole of the visible spectrum, but these colors are seldom vivid. Only one chlorophyll, chlorophyll *a*, is present. The carotenoids include beta carotene and a set of xanthophylls that are more or less unique to this group. Additionally two *phycobilin* pigments occur. One of these, *phycocyanin*, is blue while the other, *phycoerythrin*, is red. Neither of these is fat soluble and they are not extractable with such solvents as acetone, alcohol, or petroleum ether. They are considered to be water soluble but because of chemical associations with proteins they are not amenable to simple extraction techniques. Since green, yellow, yellow-brown, blue, and red pigments are mixed together in varying proportions the wide range of observable colors is understandable. Genetic influences affect the quantities of each pigment that may be produced by a given species; several environmental factors may affect the rates of production and destruction of any given pigment.

Some blue-green algae grown in deep clear waters may develop the red pigment strongly. This is apparently an aid to photosynthesis since blue light penetrates more deeply into water than red light and a red pigment is the most efficient in absorbing blue light. (The absorbed energy is transferred to the photosynthetic mechanism.)

Chlorophyll is rapidly destroyed by exposure to intense sunlight and thus algae trapped at the surface of a body of water tend to lose some of their green color. The carotenoid pigments are more stable under such circumstances; therefore blue-green algae frequently have a yellowish appearance. Under some environmental conditions both chlorophyll *a* and phycocyanin develop strongly; in such cases a blue-green

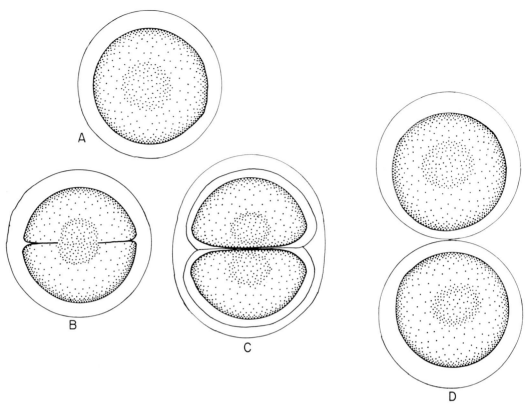

Fig. 10.2. Schematic representation of cell division in a single-celled blue-green alga. A. Undivided cell. Note slightly denser central body. B. Note inward growth of new cell wall. C. Division completed and new sheaths are forming. D. Original sheath has broken, allowing separation of daughter cells.

alga may actually have a blue-green appearance.

ADDITIONAL CELL CHARACTERISTICS

Many blue-green algal cells develop minute gas bubbles called *pseudovacuoles* or *gas vacuoles* that serve to increase cell volume without adding to cell mass. The pseudovacuoles increase the ability to float in water and frequently cause the cells to accumulate in surface layers where they become conspicuous as *algal blooms*. When viewed with transmitted light (as with the compound microscope) the bubbles appear as black granules. They give the cells a much darker appearance than is the case when they are absent.

Granules in several size categories are also observable in electron micrographs; some of them are large enough to be visible also with the light microscope. Among the smaller granules are numerous *ribosomes*. Some larger particles that have a characteristic appearance are called *structured granules*. Others are simply placed in size categories; much research remains to be done before their functional significances are understood.

In this group the type of starch characteristic of green algae and higher plants does not occur; tests with iodine solutions do not yield a typical blue-black coloration. Instead, a smaller-molecule reserve carbohydrate is formed that stains a dull nondescript red with iodine solutions. This was once thought to be glycogen but presently is designated as *cyanophycean starch*. Particles of stored food may contribute in some instances to the granular appearance of the protoplasm.

## NEGATIVE CHARACTERISTICS

At this point a general overview of the blue-green algae shows them to have many negative characteristics when compared to other algae or to the higher green plants. They do not:

- have chloroplasts
- have mitochondria
- have Golgi membranes
- have endoplasmic reticula
- have true nuclei
- undergo mitosis
- undergo meiosis
- have sexual life cycles

## ILLUSTRATION OF PLANT BODY TYPES IN THE BLUE-GREEN ALGAE

Many of the simpler types of plant bodies found in the green algae have counterparts in the blue-green algae. However, two types are particularly notable for their absence: among blue-green algae there are no motile unicells and no motile colonies. A type of motility occurs in some filamentous forms (see discussion of *Oscillatoria*) but this is not due to flagella. It is particularly helpful in this group to consider how a geometric control of the planes of cell division affects plant body types.

In some of the common genera the cells are spherical before division and divisions occur in three sets of planes, each at right angles to the others. This tends to result in a three-dimensional distribution within a common gelatinous matrix. In some of these forms the sheaths of individual cells are distinct while in others they coalesce to form a homogeneous mass. *Chroococcus* (Fig. 10.3A) demonstrates the first condition and furthermore tends to revert to the single-celled state by fragmentation after one or a few divisions. In *Microcystis*, which is an abundant nuisance form in many shallow warm-water lakes, hundreds or thousands of small spherical cells (Fig. 10.3B) are distributed uniformly throughout coalescent sheath substance to form macroscopically visible masses. (*Gloeocapsa* is frequently used in textbooks to illustrate this type of blue-green alga but many species in this genus have been reassigned to *Chroococcus* and one modern revisionist has assigned all species of *Chroococcus*, *Microcystis*, *Gloeocapsa*, and several other genera to the genus *Anacystis*.)

In the new genus *Coccochloris* (Fig. 10.3C) (which now includes *Aphanothece*) individual cells vary from being egg-shaped to rod-shaped but always divide transversely to the long axis. In this respect they are quite different from *Microcystis* but are similar in having many cells distributed at random in a common gelatinous matrix.

In *Merismopedia (Agmenellum)* (Fig. 10.3D) cell divisions occur in only two sets of planes, each at right angles to the other, to form a sheet of cells that are neatly organized in horizontal and vertical rows in the common matrix.

In *Gomphospheria* (which now includes *Coelosphaerium* according to one authority) cell divisions occur in planes perpendicular to tangents of a curved surface. Thus the cells tend to be arranged more or less in a single layer near the surface of a spherical mass of gelatinous material (Fig. 10.3E).

In *Oscillatoria* (Fig. 10.3F) each cell division occurs in a plane parallel to the wall formed by the preceding division. The cells remain attached following divisions and this results in a linear row of cells called an *unbranched filament*. In *Oscillatoria* a gelatinous sheath is present *but not evident* while in the related genus *Lyngbya* it is clearly evident. Filaments get longer because of cell division and a type of reproduction occurs when filaments fragment. The occurrence of dead cells with or without gelatinous discs within the filament often provides weak points in the filament where breakage can occur.

*Oscillatoria* demonstrates several types of movement which have fascinated microscopists ever since the development of the compound microscope. Filaments may move forward or backward and frequently reverse direction. They may rotate about the longitudinal axis and they frequently swing in wide arcs. Although no completely satisfactory explanation of these movements has been advanced, they seem to be associated with the more or less constant discarding of a delicate sheath.

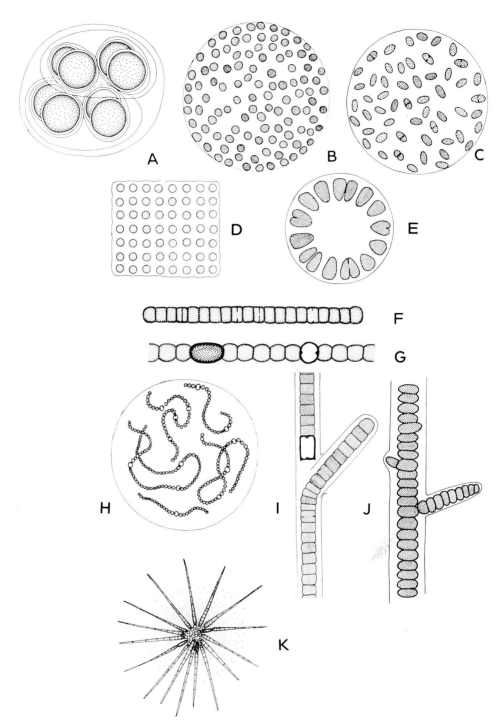

Fig. 10.3. Blue-green algae. A. *Chroococcus*. B. *Microcystis*. C. *Coccochloris*. D. *Merismopedia*. E. *Gomphospheria*. F. *Oscillatoria*. G. *Anabaena* (with akinete and heterocyst). H. *Nostoc*. I. *Tolypothrix*. J. *Stigonema*. K. *Gleotrichia*.

The cells of *Oscillatoria* are essentially all alike but in some other filamentous forms two types of special cells are common. One is a type of resting spore called an *akinete,* which is formed by enlargement of an existing cell (Fig. 10.3G). The original cell wall thickens to become the akinete wall. The protoplasm becomes much more dense and is often grainy in appearance, probably due to the formation of storage granules of reserve foods. The partial dehydration and subsequent physical modification of the protoplasm enables the akinete to withstand environmental stress better than an unmodified vegetative cell. Thus the akinete should be considered as a survival mechanism rather than a device for rapidly increasing populations. Experiments have shown akinetes may germinate after many years of dormant storage in dry soil.

The other special cell is the *heterocyst* (Fig. 10.3G), a "different cell" whose significance remained elusive for more than a century after its discovery. Heterocysts have somewhat thickened walls and a more translucent (but *not* transparent) appearance than vegetative cells. As they mature a thickening develops on the wall next to each adjacent cell. These thickenings are called *polar nodules* and give heterocysts a characteristic appearance.

Many theories concerning possible heterocyst functions have been examined and generally found inadequate; even the fascinating thought that they are vestigial evidences of a lost function has been considered. One possible significance is based on observational relationships that seem to extend beyond coincidence. The probability that blue-green algae are able to fix atmospheric nitrogen in a similar manner to that of certain bacteria was generally accepted long ago but the original experimental evidence was treated with caution because the cultures apparently were not bacteria free. In recent years it has become possible to obtain bacteria-free cultures of blue-green algae and it is now clear that some of them are in fact able to fix atmospheric nitrogen. So far those species for which convincing proof has been presented

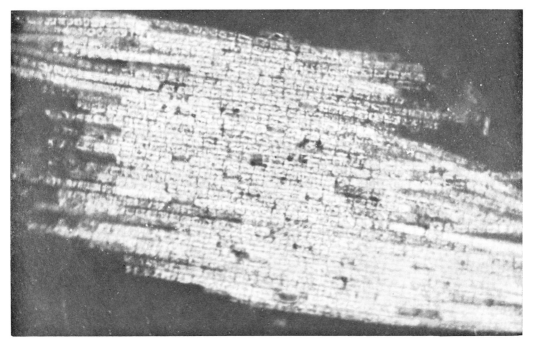

Fig. 10.4. Raftlike macrocolony of *Aphanizomenon* with many filaments lying side by side.

all seem to be heterocyst formers and thus it is possible the heterocyst may be concerned with this process. Proof of the corollary that all species with heterocysts do fix nitrogen will require many years of research.

Some common genera which produce spores and heterocysts are *Anabaena* (Fig. 10.3G), *Nostoc* (Fig. 10.3H), and *Aphanizomenon* (Fig. 10.4). All three exist as unbranched filaments and are recognized by their growth habits. *Aphanizomenon* forms raftlike macrocolonies of hundreds of filaments lying side by side, the whole appearing like short grass clippings floating in the water. Most *Nostoc* species form tough macrocolonies of twisted filaments; these masses are commonly as large as cherries but are occasionally larger than walnuts. *Anabaena* filaments may be straight or twisted and they may be solitary or clumped together in a gelatinous matrix that is soft and watery (in contrast with the firm tough matrix of *Nostoc*).

The genus *Tolypothrix* (Fig. 10.3I) demonstrates a phenomenon called *false branching* in which the two fragments of a broken filament remain in close proximity, held in place by the tough outer sheath. One fragment grows out past the other and its new sheath is contiguous with the older one; the appearance of branching is due to the sheaths, not the cells.

*True branching* occurs when an occasional cell in a filament divides in a plane not transverse to the long axis of the filament. (This event may or may not be associated with a preliminary outward bulge of the dividing cell.) Succeeding divisions of this cell occur in planes parallel to the aberrant wall and thus a filament results that is a true branch of the main filament. A genus commonly cited to demonstrate this phenomenon is *Stigonema* (Fig. 10.3J).

*Gleotrichia* (Fig. 10.3K) and a similar genus, *Rivularia*, have basal heterocysts and tapered filaments with broad flat cells at the heterocyst end and long narrow cells toward the opposite end. Species in both genera exist as macrocolonies with many radiating filaments in a gelatinous matrix. The technical difference between them is based on akinete production: *Rivularia* being the one that never forms akinetes.

Blue-green algae are generally unpopular. Dense growths are displeasing to the eye. They stick to boat pilings and pile up on beaches, particularly on the downwind shore. When they accumulate in sufficiently large masses they begin to die and decompose. The total respiration requirements for algae, bacteria, protozoa, insect larvae, microcrustaceans, fish, and other organisms may lower the available dissolved oxygen concentrations to such levels that fish kills result. In addition the decomposition products generally have foul odors. Furthermore, the last of the pigments to decompose is usually phycocyanin, so the decomposing mass tends to turn a bright blue, often causing a beach to appear as though covered with spilled blue paint.

Some species of blue-green algae have evolved toxic strains that can kill those warm-blooded animals that ingest them. Unfortunately the toxic strains cannot be told from the nontoxic strains of the same species by microcsopic examination. Thus when dense growths of blue-green algae appear in farm ponds or other bodies of water it is a wise precaution to keep cattle from drinking the water.

RED ALGAE

The *red algae* are primarily marine but a few species exist in freshwater habitats. The colors are predominantly shades of red; in some forms a violet-gray-green color is evident. The pigments have long been said to include chlorophylls *a* and *d*, carotenes, xanthophylls, and two phycobilin pigments (phycocyanin and phycoerythryn). The last two are blue and red respectively and are only slightly different chemically from similar pigments in the blue-green algae. As was noted also in the discussion of the blue-green algae, the red pigment tends to develop more strongly in deeper waters, increasing the ability of these plants to use blue light in photosynthesis. Recent evidence suggests that chlorophyll *d* is a degradation product of chlorophyll *a*.

The most common food storage product is a carbohydrate, *floridean starch,* that is more closely allied to the cyanophycean starch of blue-green algae than it is to the starch of the green algae.

The number of chloroplasts per cell is quite varied. Some forms have many small chloroplasts per cell. Others may have only one; this is apt to have a lobed stellate appearance with a distinct pyrenoid in the center. A single nucleus is present in each cell and this may be haploid or diploid depending on the stage of the life cycle.

Most species are macroscopic but do not approach the size of the large kelps in the brown algae. Some forms are microscopic; one freshwater genus consists of pseudofilaments that frequently break apart into the single-celled condition.

A few forms of red algae are used in the human diet; some of them are cultivated in the Orient. The familiar substance *agar,* widely used in biology laboratories as a culture medium, is derived from a red alga. "Irish Moss" is a red alga that has many pharmaceutical and industrial uses (Fig. 10.5). The ability of some species to precipitate coatings of lime on their bodies is of major significance, particularly in tropical seas. These coralline algae contribute greatly to the formation of coral reefs and island beaches.

Red algae are more common than brown algae in tropical waters; they add much to the underwater beauty of shoals and reefs. A minor hobby that has waxed and waned in popularity through many generations is the drying of red algae on cards. They seem to arrange themselves in intricate patterns and stick to the paper as they dry, thus preserving their natural artistry (Fig. 10.6).

No flagellated cells occur in any members of this group but some reproductive units achieve a kind of passive mobility by floating about in water. In certain species asexual reproduction is achieved by *monospores,* single-celled units that become detached, float away, and eventually become attached elsewhere to begin formation of new plants without any change in genetic makeup. Sexual reproduction is a complex matter and may or may not involve alternation of generations. Only one of the simpler life cycles is presented here as an example. Interested students are referred to advanced level texts for details of the more complex cycles.

LIFE CYCLE OF *Nemalion*

The haploid plant body of *Nemalion* consists of a branching system of wormlike threads, each of which is made of compacted filaments held in a gelatinous matrix. The female gametangium usually consists of a cell at the tip of a filament that is modified by enlarging somewhat and developing a slender hairlike outgrowth. The swollen cell is called a *carpogonium;* its protoplasmic content functions as a female gamete. The hairlike outgrowth is the *trichogyne* (Fig. 10.7A).

Individual antheridia are small and single celled; they occur at the ends of branching filaments. The protoplast of each antheridium escapes and floats about as a naked protoplast (Fig. 10.7B). Being nonflagellate it is not a sperm in the strict definition of that term, but it is a male gamete and is usually called a *spermatium.* Eventually a floating spermatium comes in contact with a receptive trichogyne and syngamy occurs. The zygote remains inside the carpogonium and does not become a resting cell. Meiosis occurs shortly after

Fig. 10.5. Irish moss *(Chondrus crispus).*

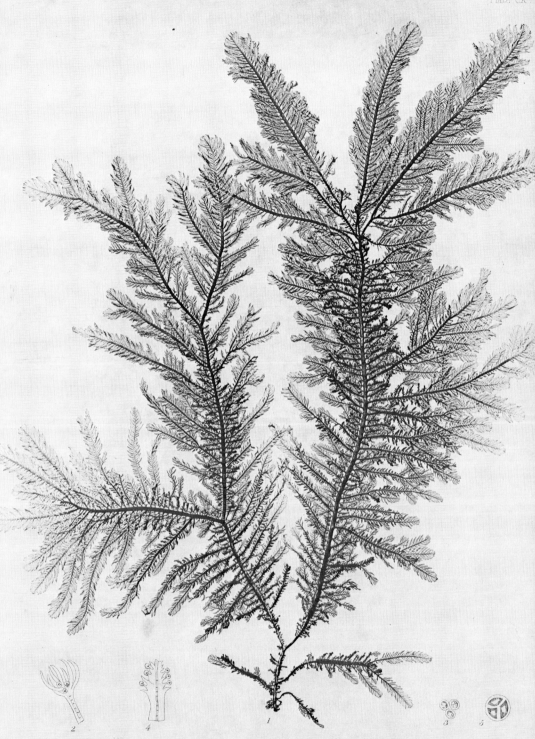

PTILOTA plumosa, AG.

Nature Printed by Henry Bradbury.

Fig. 10.6. (left) Specimen of a red alga decoratively arranged and dried on the page of a book.

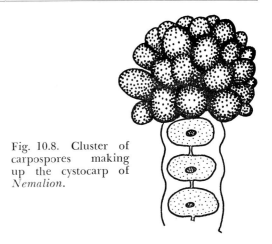

Fig. 10.8. Cluster of carpospores making up the cystocarp of *Nemalion*.

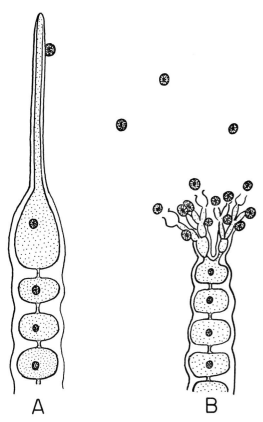

Fig. 10.7. A. Schematic illustration of a filament of *Nemalion* with terminal carpogonium. Note spermatium lodged against trichogyne. B. Schematic illustration of a filament of *Nemalion* with antheridia releasing spermatia.

syngamy and several mitotic divisions then serve to increase the number of haploid nuclei. Ultimately the protoplasm of the carpogonium is divided into several cells with one nucleus in each. These cells are known as *carpospores* and the mass of carpospores resulting from one carpogonium is called a *cystocarp* (Fig. 10.8). When the carpospores are disseminated, each one is capable of developing into a new haploid plant.

The red algae are considered to have achieved a very high level of evolutionary advance even though they possess some characteristics in common with the blue-green algae, which are primitive and (along with bacteria) the oldest of living things.

## BROWN ALGAE

The *brown algae* are almost exclusively marine. Most of them are macroscopic and are generally familiar as seaweeds. The group includes the *giant kelps* that grow so profusely in the ocean along the western shores of North America, as well as the familiar *rockweeds* that are exposed twice daily by ebbing tides in many parts of the world. They are mostly brown in color due to a masking of chlorophylls *a* and *c* by the carotenoid pigments, notably certain brownish xanthophylls. They do not store starch but a large molecule sugar, *laminarin*, is a common reserve carbohydrate.

A gelatinous hydrophilic material in the cell walls gives these plants a slippery feeling and is undoubtedly of value in protecting them from excessive dehydration when the tide is out. An important item of commerce, *algin*, is derived from this wall material and is useful in maintaining a desirable texture in such widely diverse products as chocolate milk, soup, paint, and ice cream. Some brown algae are used as food in the human diet, particularly in the Orient, and as additives to animal feeds. Brown algae are particularly efficient in extracting potassium from sea water; in the past large quantities were burned to provide a source of potash. Iodine, bromine, and some metal elements may be retrieved from the oceans by brown algal harvests. Kelp beds are an important resource that may be damaged by too intensive harvest It is a matter of considerable public interest that this resource be treated and managed as a renewable one.

No adequate concept of the brown algae can be obtained in a brief introductory course, but an examination of three common genera, *Ectocarpus, Laminaria,* and *Fucus,* can lead to an appreciation of their diversity. These genera are widely used for teaching purposes since they provide additional examples to reinforce the concept of alternation of generations.

### ECTOCARPUS

*Ectocarpus* is in many ways the simplest of the brown algae. Its plant body is a branched filament that may be found attached to rocks or to other algae. In general appearance (Fig. 10.9A) it resembles the green alga *Cladophora*. However, the individual cells are quite different from *Cladophora* cells since they have several small discoid chloroplasts and contain a single nucleus.

Species in this genus demonstrate an *isomorphic* alternation of generations in the sense that both generations are similar in appearance even though one is haploid and the other diploid. The haploid generation is the *gametophyte*. When mature it produces reproductive structures that are many-celled clusters, with each cell giving rise to a single biflagellate isogamete. The many-celled structure is called a *plurilocular gametangium* (Fig. 10.9B). *Ectocarpus* species are heterothallic and the isogamous union of gametes occurs only when plus and minus mating types intermingle. (It should be noted also that a direct germination of unfertilized gametes may occur, resulting in asexual propagation of the haploid generation.)

The zygotes do not become resting cells and germinate shortly after their formation. They do not undergo meiosis but undergo repeated mitotic divisions instead.

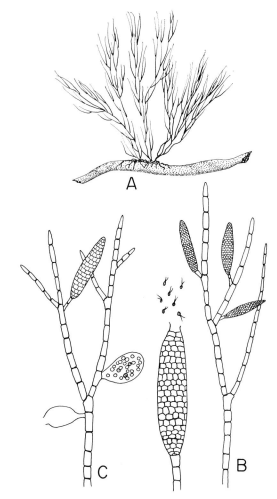

Fig. 10.9. *Ectocarpus*. A. Habit sketch of plant growing attached to a larger alga. B. Gametophyte with plurilocular gametangia producing gametes. C. Sporophyte with plurilocular zoosporangia producing diploid zoospores and unilocular zoosporangia producing haploid zoospores.

This results in the formation of a diploid plant body that, as noted above, resembles the haploid plant body. This *sporophyte* generation may produce two types of reproductive structures, *plurilocular zoosporangia* and *unilocular zoosporangia* (Fig. 10.9C). The former appear exactly like the plurilocular gametangia of the gametophyte and each cell gives rise to 1 zoospore. The zoospore is diploid like the cell from which it was derived; since it gives rise to a plant that is genetically the same as the parent plant this is an asexual reproductive process.

The unilocular zoosporangium is quite different. It consists of a single cell that enlarges without cell division. Ultimately its diploid nucleus undergoes meiosis to form 4 haploid nuclei and the number of nuclei is then increased to 32–64 by repeated mitotic divisions. Each of the haploid nuclei is incorporated into a unit of cytoplasm and these uninucleate cells are modified into zoospores. Upon release from the unilocular zoosporangium they swim about and become dispersed. Each one that survives eventually becomes attached to a suitable substrate and gives rise to a new gametophyte.

It would be convenient to assert that these particular zoospores are meiospores and the unilocular zoosporangia are meiosporangia. However, a semantic quibble arises from the fact noted above that the number of haploid nuclei is increased by mitosis before the spores are actually formed. If the general concept of alternation of generations is well understood this should be a relatively minor source of confusion.

## LAMINARIA

Of the several genera that are called *kelps*, *Laminaria* is most commonly used for teaching purposes to demonstrate a type of *heteromorphic* alternation of generations, in which both generations are separate and independent but totally different in appearance. The large familiar plant is the sporophyte and consists of a long *blade* portion, which may or may not be subdivided, and a short tough *stipe* that serves as a stalk to connect with the *holdfast*, a rootlike system of rhizoids that anchors the plant to a rock (Fig. 10.10A).

The tissue systems of the blade are complex and have a basic separation into two regions that are reminiscent of the pith and cortex of higher plants. Apparently these tissues evolved from modifications of compacted and intertwined branched fila-

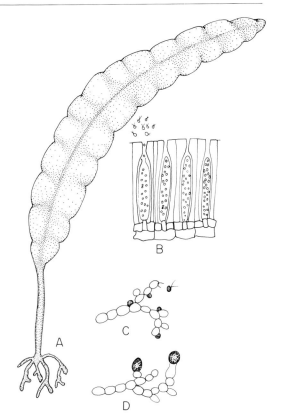

Fig. 10.10. *Laminaria*, one of the kelps. A. Habit sketch drawn from a plant about 60 cm long. Note basal attaching rhizoidal system, stipe, and blade. B. Portion of a sorus seen in section showing unilocular zoosporangia producing haploid zoospores. C. Microscopic few-celled male gametophyte. Note sperm cell being released from an antheridium. D. Microscopic few-celled female gametophyte. Note egg perched on lip of oogonium.

ments. In some kelps cells recognizable as sieve tube elements occur and function in the transport of organic foods to the basal holdfast. However, no cells recognizable as xylem elements have been observed.

Sporangia occur in clusters *(sori)* on the surface of the blade. Each sporangium is unilocular (Fig. 10.10B), and produces 32–64 haploid zoospores by the same sequence of events as occurs in *Ectocarpus* (meiosis followed by several mitotic divisions). Due to a segregation of sex-determining genes during meiosis, half of the zoospores give rise to female gametophytes and the other half to males. Both gametophytes are minute consisting only of few-celled branching filaments that grow independently of each other. The casual observer, if noting them at all, would have no basis for relating them to the sporophyte generation.

Terminal cells of the male plants function as *antheridia;* each one releases a single biflagellate sperm at maturity (Fig. 10.10C). In a somewhat similar way a few cells of the female filaments function as *oogonia*. The entire protoplast of each oogonium is released as a single nonmotile egg that may remain perched on the outside of the oogonium (Fig. 10.10D). One of the swimming sperms unites with the egg to form a zygote, which begins development almost immediately to form a new sporophyte.

Fucus

Species of *Fucus* and several closely related genera comprise that segment of the brown algae commonly called *rockweeds* because they occur in such vast numbers attached to rocks in the intertidal zones. The life cycle of *Fucus* is comparable to that of *Ectocarpus* and *Laminaria* but is of special interest because the independent gametophyte generation is essentially eliminated.

Individual sporophytes are attached by holdfasts and the stipe continues into the blade as a kind of midrib (Fig. 10.11). Dichotomous branching is common, giving the plant body a characteristic shape. Many species have air bladders in the blade on either side of the midrib; these are useful as flotation devices during parts of the day when the plants are submerged. (The odors associated with the rockweeds sometimes seem unpleasant to newcomers on the beach, but they are nostalgic remembrances to former beachdwellers.)

Zoospore formation does not occur in *Fucus;* except for fragmentation there are no methods of asexual reproduction. With the onset of the sexual cycle terminal portions of the branches become modified into *fruiting tips*. There is a slight swelling, a change in coloration, and an appearance of

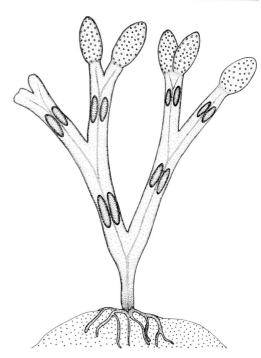

Fig. 10.11. *Fucus*, one of the rockweeds. Habit sketch drawn from a specimen about 12 cm long. Note attaching rhizoidal system, dichotomously branched blade with midribs, and air bladders. The several fruiting tips show the positions of numerous conceptacles.

numerous small eruptions on the surface of such tips (Fig. 10.11).

Each of the eruptions is associated with a distinct subepidermal chamber called a *conceptacle* (Fig. 10.12A,B,D). The *ostiole* is a minute opening of the chamber to the surface. Depending on the species a given conceptacle may contain eggs or sperm or both. Sometimes they are formed in different conceptacles on the same plant but in many species they occur on separate (male and female) plants. The illustrations used here show them in separate conceptacles (Fig. 10.12B,D).

The *antheridia* are borne at the tips of branching filaments that arise from the inner surface of the conceptacle (Fig. 10.12B). Each of these filaments may be likened to a branching filament of *Ectocarpus* and the antheridia are essentially like the unilocular zoosporangia of that genus (Fig. 10.12C). Each antheridium be-

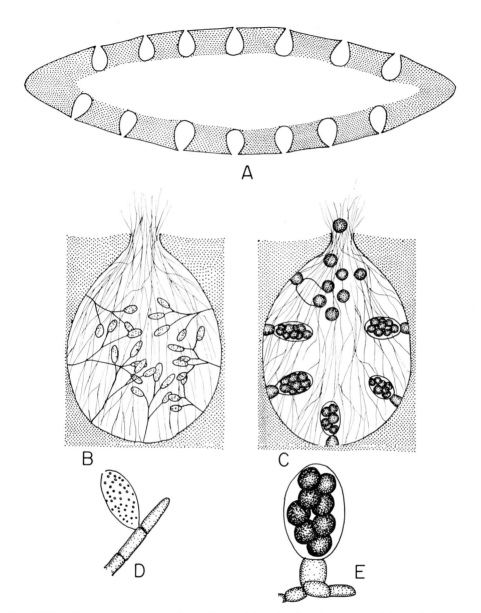

Fig. 10.12. *Fucus*. A. Cross section of a fruiting tip showing the positions of the conceptacles. B. Detail of male conceptacle showing numerous antheridia attached to branching stalks. C. Detail of female conceptacle showing several oogonia and the release of eggs from one of them. (Both B and C show numerous paraphyses.) D. Detail of single antheridium containing numerous sperm cells. E. Detail of single oogonium showing production of 8 eggs.

gins as an enlarged cell with a single diploid nucleus. Meiosis results in 4 haploid nuclei and additional mitotic divisions bring the total to 32–64. Each nucleus becomes incorporated in a cell that becomes a laterally biflagellate sperm. (It should be recalled that in *Ectocarpus* and in *Laminaria* also such cells would function as zoospores.)

In female conceptacles the reproductive filaments are reduced to a two-celled stage, with one cell functioning as a stalk and the other as an *oogonium*. This too is similar to a unilocular zoosporangium in that its single diploid nucleus undergoes meiosis first and then a single mitosis to produce a total of 8 haploid nuclei, each of which becomes incorporated into a cell (Fig. 10.12E). Observations show that these 8 cells are released into the surrounding seawater where they function as nonmotile female gametes or eggs.

In both types of conceptacles sterile filaments called *paraphyses* intermingle with the fertile filaments and extend outward through the ostioles. They seem to serve as guides when the mature gametes are squeezed out of the conceptacles by contractions induced by drying during periods of low tide. The jellylike masses are washed off the plants with the flooding tide and the gametes are mixed together in enormous numbers.

Each free-floating egg is quickly surrounded by a swarm of male gametes and is caused to whirl about rapidly by sperm activity. Fertilization occurs when one sperm enters an egg; a wall forms immediately about the new zygote. The chemical attractant secreted by the unfertilized egg dissipates; the sperm cloud vanishes; and the zygote no longer spins. It settles to a rock surface and becomes attached. Germination of the zygote occurs promptly and the new sporophyte grows rapidly.

In this cycle we see a curious evolutionary advance in that the potential spores produced in the unilocular organs are converted into gametes directly without the intervention of a separate gametophyte plant body.

## DIATOMS

*Diatoms* generally have a yellow-brown color which is so characteristic that many earlier observers were not aware of the presence of green pigments in this group. The carotenoid pigments include beta carotene, which is bright golden yellow, and several xanthophylls, at least one of which is brown. The two chlorophylls, *a* and *c*, are very effective in photosynthesis even though they are masked by the carotenoids.

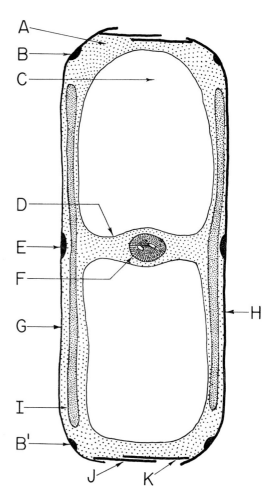

Fig. 10.13. Longitudinal section of a representative pennate diatom, *Navicula*. A. Cytoplasm. B, B'. Polar nodules. C. Central vacuole. D. Cytoplasmic bridge. E. Central nodule. F. Nucleus. G. Epivalve. H. Hypovalve. I. Edge view of a large pancake-shaped chloroplast. J. Girdle band attached to epivalve which overlaps girdle band (K) attached to hypovalve.

Starch formation does not occur in diatoms; the main storage product is oil.

The cytoplasm is bounded by a plasma membrane on the outer surface and a vacuolar membrane around the large central vacuole. There may be many small chloroplasts per cell or a few large ones that are like elaborately lobed pancakes in some species. Each chloroplast is bounded by a chloroplast membrane and contains several thylakoids. Pyrenoids are present in some forms. The nucleus is a typical true nucleus in that it has a nuclear membrane and nucleoli and undergoes a normal mitosis. It usually lies in a cytoplasmic bridge across the vacuole (Fig. 10.13F). Mitochondria, Golgi membranes, endoplasmic reticula, and ribosomes are clearly evident in electron micrographs. The main substance of the wall is silica distributed in a matrix of pectic materials. Cellulose is completely lacking. The wall is not degradable by enzymatic action; the walls of dead cells sink to the bottom. Accumulations of diatom remains, especially those deposited in ancient seas whose beds have since been elevated as part of land masses, are called diatomaceous earth. This material is an item of commerce, being important as a filtering material, as an insulating material, as a fine polishing substance, and in other ways.

Diatoms are major components of the biota of both freshwater and marine habitats; even though some estimates of their total contribution are perhaps too high it seems probable that at least half of the total productivity of the oceans is due to diatoms. And since the contribution of the oceans is greater than that of the land masses diatoms quite possibly exceed all other plant groups in the primary production of organic substances and oxygen. Directly and indirectly they contribute significantly to seafood harvests. Certain types of petroleum deposits may owe their creation to the diatoms of ancient seas.

A characteristic feature of diatoms is their cell wall construction, which is variously likened to a box and its lid, or the two halves of a Petri dish (Figs. 10.13, 10.14). The entire cell wall is termed a *frustule* but the top and bottom surfaces are called *valves* (the *hypovalve and epivalve* respectively). In most forms the two valves are alike in appearance but in a few genera they are distinctly different. The margin of each valve usually curves somewhat and the curved portion is referred to

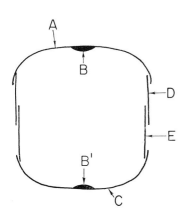

Fig. 10.14. Schematic median transverse section of the frustule of *Navicula*. A. Epivalve. B, B'. Central nodules. C. Hypovalve. D. Girdle band attached to epivalve which overlaps girdle band (E) attached to hypovalve. (The downwardly curved portions of each valve are the valve mantles.)

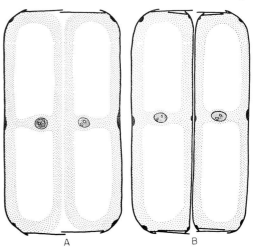

Fig. 10.15. Cell division in *Navicula*. A. Mitosis and cytokinesis have been completed and 2 protoplasts now exist within old frustule. B. New valves and girdle bands have formed to complete the division process.

as a *valve mantle* (Fig. 10.14). A narrow siliceous band, the *girdle band,* is attached to the inner surface of each valve. It is quite like a barrel hoop in some respects. The girdle band of the epivalve overlaps that of the hypovalve and in this way the similarity to a box and lid is created (Figs. 10.13, 10.14, 10.15).

Prior to cell division the protoplasm swells until the overlapped girdle bands almost separate. A normal mitosis of the nucleus occurs and is followed promptly by cytokinesis in a plane parallel to the valve surfaces (Fig. 10.15A). Two new valves then form, lying opposed to each other between the two newly formed protoplasts. When the new valves are completed a new girdle band is formed for each of them (Fig. 10.15B).

One of the new cells has the original epivalve and a new hypovalve and is identical in size to the original cell (Fig. 10.15A). In the other cell the original hypovalve becomes the epivalve and a new hypovalve is formed. This cell is slightly smaller than the other one (Fig. 10.15B); with each generation of new cells the average size of cells in the population decreases. After many generations the size of cells at the small end of the population may be less than half those of the larger end. When this disparity begins to become extreme the population becomes sensitive to an unknown environmental trigger that sets in motion a process called *auxospore formation.*

## Auxospore Formation in Diatoms

At first glance, this process appears to involve merely a shedding of existing walls and rapid growth of the naked protoplast to maximum size for the species, whereupon new valves and girdle bands are formed. These first-formed new walls may be somewhat atypical for the species but after one or two normal vegetative divisions the typical appearance is achieved. The cell unit which grows as described above is called an *auxospore* and its formation is by no means as simple as it appears. In those cases where it has been investigated carefully auxospore formation has been shown to be part of a sexual cycle. Normally the vegetative cell nuclei of diatoms are diploid and meiosis occurs at the onset of the sexual process (Fig. 10.16B,B'). The 4 haploid nuclei that usually result from this process do not always materialize due to breakdowns in the mechanism at some point. It is quite common apparently

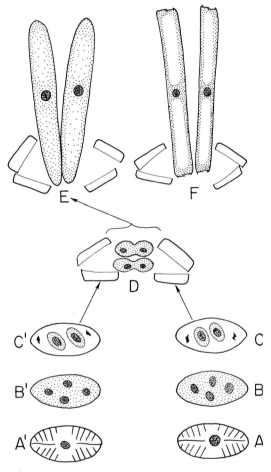

Fig. 10.16. One of several methods of auxospore formation in diatoms. A and A'. Diploid vegetative cells in a premeiotic condition. B and B'. Cells with 4 haploid nuclei following meiosis. C and C'. Degeneration of 2 of the 4 haploid nuclei in each cell and incorporation of surviving nuclei into gametes. D. The valves of 2 conjugating cells have separated allowing the gametes to undergo syngamy. (Karyogamy will occur shortly to reestablish the diploid condition.) E. The 2 zygotes have enlarged as auxospores to the maximum size for the species. F. New valves and girdle bands have formed to complete the process.

for only 1 or 2 of the haploid nuclei to survive (Fig. 10.16C,C'). (It is also possible for several haploid nuclei to result from additional mitotic divisions after the completion of meiosis.) Each surviving haploid nucleus becomes part of a gamete when cytokinesis occurs; depending on the species there may be 1, 2 (Fig. 10.16C, C'), 4, or several gametes per cell.

(This process is observed infrequently in nature and the cells in which it happens do not have a modified appearance. Furthermore it is not readily induced in laboratory cultures. Thus, far fewer of the details are known than for other groups of algae.)

In those species where conjugation occurs the paired cells lose their walls and the gametes approach each other by amoeboid movements (Fig. 10.16D). They unite in pairs to form zygotes; from the above discussion it is evident that conjugating cells may give rise to 1, 2 (Fig. 10.16E), or 4 zygotes, depending on the numbers of gametes formed. In species where many gametes are formed these tend to become motile by means of flagella and function in the manner of sperm. In other cells of the same species 3 of the 4 meiotic nuclei degenerate and the surviving cell unit remains in place to function in the manner of a female gamete (egg).

It is quite common for some diatom species to undergo a process called *autogamy* in which only 2 of the meiotic nuclei in a given cell survive and these 2 unite promptly to restore the diploid condition. The cell walls are then shed and the single naked protoplast is essentially a zygote.

Regardless of variations in the mechanism of formation *each zygote becomes an auxospore which undergoes the growth process described above to establish the maximum cell size in the population range* (Fig. 10.16F).

A SIMPLE PROCEDURE FOR
EXAMINING DIATOMS

A laboratory exercise which usually generates considerable student interest involves an unstructured examination of living collections of algae gathered locally, particularly if the collections are rich in species. It is not a reasonable aim to expect identifications except for a few very common forms, but morphological diversity can be appreciated and considerable satisfaction can be obtained from matching observed forms against available illustrations. Despite their overall significance, the numerous species of diatoms in such collections are usually ignored because many features essential to identification are not clear in fresh mounts.

The *burned mount technique* is a relatively simple procedure for preparing diatoms so they may be examined microscopically within one laboratory period. A drop of water containing diatoms (but free of large sand particles) is placed on a glass cover slip that is heated on a hot plate to approximately 500 C for 20–30 minutes. (It is considered a superior procedure to allow the material to air dry first on the cover slip but this takes more time than is available in one laboratory period.) The heat treatment causes organic matter to oxidize and the frustules of diatoms are cleaned of their protoplasm in this way. The frustules are not modified as they are composed of silicon dioxide.

The cover slip is then inverted and placed in a drop of a mounting medium (such as Hyrax) that has a refractive index suitable to making the wall markings clearly visible under the microscope. Usually heating is needed to drive off the solvent of the mounting medium.

A number of sketches and photomicrographs of common genera of freshwater diatoms have been included in the text to facilitate this exercise but it is to be hoped they will not become a memorization chore. The list is by no means complete; there are more than 50 genera of diatoms in central North America alone.

RECOGNITION FEATURES OF SOME
COMMON FRESHWATER DIATOM GENERA

The diatoms may be divided into two large groups based on considerations of frustule shape. In one group, the *centric diatoms* (Fig. 10.17), the markings on the valve surface are arranged with respect to

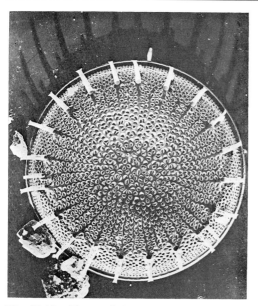

Fig. 10.17. Transmission electron micrograph of a metal-coated carbon replica of a valve of *Stephanodiscus*. Note the radiating lines made of punctae and the marginal spines.

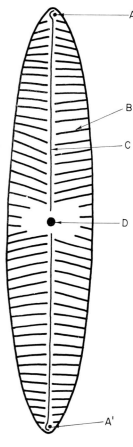

Fig. 10.18. Surface view of one valve of *Navicula*. A, A'. Polar nodules. B. One of the striae. C. Raphe. D. Central nodule.

a point. Many of them have valves that are circular in outline and the valve markings are radiately distributed. *Pennate diatoms* on the other hand have markings on the valve surface arranged laterally with respect to a real or imagined longitudinal line (Fig. 10.18). Many members of this group are boat-shaped. Both groups occur in marine and freshwater habitats but species diversity of centric forms is greater in marine habitats. The number of pennate species in freshwater habitats is much larger than the number of species of centric forms but it is not uncommon for certain centric forms to be dominant in the plankton of freshwater habitats.

The radiate or lateral markings referred to often appear as lines. If they are simple scratches in the surface or if they may be resolved microscopically into rows of dots *(punctae)* they are called *striae*. If they are riblike thickenings of the wall or if they are tubules in the wall they are referred to as *costae*.

Two of the more common centric diatoms in freshwater habitats are *Cyclotella* (Fig. 10.19V) and *Stephanodiscus* (Fig. 10.17). In the former the radiate markings are costae while in the latter they are striae. *Stephanodiscus* also has a ring of spines around the margin. The cells which make up the filaments of *Melosira* are individually much like those of *Stephanodiscus*.

Symmetry is a feature of particular value in describing the morphology of diatom frustules. If the top half matches the bottom half the valve is transversely symmetric. If it doesn't the valve is transversely asymmetric. If the right half matches the left half the valve is longitudinally symmetric. If it does not the valve is longitudinally asymmetric. Figure 10.19A illustrates a valve which is symmetric both ways. Figure 10.19G illustrates a valve which is transversely asymmetric while

Fig. 10.19H illustrates one which is longitudinally asymmetric. The sigmoid shape of the diatom illustrated in Fig. 10.20 is not in a strict sense symmetric in any dimension.

Valve surfaces are often flat but they may be highly arched. In one very common form the valve surface undulates, taking the appearance of a rolling meadow (Fig. 10.21).

Several genera have internal partitions, or *septa*, that may be transversely or longitudinally oriented. In the genus *Tabellaria* for instance there may be 2 to several septa which lie in planes parallel to the valves. Each of these septa has a large hole or *locule* in it (Fig. 10.22). *Epithemia* has both longitudinal septa and transverse septa that interlock. The latter appear as costae in valve view (Fig. 10.19M).

In many pennate diatoms a longitudinal slit occurs in the valve surface, appearing much as a crack in a pane of glass. This is the *raphe;* it is usually interrupted at the midpoint by a thickening of the valve surface called a *central nodule* (Fig. 10.18). Similar thickenings, the *polar nodules*, occur at the extremities of the raphe. In many genera the raphe lies in a median position, but in certain genera it occupies a lateral position (Fig. 10.19I,J).

The characteristic back and forth gliding movements of many diatom species are evidently associated with the raphe since countless observations have clearly shown that only those diatoms with true raphes are capable of such motility. In some of the pennate diatoms, however, an apparent line may exist where in actuality there is no true raphe. This *pseudoraphe* is an optical illusion of sorts resulting from the longitudinal alignment of spaces between the rows of striae on each side of the valve (Fig. 10.19Q',R).

In most diatoms the two opposing valves are alike, but in at least three common genera the two valves are dissimilar, one having a true raphe and the other a pheudoraphe (Fig. 10.19E,E',F,F').

Several pennate forms have the raphes in the bottoms of canals and the canals themselves may be in elevated ridges or *keels*. These are usually eccentric and marginal (Fig. 10.19I,J).

## YELLOW-GREEN ALGAE

The *yellow-green algae* are so called because chlorophyll does not mask the carotenoid pigments but rather blends with them to give these algae their characteristic pale yellow-green coloration. Since under some circumstances the chlorophylls of green algae may break down rapidly, it is possible for green algae to be similarly yellow-green in appearance and much confusion may result if identifications are attempted on the basis of color alone. The yellow-green algae *have chlorophyll a but not chlorophyll b* and also have been reported in a few instances to have chlorophyll e. The carotenoid pigments include beta carotene and a set of xanthophylls. Starch formation is not characteristic of the group as a whole; oils and *chrysolaminarin (leucosin)* are common reserve foods. In the cells of most forms the chloroplasts are numerous, small, and disclike. Most forms lack cellulose also and the main wall substance is pectic in nature. Silica may be deposited in the walls of some species, particularly in certain spore types adapted as survival mechanisms.

A widespread characteristic in the group is the presence of overlapping layers in the cell walls. This frequently results in the dissociation of the walls of dead cells into two parts. (See illustrations of *Tribonema* and *Ophiocytium*, Fig. 10.23F,G.)

Most of the plant body types in this group are relatively simple, single cells (motile or nonmotile), small colonies, and a few filamentous types. Those cells that are motile usually have two flagella that are of different types and different lengths.

*Vaucheria,* which is the best known genus, was long classified as a green alga but was transferred ultimately as the result of careful studies of its pigments, its food reserves, and the flagella of its motile cells. This common inhabitant of springs, cool

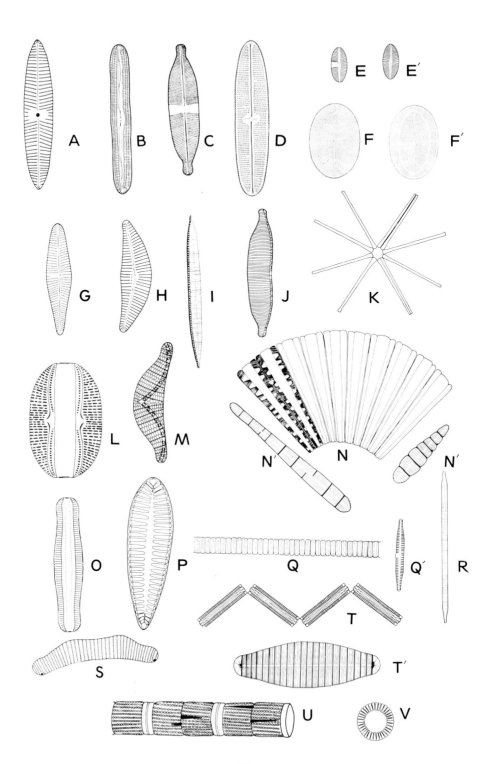

Fig. 10.19. Some common genera of freshwater diatoms.
A. *Navicula*. Symmetric both ways in both views; true raphe in both valves; striae are punctate or linear but not costate.
B. *Pinnularia*. Like *Navicula* except for presence of costae instead of striae; costae usually traversed by paired lateral lines.
C. *Stauroneis*. Like *Navicula* but central nodule is expanded laterally across valve to form a *stauros*.
D. *Neidium*. Like *Navicula* but has marginal lateral lines due to missing punctae in the striae. Raphe ends may be oppositely hooked in the middle next to central nodule.
E, E'. *Achnanthes*. Much like *Navicula* but opposing valves differ in that one has a true raphe while other has a pseudoraphe.
F, F'. *Cocconeis*. Like *Achnanthes* in having one valve with true raphe and other with pseudoraphe. Main difference is that *Cocconeis* is transversely bowed, while *Achnanthes* is longitudinally bent.
G. *Gomphonema*. Like *Navicula* except it is transversely asymmetric.
H. *Cymbella*. Like *Navicula* except it is longitudinally asymmetric, often shaped like crescent moon.
I. *Nitzschia*. Essentially symmetric both ways but the canal raphe is eccentric. The keel punctae are internal openings into the canal. The raphes of opposing valves are diagonally opposite.
J. *Hantzschia*. Similar to *Nitzschia* except that raphes of opposing valves are directly superimposed, one above the other.
K. *Asterionella*. Wheellike colonies of radiating cells more or less in one plane. Individual cells are like those of *Fragilaria* (below) except they are slightly inflated at one end.
L. *Amphora*. Individual valves are *Cymbella*-like but frustule has one broad girdle and one narrow girdle, causing the whole structure to have a wedge shape.
M. *Epithemia*. Longitudinally asymmetric (like a *Cymbella*) but has both transverse and longitudinal septa. The canal raphe has a conspicuous V-shape.
N. *Meridion*. Individual cells are *Gomphonema*-like in being transversely asymmetric. However, they lack raphes completely and pseudoraphes may be distinguished. Costae are prominent. Cells usually form colonies which are fan-shaped.
O. *Rhopalodia*. Valves somewhat like those of *Epithemia*, but the canal raphe is not V-shaped. Frustules are like *Amphora* in having one broad and one narrow girdle.
P. *Surirella*. Usually large and elegant diatoms which may be transversely symmetric or transversely asymmetric. Each valve has two marginal keels which are usually prominent. Along the edge of each keel is a canal raphe. (Note: *Cymatopleura* [Fig. 10.21] is quite similar to *Surirella* except that its valves have undulating surfaces.)
Q. *Fragilaria*. Linear colony or filament of pennate cells joined valve to valve.
Q'. *Fragilaria*. Individual cell seen in valve view showing presence of a pseudoraphe.
R. *Synedra*. Usually exists as long, slender, single cells. Valves are much like those of *Fragilaria* in having pseudoraphes.
S. *Eunotia*. Longitudinally asymmetric *Cymbella*-like shape but with only two minute, vestigial (or rudimentary), raphes, one near each pole.
T. *Diatoma*. Zigzag colony of cells joined at the corners by gelatinous pads secreted through small jelly pores. (Note: *Tabellaria* frequently forms zigzag colonies also, but the cell structures of the two genera are quite different. Compare T' below, with Fig. 10.22).
T'. *Diatoma*. Valve view of single cell. The transverse bars are costae (lacking in *Tabellaria*). No internal septa are present; the longitudinal central line is a pseudoraphe but the striae which in a sense create the pseudoraphe are difficult to see.
U. *Melosira*. A filamentous centric diatom with valve faces joined and thus not ordinarily visible. The seemingly transverse striae are continuations of the radiating striae of the valve on to the valve mantles. The faint lines in the cell middles represent the overlapping girdle bands.
V. *Cyclotella*. Usually single-celled centric diatom with costae instead of striae. This should be compared with *Stephanodiscus* where the radiating lines are striae.

Fig. 10.21. A common species of *Cymatopleura* with an undulating valve surface.

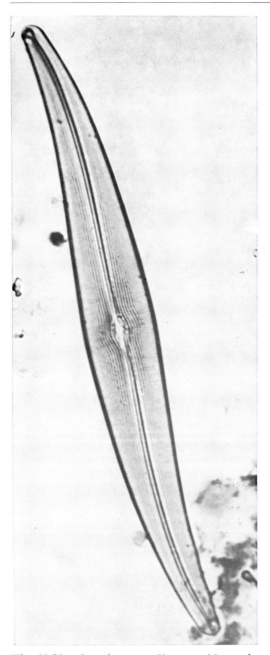

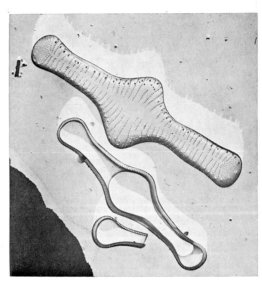

Fig. 10.22. Electron micrograph of a metal-coated replica of *Tabellaria* illustrating the internal septum with large locule.

Fig. 10.20. *Gyrosigma*, a diatom with a sigmoid valve.

slow-moving streams, and moist soils has a tubular branching plant body without cross walls except where reproductive structures occur (Fig. 10.23A). Occasionally, for example, a cross wall forms to cut off the tip of a branch and from this a large multiflagellate zoospore escapes to function in asexual reproduction. (A pair of flagella is associated with each of the numerous nuclei in this structure.) The chloroplasts are exceedingly numerous small discs with a decidedly green coloration.

*Vaucheria* is oogamous with a single egg maturing inside each oogonium. The oogonium is a rounded structure with a beak at one end (Fig. 10.23B). It is separated from the filament by a cross wall. One or more short coiled branches arise in the general vicinity of oogonia and a cross wall separates the tip region of each one of them

# THE NONGREEN ALGAE

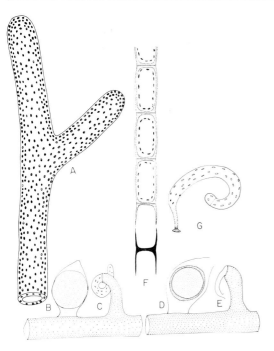

Fig. 10.23. Some common yellow-green algae. A. Vegetative filament of *Vaucheria* showing its branching tubular habit. B. Oogonium of *Vaucheria* with single large egg. C. Coiled tip of a filament has been cut off by a cross wall to form an antheridium containing several sperm cells. D. Mature oospore. E. Empty and collapsed antheridium. F. Portion of a filament of *Tribonema*. Note formation of H-pieces as dead end-cells disintegrate. G. *Ophiocytium*, a single-celled form with attaching holdfast.

into a male gametangium *(antheridium)* (Fig. 10.23C). Mature antheridia break open at their tips releasing numerous sperm which swim rapidly and are attracted to mature oogonia by a chemical secretion. A sperm enters the open beak of an oogonium and unites with the egg to form the zygote, which then develops into a thick-walled resting *oospore* (Fig. 10.23D).

Another fairly common representative of the yellow-green algae is *Tribonema*, an unbranched filament with several chloroplasts per cell. It is characteristic of this plant that when cells die and begin to disintegrate a break occurs in the cell walls where the layers of this structure overlap. Two adjacent half-cells and the cross wall between them usually remain associated for some time in the form of H-pieces (Fig. 10.23F).

*Ophiocytium* is a single-celled form often having a coiled appearance and disc-shaped holdfast (Fig. 10.23G). When asexual reproduction occurs the zoospores are released as the overlapped layers of the wall separate and the tip portion of the cell falls off.

## GOLDEN-BROWN ALGAE

The *golden-brown algae* also owe their characteristic color to a masking of chlorophyll *a* by carotenoid pigments that have more pronounced color values than those of the yellow-green algae. In many species each cell has two large opposing disc-shaped chloroplasts.

Very few complex body types are known in this group but the existing types are quite variable in appearance. *Mallomonas* (Fig. 10.24A) is a single-celled motile form with numerous siliceous bristles extending into the water from siliceous scales firmly attached to the cytoplasm. *Synura*, which forms motile colonies of *Mallomonas*-like cells, can be abundant enough in the cooler seasons to cause taste and odor problems in public water supplies. The dendroid (treelike) motile colonies of *Dinobryon* (Fig. 10.24C) are composed of open vase-shaped containers attached base to lip in branching series. Each container is called a *lorica* and holds a flagellated cell within. Motility of the whole colony is due to the concerted action of the component cells.

Modern investigations have shown two groups of ocean-dwelling chrysophytes to have immense significance as part of marine food chains. They are the *coccolithophorids*, named for the production of vast numbers of minute calcite crystals called *coccoliths*, and the *silicoflagellates*, whose cells form internal silicified skeletons on which the protoplasm is supported in a thin delicate layer.

The golden-brown algae are especially noted for the production of silicified *statospores*, which are often elaborately sculp-

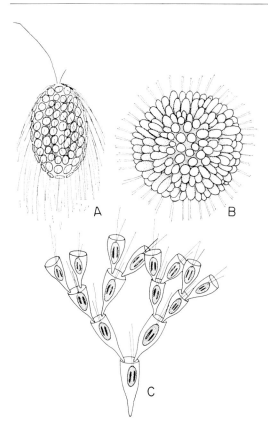

Fig. 10.24. Some common golden-brown algae. A. *Mallomonas*. B. *Synura*. C. *Dinobryon*.

tured. A small opening in each statospore is closed with a siliceous plug.

## DINOFLAGELLATES

*Dinoflagellates,* which tend to rival the diatoms in importance in marine habitats, are abundant also in many freshwater habitats but usually are somewhat less significant than the diatoms, blue-green algae, and green algae. Members of this group are frequently treated as protozoans but the presence of photosynthetic pigments, the ability to form starch as a storage product, and the occurrence of cellulose in their walls are suggestive of algal affinities. The normal color is brownish due to the masking of chlorophylls $a$ and $c$ by carotenoid pigments. A conspicuous reddish eyespot is usually evident.

The great majority of species are flagellated and actively motile although a few sedentary forms are known. Each motile cell has two flagella, one of which lies in a characteristic transverse groove. Its undulatory movements seem to cause cell rotation. The other is attached laterally and trails behind, serving to propel the cell forward. Many species have characteristic polygonal plates in their walls; these forms are known as the armored dinoflagellates (Fig. 10.25A).

*Ceratium* (Fig. 10.25B) is a very common freshwater form which may become so abundant at times that it causes an algal bloom. Certain marine dinoflagellates contain toxins that are accumulated by filter-feeding animals such as oysters and other shellfish. In seasons when these toxic species are abundant consumption of these animals by humans can have fatal consequences. Certain other marine forms exhibit a dramatic phosphorescence involving a luciferin-luciferase enzyme system. A phenomenon known as the *red tide* is the formation of a bloom of certain marine dinoflagellates which are poisonous to fish.

## EUGLENOIDS

The *euglenoids* comprise a group of organisms that have many characteristics in common with protozoans. Many of them have a bright grass-green coloration similar to that of the green algae. Pigment analyses have shown they contain both chlorophyll $a$ and chlorophyll $b$, which may be interpreted as suggesting a strong alliance with the green algae. Their main food reserve product is a starchlike carbohydrate called *paramylum* that does not stain with iodine solutions in the same manner as the starch of green algae. They do not produce cellulose and generally lack a cell wall. In *Euglena* and some other genera the cell demonstrates an ability to change shape constantly. A single flagellum moves the cell about rapidly in a forward direction. The flagellum is attached at the base of an invagination of the protoplasm referred to as a *gullet* (Fig. 10.25C). A large brilliantly

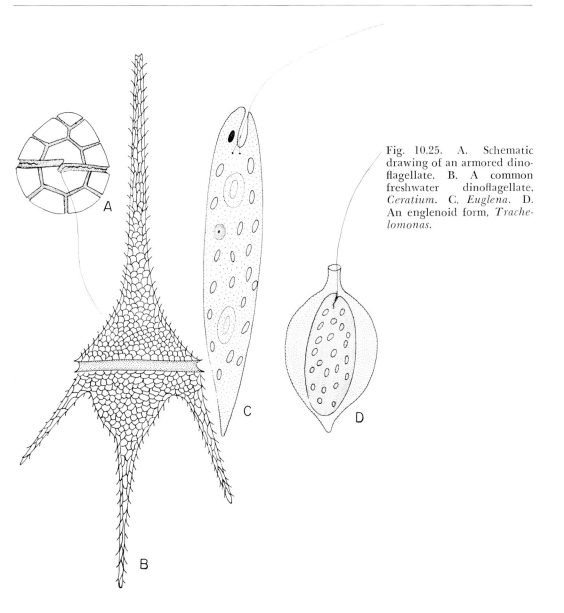

Fig. 10.25. A. Schematic drawing of an armored dinoflagellate. B. A common freshwater dinoflagellate, *Ceratium*. C. *Euglena*. D. An englenoid form, *Trachelomonas*.

colored eyespot is usually present and most forms have many small chloroplasts.

*Euglena* species are not perfectly autotrophic; many require the presence of specific organic compounds (such as certain vitamins) in the environment. Thus they tend to be found most abundantly in habitats supplied with organic waste materials. Some species regularly form resting cysts which may become attached to the surface water film of a pond. Occasionally these cysts accumulate dense quantities of a reddish pigment called *hematochrome* and the net effect is to give the pond a bloodred appearance.

Species in the genus *Phacus* are among the most beautiful of microscopic objects (Fig. 10.26). They are basically euglenoid but have become flattened and rigid in shape. As they swim through the

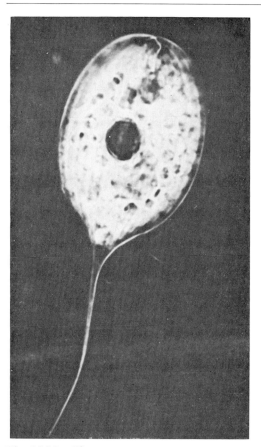

Fig. 10.26. *Phacus*.

in an outer coating, the *lorica*, which has a hole at one end through which the flagellum protrudes (Fig. 10.25D). The lorica darkens with age, ultimately becoming almost black in appearance. This is due to the adsorption of iron from the water; dense growths of *Trachelomonas* may be of great environmental significance if they affect the availability of iron to other organisms in this way.

Many aquatic animals contain minute green algae within their bodies and the algae are given the general name of *Zoochlorellae*. It is a possibility to be considered that the chloroplasts of *Euglena* originated as *Zoochlorellae* but have since lost many of the cell organelles by which they might be clearly recognized as such. Such theories are extraordinarily difficult to substantiate and as things stand, the affinities of the euglenoids are still a matter of debate.

water they constantly flip over and over in a manner resembling falling leaves. Species of *Trachelomonas* are also *Euglena*-like, but the individual cells are loosely encased

*Ad Astra ad Anima*

*Euglena* gazed into the sky
    And seeing there a human eye
Bowed with metamorphic grace
    And spoke from microcosmic space,
"Kind sir, there is an ancient riddle
That leaves me always in the middle.
    Pray, do you know
    If Father Noah
Called me plant,
    Or protozoa?"

# ELEVEN

LIVING organisms are generally classified in an ascending series of categories, each of which may be further subdivided. The major categories are listed below and a plant that has been used in the text is classified as an example.

| Plant Kingdom | Example |
|---|---|
| Division (or Phylum)* | Bryophyta |
| Class | Hepaticae |
| Order | Marchantiales |
| Family | Marchantiaceae |
| Genus | *Marchantia* |
| Species | *polymorpha* |

* The use of the term phylum in place of division is considered desirable by some authorities but is not permissible under the *International Code of Botanical Nomenclature* which outlines the rules for plant nomenclature.

When a specific kind of living organism is discussed it may be referred to either by a common name or by a scientific name. However, common names vary from place to place and from language to language; furthermore the same common name often applies to different plants in different localities.

On the other hand, the scientific name of an organism is the same in all lands and in all languages. Scientific names are written in Latin and when a new species is described a Latin translation of the description is included as a part of the publication no matter what the native language used in the publication may be. Thus a dead language has an important role in the transmission of scientific information across existing language barriers.

The scientific name is written as a binomial in which the generic name *(genus)* is listed first followed by the species name. For instance, the generic name of the maples is *Acer*, and the full binomial of the common silver maple is *Acer saccharinum* L. It is standard practice to include with the binomial an abbreviation of the name of the man who named the species. In the binomial cited above, the initial L. stands for Linnaeus, one of the great biologists of the eighteenth century.

As far as the larger categories in the system of classification are concerned major changes are being made and others contemplated by plant scientists. As a result the student who compares textbooks will find variations among them with respect to the systems of classification used.

For many years a standard classification was used that divided the plant kingdom into the following four divisions:

I—Thallophyta—Algae and Fungi
II—Bryophyta—Mosses and Liverworts
III—Pteridophyta—Ferns and Fern Allies
IV—Spermatophyta—All Seed Plants

This very simple and convenient classification had to be abandoned for several reasons, among them being:

1. The various groups of algae are no longer considered to be closely related to one another.
2. The fungi are no longer considered to be degenerate algae. Also the various groups of fungi may have had independent origins.
3. The discovery of the fossil psilophytes resulted in a reevaluation of the relationships of the vascular plants. One significant result of this process has been the realization that ferns are more closely related to the seed plants than they are to the club mosses and horsetails.

Many of the current classification changes are being made at the levels of the

# The Classification of Plants

divisions, subdivisions, and classes of the plant kingdom. Despite these changes the major groups of plants remain well defined, and it is possible to discuss these groups without confusion by means of common names. In the following paragraphs each group will be discussed briefly. Then the common names will be linked to the scientific names as used in two of the contemporary classification systems.

*Green algae* contain chlorophylls $a$ and $b$ in addition to carotenoid pigments. They store starch and usually have cellulose in their cell walls. The plant bodies show a wide range from flagellated unicellular types to complex, nonmotile, branching systems. When flagellated cells occur the flagella are alike and usually of equal length. Sexual reproduction ranges from isogamy to oogamy and from homothallism to heterothallism. Alternation of generations occurs in some species.

*Stoneworts* are algae of macroscopic size which are characterized by definite nodes and internodes. The major cell of each internode extends from one node to the next. The pigmentation and food storage products are similar to those of the green algae. Sexual reproduction is oogamous and involves unique multicellular fruiting bodies. Some species, as the name implies, become encrusted with lime.

*Euglenoids* are a group of animallike plants (or plantlike animals) that have a pigmentation similar to that of the green algae. They lack cellulose, however, and their reserve food is a carbohydrate called *paramylum* rather than true starch. Some forms are incompletely autotrophic and require organic nutrients from the environment. Most of them are unicellular and motile.

*Yellow-green algae* contain chlorophylls $a$ and $e$ but the carotenoid pigments develop strongly to give the characteristic yellow-green color. Food reserves are *chrysolaminarin (leucosin)* and oil rather than starch. The plant body types vary from motile unicellular forms to branching filaments. Gametic reproduction is not common, involving only motile gametes in all known cases except *Vaucheria,* which is oogamous. In some genera the wall consists of overlapping halves. Flagellated cells have two different kinds of flagella, one of which is usually shorter than the other.

*Golden-brown algae* owe their characteristic color to a strong development of carotenoid pigments that tend to mask the presence of chlorophyll $a$. Commonly the pigments are located in two large chromatophores in each cell. There are only a few genera with the filamentous type of organization; most forms are motile. A number of species are amoebalike and may ingest solid food. Food reserves are chrysolaminarin and oil. A special type of spore, the *statospore,* is common in the group. This structure has a silicified outer wall.

*Diatoms* also have a strong development of carotenoid pigments that tend to mask the chlorophylls $a$ and $c$. Their wall structure is highly characteristic, being composed of two overlapping halves that fit together like the two halves of a petri dish or a pillbox. The walls are silicified and ornamented in various ways. In one of the two major groups, the centric diatoms, these ornamentations are grouped about a point; in the other group, the pennate diatoms, they are organized with respect to a longitudinal line. One of the chief food reserves of diatoms is oil. Diatoms grow in soil, creeks, rivers, ponds, lakes, and in the oceans where they are the most important primary producers of organic food.

*Dinoflagellates* compose another group of organisms that are sometimes classified in the animal kingdom. Chlorophylls *a* and *c,* carotene, and certain unique xanthophylls give the chromatophores a brownish color. Food reserves are stored as starch or oil, and cellulose may be present in the cell wall. All motile cells of this group are characterized by a transverse groove and two flagella, one of which lies in the groove while the other trails.

*Brown algae* owe their color to carotenoid pigments, especially certain brownish xanthophylls, that mask the presence of chlorophylls *a* and *c*. Food reserves in this group are polysaccharides that are more sugarlike than starchlike. Brown algae are almost exclusively marine; most of them are macroscopic. The plant bodies of the more complex forms are parenchymatous or are made up of compact and intertwined filaments while the simpler forms are branched filaments resembling the green alga *Cladophora*. The occurrence of flagellated cells is restricted to zoospores and gametes. Gametic reproduction varies from isogamy to oogamy and alternation of generations is widespread in this group.

*Red algae* contain red and blue pigments in addition to carotenoids and chlorophylls *a* and *d*. Most of them are macroscopic but they are more delicate in appearance than the brown algae. Frequently they are gelatinous to the touch. A few of them occur in fresh waters but they are most abundant in marine habitats. Their life cycles are complex and alternation of generations is a common feature. No motile cells are produced in this group. Even the male gametes are without flagella and are carried by water currents to the female gametic structures.

*Blue-green algae* have red and blue pigments in addition to carotenoids and chlorophyll *a* only. As a group they are variously colored but many of them are actually blue-green. The more common food reserve is a carbohydrate that is sometimes referred to as *cyanophycean starch*. The photosynthetic membranes (thylakoids) are distributed in the cytoplasm rather than segregated in plastids. The chemical substances characteristic of nuclei in general are present in these plants but are not organized as in most other organisms; DNA fibrils can be demonstrated by appropriate chemical methods, either diffusely arranged or densely clustered in the central region of the cell. True mitosis does not occur and cell division is accomplished by the inward growth of a ringlike furrow. There is no gametic reproduction of any kind; there are no flagellated cells. The most complex body type achieved in this group is the branching filament.

*Bacteria* are for the most part minute single-celled organisms whose internal structure is difficult to ascertain with the light microscope. Most of them are unable to manufacture their own basic foods and must function as either saprophytes or parasites. A few are able to carry on a type of photosynthesis and some others are chemosynthetic. Present-day research with the structure and physiology of bacteria is so intensive that it is difficult to make definitive statements concerning the organization of nuclear substances and the nature of the life cycle.

*Actinomycetes* are often referred to as higher bacteria. The cellular organization is bacterialike but the cells are joined in a filamentous organization. They are common inhabitants of the soil. Several important antibiotics including streptomycin are derived from species in this group.

*Slime molds* have a vegetative stage that consists either of a mass of separate amoeboid cells or of a multinucleate plasmodium. Details of the gametic life cycle vary considerably, but an intricate fruiting structure usually is formed in which spores adapted to wind dispersal are produced. These organisms have many animallike characteristics and are classified as animals by most zoologists.

*True fungi* include the three major groups listed below. All true fungi lack chlorophyll and exist either as saprophytes or as parasites.

*Algalike fungi* exhibit a marked tendency towards the coenocytic growth habit, i.e., the plant body is generally without septae to divide it into cellular units. This is the only group among the true fungi in which flagellated reproductive cells occur. Sexual reproduction varies from isogamy to oogamy and from homothallism to heterothallism. In some forms the spores produced by asexual processes are adapted to wind dispersal while in more primitive ones zoospores are produced. Many of the aquatic fungi (water molds) belong in this group.

*Sac fungi* have plant bodies composed of cellular units. Their gametic reproduction is complex and frequently involves a time lapse between plasmogamy and karyogamy during which the paired but not united nuclei from the male and female source are increased. Meiosis follows immediately after the eventual nuclear fusions. Then a mitosis occurs that increases the number of nuclei resulting from each diploid nucleus to eight. This group of eight nuclei is enclosed characteristically in a special cell called an ascus. The protoplasm of the ascus then becomes divided into a row of eight ascospores within the ascus. Asexual reproduction is common in this group and frequently results in the production of such highly efficient wind-dispersed spores as conidia.

*Club fungi* also have plant bodies composed of cellular units. They further resemble the preceding group in that the nuclei of the male and female gametes do not fuse immediately after plasmogamy. Commonly the binucleate (dikaryotic) condition lasts for considerable time. In fact there are often two different types of plant bodies formed by the same species, one with mononucleate cells and the other with binucleate cells. Nuclear fusions occur within specialized cells called basidia and are followed by meiosis. Each of the four haploid nuclei within a basidium becomes incorporated into a basidiospore that is formed by an extrusion process on the outside of the basidium, rather than internally as in the ascus.

*Lichens* are organisms consisting of an algal component and a fungal component. Each lichen has a characteristic appearance and structure which permits classification as a specific entity. The fungus and the alga can be grown separately and have specific identities. However, these separate cultures bear no resemblance to the lichen from which they were isolated. The fungus is most commonly a member of the sac fungi, while species of both green algae and blue-green algae are found in various lichens.

*Imperfect fungi* include most of the fungi for which no sexual stage has been observed. This artificial group is necessary because the vegetative distinctions between the major groups of fungi, particularly the sac fungi and the club fungi, are not clear-cut. Without the sexual or perfect stage a proper assignment of the fungus cannot be made. As research with the fungi continues and observations of sexual stages are obtained species in this group are transferred to the appropriate category.

*Liverworts* are a group of nonvascular, green, land plants in which the gametophyte is the dominant and independent generation. The sporophyte is epiphytic on the gametophyte and at least partially dependent on it. There are three series (among others) in the group: (1) the *Marchantia* series in which the plant body of the gametophyte has internal differentiation of tissues, (2) the *thallose* series in which the plant body is a simple ribbon-like thallus without internal differentiation, and (3) the *leafy* series in which the plant body has an external differentiation into leaflike and stemlike parts. In all three series the sporophyte consists of a foot, a stalk, and a capsule. The capsule is simply constructed and the stalk does not elongate until after the spores have matured. As a significant exception it should be noted that the sporophyte of Riccia consists only of a capsule.

*Hornworts* are similar in many respects to the liverworts. The gametophyte is an independent land plant and is similar in structure to the gametophytes of the thallose liverworts. The sporophyte is also similar to the liverwort sporophyte in that it is epiphytic on the gametophyte. The major part of the sporophyte is a slender green column that has true stomates in the epidermis. At the base of this cylindrical capsule there is a meristematic region that adds new tissues to the capsule. As the terminal part of the capsule matures it splits open in two valves and the spores are released. Between the terminal portion and the base of the capsule it is usual to encounter various stages in spore development.

*Mosses* are also nonvascular, green, land plants in which the sporophyte is epiphytic on the independent gametophyte. The plant body of the gametophyte is leafy and bears some resemblance to that of a leafy liverwort. However, it is more robust and complex in its development. The moss sporophyte is also more complex in structure than the liverwort sporophyte. Furthermore, in contrast with the liverworts, the stalk elongates early, before the capsule tissues are fully differentiated. The whole sporophyte is green and true stomates may occur in the basal region of the capsule. The capsule tissues are organized as a series of concentric cylinders with one of the cylinders forming sporogenous tissue. The mouth of the capsule is covered by a lidlike structure, the operculum.

*Peat mosses* include all species in the genus *Sphagnum*. They are classified with the true mosses rather than the liverworts even though certain vegetative and reproductive structures are comparable with those of the liverworts. This is done primarily because the complexities of the sporophyte structure make it more readily comparable with the moss sporophyte. The occurrence of large water-storage cells intermixed with the smaller photosynthetic cells in the leaf is a characteristic feature of *Sphagnum*.

*Psilophytes* comprise a group of primitive vascular plants in which the plant body of the sporophyte consists primarily of a branching stem system. Psilophytes are rootless and for the most part leafless. Sporangia are borne terminally on the stems. Most members of this group are fossils dating back to the Lower Devonian and Upper Silurian. Two living genera, *Psilotum* and *Tmesipteris*, fit the above description with minor exceptions and may be classified as psilophytes. The gametophytes of the fossil forms are unknown while those of the living forms resemble fragments of rhizomes of the sporophyte.

*Club mosses* and *quillworts* are vascular plants whose leaves are classified as microphylls. The sporangia are associated with modified leaves called sporophylls and these are often segregated from the vegetative leaves in cones (strobili). The gametophytes are reduced in size and inconspicuous. They are rarely green and usually depend on food stored in the spore or on an association with a fungus for their basic food supplies. This group contains many fossil genera that were conspicuous elements of the Coal Age swamp forests. Of the few living genera of club mosses, *Lycopodium* and *Selaginella* are best known. *Isoetes* is the generic name of the living quillworts. Club moss leaves are rather small; quillwort leaves are longer and taper to a point. *Lycopodium* species are homosporous and their leaves are without ligules. Species of *Isoetes* and *Selaginella* are heterosporous and their leaves possess ligules. The quillworts differ from the club mosses in being aquatic plants. Also their stems are very short and have a structural organization that is unique among living vascular plants.

*Horsetails* are sometimes called scouring rushes or snake grass. The sporophyte is vascular and is characterized by stems that are jointed and ridged. The leaves are essentially without function and occur in definite whorls at the nodes of the stem. True roots arise from the nodes of underground portions of the stem. The spore-bearing structures are segregated into terminal cones. The spores give rise to minute, green, independent gametophytes. The single living genus, *Equisetum*, had many relatives in the Coal Age swamp forests.

*Ancient ferns* and *fernlike plants* were similar in many respects although it is unlikely that only one evolutionary line is involved. The dominant vegetative feature of such plants was the large macrophyllous leaf. The macrophyll probably originated more than once as a result of the flattening and webbing of a psilophyte branch system. In some lines there was a segregation of fertile (sporangia-bearing) branch systems from vegetative branch systems before the evolution of the leaf occurred. In other lines this prior segregation did not occur. In some of these ancient lines, now extinct, the evolution of seed production occurred. The fossil remains of such plants are often fragmentary and it is difficult for a nonspecialist to distinguish between the leaf of a fossil fern and the leaf of a fossil seed plant. In this brief discussion, the primitive macrophyllous plants are grouped together under the artificial heading used above. From this group arose the several separate evolutionary lines discussed below.

*Leptosporangiate ferns* include the vast majority of living ferns. In this group the sporangia are often grouped in small clusters (sori) on the under surface of the leaf. The sporangia are delicate, thin-walled, and stalked. The number of spores in each sporangium is small (rarely more than 64). The leaves are the dominant vegetative organs and in most cases the stems are inconspicuous rhizomes from which the leaves as well as the roots arise. Fern gametophytes are small, green, and independent.

*Eusporangiate ferns* comprise a rather small group of ferns that are older (in the evolutionary sense) than the group described above. Most of them have large compound leaves. One essential difference between the two groups of ferns has to do with sporangium development. In this group the sporangium is a massive structure that gives rise to an indefinitely large number of spores.

*Gymnosperms* include all seed plants in which the seed is not completely enclosed within the seed-bearing structure. In all gymnosperms the microspores mature into pollen grains that are wind-disseminated. Each pollen grain is able to form a pollen tube that penetrates the tissues of the nucleus and transports the sperms to the vicinity of the eggs. The whole group is admittedly a convenient but artificial complex of several independent evolutionary lines.

*Cycads* are the oldest living group of seed plants. Their leaves are large and fernlike. Their stems are thick, short, erect, and fleshy, with relatively poor development of vascular tissue. Pith and cortex tissues make up the larger part of the stem. The stems are clothed with the bases of dead leaves. Seeds are produced on megasporophylls that are segregated into strobili. Microsporangia are numerous on the surface of the microsporophylls that also are grouped into strobili. The two kinds of cones occur on separate plants.

*Conifers* are, for the most part, treelike forms with an extensive development of secondary vascular tissue and an outer corky bark. The pith and cortex tissues are relatively small. In most of them the leaves are highly modified as scales or needles. Conifers frequently retain their leaves during the winter season and are known commonly as evergreens. The strobilus (cone) is a basic reproductive structure in the group but is sometimes so highly modified that it no longer is identifiable as such. In many conifers the ovuliferous scales bear two ovules apiece and the microsporophylls have two microsporangia apiece.

*Ginkgo* is a large profusely branched tree that bears simple deciduous leaves with dichotomously branched veins resembling the vein system of the maiden-hair fern leaf. The stem structure is more like that of a conifer than a cycad for it has extensive development of secondary xylem and a small pith and cortex. The stem has a characteristic arrangement of long shoots and spur shoots. The microsporangia develop in simple strobili while the ovules develop in pairs on slender stalks. As the ovule develops into the seed the outer layer of the integument becomes fleshy and malodorous.

*Ephedra, Gnetum,* and *Welwitschia* are three genera with gymnospermous features that seem to be quite separate from other gymnosperms and quite probably are not closely related to one another. Among their characteristics are compound reduced microstrobili and megastrobili and the presence of vessels in the xylem.

*Angiosperms* include all the flowering

Table 11.1. A Comparison of Two Contemporary Classification Systems of the Plant Kingdom

| Classification of Tippo (1942) | Common Names | Classification of Bold (1973) |
|---|---|---|
| Subkingdom .... Thallophyta | Thallus plants | ************† |
| Phylum 1* ...... Cyanophyta | Blue-green algae | Division 8 .... Cyanochloronta |
| Phylum 2 ....... Chlorophyta | Green algae | ************ |
| Class ........ Chlorophyceae | Green algae | Division 1 .... Chlorophycophyta |
| Class ........ Charophyceae | Stoneworts (Charophytes) | Division 3 .... Charophyta |
| Phylum 3 ....... Euglenophyta | Euglenoids | Division 2 .... Euglenophycophyta |
| Phylum 4 ....... Phaeophyta | Brown algae | Division 4 .... Phaeophycophyta |
| Phylum 5 ....... Rhodophyta | Red algae | Division 5 .... Rhodophycophyta |
| Phylum 6 ....... Chrysophyta | Chrysophytes | Division 6 .... Chrysophycophyta |
| Class ........ Xanthophyceae | Yellow-green algae | Class ........ Xanthophyceae |
| Class ........ Chrysophyceae | Golden-brown algae | Class ........ Chrysophyceae |
| Class ........ Bacillariophyceae | Diatoms | Class ........ Bacillariophyceae |
| Phylum 7 ....... Pyrrophyta | Dinoflagellates | Division 7 .... Pyrrhophycophyta |
| Phylum 8 ....... Schizomycophyta | Bacteria and Actinomycetes | Division 9 .... Schizonta |
| Phylum 9 ....... Myxomycophyta | Plasmodial slime molds | Division 10 ... Myxomycota |
|  | Cellular slime molds | Division 11 ... Acrasiomycota |
| Phylum 10 ...... Eumycophyta | True fungi | ************ |
| Class ........ Phycomycetes | Algalike fungi | ************ |
|  | Chytrids | Division 12 ... Chytridomycota |
|  | Water molds | Division 13 ... Oomycota |
|  | Black bread molds | Division 14 ... Zygomycota |
| Class ........ Ascomycetes | Sac fungi | Division 15 ... Ascomycota |
| Class ........ Basidiomycetes | Club fungi | Division 16 ... Basidiomycota |
| ............ ************ | Imperfect fungi | Division 17 ... Deuteromycota |
| ............ (no ranking given) | Lichens | (classified among fungi) |
| Subkingdom .... Embryophyta | Embryo plants | ************ |
| Phylum 11 ...... Bryophyta | Bryophytes | ************ |
|  | ************ | Division 18 ... Hepatophyta |
| Class ........ Hepaticae | Liverworts | Class ........ Hepatopsida |
| Class ........ Anthocerotae | Hornworts | Class ........ Anthocerotopsida |
| Class ........ Musci | Mosses | Division 19 ... Bryophyta |
| Order ........ Sphagnales | Peat mosses | Class ........ Sphagnopsida |
| Order ........ Bryales | True mosses | Class ........ Mnionopsida |
| Order ........ Andreaeales | Granite mosses | Class ........ Andreaeopsida |

242

A Comparison of Two Contemporary Classification Systems of the Plant Kingdom (Continued)

| Classification of Tippo (1942) | Common Names | Classification of Bold (1973) |
|---|---|---|
| Phylum 12 .............. Tracheophyta | Vascular plants | ********** |
| Subphylum ............. Psilopsida | Psilophytes | Division 20 .. Psilophyta |
| Subphylum ............. Lycopsida | Club mosses and Quillworts | Division 21 .. Microphyllophyta |
| Subphylum ............. Sphenopsida | Horsetails | Division 22 .. Arthrophyta |
| Subphylum ............. Pteropsida | Plants with macrophylls | ********** |
| Class ............. Filicinae | Ferns | Division 23 .. Pterophyta |
| | Eusporangiate ferns | Class ......... Eusporangiopsida |
| | Leptosporangiate ferns | Class ......... Leptosporangiopsida |
| Class ............. Gymnospermae | Gymnosperms | ********** |
| Subclass ......... Cycadophytae | Cycads | Division 24 .. Cycadophyta |
| Subclass ......... Coniferophytae | ********** | ********** |
| Order ......... Ginkgoales | Ginkgo | Division 25 .. Ginkgophyta |
| Order ......... Coniferales | Conifers | Division 26 .. Coniferophyta |
| Order ......... Gnetales | Gnetum | Division 27 .. Gnetophyta |
| Class ............. Angiospermae | Flowering plants | Division 28 .. Anthophyta |

* The term Phylum has essentially the same significance as Division but its use is not acceptable under the International Code of Nomenclature that applies to plants. Blue-green Algae and Bacteria are frequently placed in a separate kingdom called the Monera, or the Procaryota. ÷ The use of several asterisks indicates no term is used in the particular slot in the classification system.

plants. The basic flower type consists of a shortened axis on which are borne sepals, petals, stamens, and pistils. The stamen is considered to be a microsporophyll and the simple pistil is a modified megasporophyll. A basic angiosperm feature is the enclosure of the ovules within the ovary portion of the pistil.

In Table 11.1 two contemporary systems for the classification of living plants are outlined for purposes of comparison with the common names of each of the major groups listed in a central column.

The column to the left is a classification system based mainly on one published by Oswald Tippo in 1942. It is used in many current textbooks including *College Botany* (Henry Holt & Co., 1949) written by Tippo in collaboration with Harry Fuller.

The classification system in the right-hand column is one presented by Harold C. Bold in his text *Morphology of Plants* (Harper and Bros., 1973). This system represents a more extensive revision of the major categories than many botanists are ready to accept, even though the proposed changes are based on cogent arguments.

The chart on p. 245 shows possible evolutionary relationships between the major groups of living plants. The reader is cautioned that charts such as these are synthesized from many hypotheses and are speculative in nature.

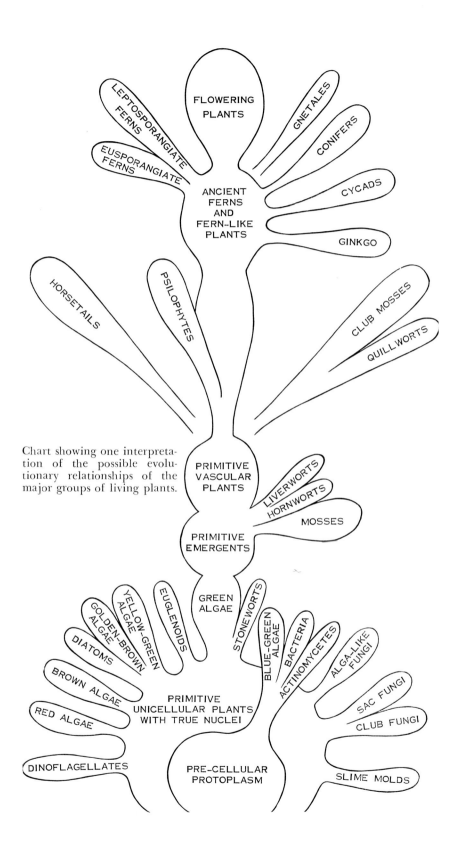

Chart showing one interpretation of the possible evolutionary relationships of the major groups of living plants.

# INDEX

[Grateful acknowlegment is due my wife, Jeanne N. Dodd, without whose help this index might never have been accomplished]

Abscissic acid, 133
Abscission layer, xiv
Accessory cell, 4
Accessory fruit, 122
Acetic acid, 177
Acetone, xiii
Achene, 122, 123
*Achnanthes*, 228, 229
Acid production, bacterial, 202
Actinomycetes, 178, 200, 238
Adenine, xxxii
Adenosine diphosphate (ADP), xix
Adenosine triphosphate (ATP), xix, xxiii. See also Photosynthesis; Respiration
Adhesive force, xxvi
Adventitious bud, 51
Adventitious root, 56, 94
Aecial initial, *Puccinia graminis*, 194
Aeciospore, *Puccinia graminis*, 194, 195
Aecium, *Puccinia graminis*, 195, 196
Aerial dispersal, fungus spores, 180-83
Aerial root, 56, 58
Aerobic bacteria, 201
Aerobic respiration, 204, 205
After-ripening, 128
Agar, 215
Aggregate fruit, 122
*Agmenellum*, 211
Air bladder, *Fucus*, 220
Air chamber, *Marchantia* thallus, 83, 84
Airplane plant, 54
Akinete, 212, 213
Albino condition, xvi
*Albugo*, 174, 175
Alcohol, xiii, xxiii, 177
Aleurone layer, 122
Algae, xiii, 136. See also individual algal groups, i.e., Green algae
Algal bloom, blue-green algae, 210
Algal filaments, *Marchantia* thallus, 83, 84

Algalike fungi, 238. See also Phycomycetes and such genera as *Rhizopus, Phycomyces,* and *Saprolegnia*
Algal mat, xvii
Algin, 217
*Alternaria*, 182
Alternate hosts, *Puccinia graminis*, 193
Alternation of generations, xxxiii, 74, 157-60
  origins, 159-60
  theory, 159
Amino acids, xxii
Ammonia, 2
Ammonification by bacteria, 203, 204
*Amphora*, 228, 229
Amyloplast, xii
*Anabaena*, 212, 214
*Anacystis*, 211
Anaerobic bacteria, 201
Anaphase
  meiosis I (anaphase I), xxxvi
  meiosis II (anaphase II), xxxvii
  mitosis, xxix-xxxi
Angiosperm, 72, 241, 244. See also Flowering plants
  vs. gymnosperm, 104
  sperm, 104
Anisogamy, 140, 141
Annual, 53
Annual ring, 31-35
Annular thickening, 22
Annulus, fern sporangium, 95
Anther, stamen, 113
Antheridia
  discharge of sperm, 76-77
  fern, 92
  *Fucus*, 220, 221
  *Laminaria*, 219, 220
  *Marchantia*, 83, 87
  moss, 75
  *Oedogonium*, 151
  red algae, 215, 217
  sac fungi, 185, 186
  *Saprolegnia*, 183, 184
  *Selaginella*, 100
  *Vaucheria*, 231
Antheridial branch, *Marchantia*, 83, 87, 88
Antheridial filament, charophytes, 155
*Anthoceros*
  gametophyte, 89-91

  opening of capsule, 90
  spore mother cells and meiosis, 90
  sporophyte, 89-91
Anthocyanin, x, xiii-xiv
  concentration in plasmolyzed cell, xxvi
  gene for production, xxii
Antibiotics, 178
Antipodal cell, embryo sac, 117-20
*Aphanizomenon*, 213, 214
*Aphanothece*, 211
Apical cell, 12
  fern gametophyte, 91-93
  moss, 75
Apical dominance, 132
Apical initials, 12-13
Appendages ("wings"), pine pollen, 107
Apple
  fruit (pome), 124
  leaf cell, xii
  scab disease, 182
Archegonia (archegonium)
  fern gametophyte, 92, 93
  *Marchantia*, 83, 87, 88
  moss, 75
  pine (in female gametophyte) 109-12
  *Selaginella*, 101
Archegonial branch, *Marchantia*, 83, 86, 87
Ascocarp, 187
Ascogonium, 185, 186
Ascomycetes, 185-87, 242. See also Sac fungi
Ascospore, 185-89
Ascus (asci), 185-87
Asexual reproduction
  *Chlamydomonas*, 138
  fungi, 178
Asparagus root, 39
*Aspergillus*, 182
*Asterionella*, 228, 229
Atmosphere (internal), leaf, 4
ATP, xix, xxiii
Austrian pine, 58
Autogamy, 225
Autotrophic bacteria, 201
Autumn coloration, xiv
Auxins, 132
  fruit drop, 133
  rooting of cuttings, 133
  seedless fruit, 133

# INDEX

Auxins *(continued)*
 sprouting of potatoes, 133
Auxospore of diatom, 224–25
Axial plate chloroplast, 143, 144
*Azolla*, 96
*Azotobacter*, 203

## [B]

Bacillus, 199
Bacteria, 198–206, 238
 allied to blue-green algae, 198
 bacterial chromosome, 198
 cell structure, 198
 identification, 200, 201
 nutrition, 201
 sizes, 198
 spore formation and survival, 201
 sulfur, 201–2
Bacteriopurpurin, 201
Barberry eradication, 195–96
Bark, 31–34
Basal cell, 120–21
Basal granule, 138
Base pair, xxxii
*Basicladia*, 152
Basidiomycetes, 242. *See also* Club fungi
Basidiospore formation, 189–90
 *Puccinia graminis*, 193–94
Basidium (basidia), 189
 karyogamy, 189
 meiosis, 189
 mushroom sporophore, 190
 *Puccinia graminis*, 193–94
 *Ustilago maydis*, 191
 types, 189–90
Basswood, 32–34
Bean seedling, 130–31
Beet petiole, 18
Beggar's tick, 127
*Beggiatoa*, 202
*Berberis vulgaris* (common barberry). *See* Black stem rust
Beri beri, 133
Berry, 124
Bicarbonate, 2
Biennial, 53
Binary fission, 199
Binomial, 236
Biochemical deficiencies, fungi, 176
Biological clock, 135
Bird's nest fungi, 191
Bisexual, 142
Bisporangiate strobilus, 164
Black bread mold
 mycelium, 173–74
 sexual reproduction, 184–86
 sporangia and spores, 180
Black stem rust, cereal grains, 182, 191–82. *See also Puccinia graminis*, life cycle
Blade
 *Laminaria*, 219
 leaf, 46
Blue-green algae, 208–14

cell structure, 208
classification, 238
color variations, 209–10
negative characteristics, 211
plant body types and selected genera, 211–12
similarities to bacteria, 208
unfavorable features, 214
Bold, Harold C., 242–44
Border parenchyma, 5–7
Boron, 42
Boston ivy, 52
Bound water, xxvi
Bract, pine seed cone, 104–5
Branched filament, 144, 152, 153
Branch primordium, 14
Brazil nut, 123
Bread mold spores, 180–81
Brittlewort, 154
Brown algae, 217–22
 classification, 238
 pigments, 217
Brown rot, stone fruit, 182, 187–89
 ascocarp, 189
 ascospore, 188–89
 conidia, 188
 control measures, 189
 destruction of infected fruit, 188, 189
 mycelial growth, 188
 "mummies," 189
Bryophytes, 74
 classification, 236
 divergence from prevascular plants, 159–60
Budding
 bacteria, 199–200
 yeast cells, 172
Buds
 active, 50
 axillary, 46, 50, 51
 inhibition, 132–33
 mosses, 74–75
 vascular plants, 14–16, 50–51
Bud scale, 15–16, 50–51
Bulliform cell, 57
Bundle cap, 28
Buttercup family, 165
Buttercup root, 38
Button stage, mushroom sporophores, 190
Butyric acid, 177

## [C]

*Calamites*, 68
Calcium, x, 42
Calyptra, moss, 78
Calyx, 113
Cambial zone, 28–29
Cambium, 27–30, 40–41
Canal raphe, 227
Capillaries, 6
*Capsella*, 120–21
Capsule
 bacterial cell, 198
 *Marchantia* sporophyte, 83–87

moss sporophyte, 78–79
seed dispersal by shattering, 126
*Sphagnum* sporophyte, 82
type of fruit, 124
Carbohydrate, xv, xx
Carbonates, 2
Carbon cycle, 43, 177
Carbon dioxide ($CO_2$), xv, 2
 photosynthesis, xvi
 reservoir, atmosphere and ocean, 2
 in soils, 43
 solubility, 3
"Carbon water," ($CH_2O$), xv
Carboxyl group, xxii
Carotene, xii, 136, 209, 222
Carpel, 116, 166
Carpogonium, of a red alga, 215, 217
Carpospore, *Nemalion*, 217
Caryopsis, 123
Casparian strip (of endodermis), 38, 39
Castor bean, embryo, 130
Celery, x
Cell
 enlargement, x
 environment, gene interplay, xxxi, xxxii
 generalized plant type, 8
 membranes, xiii, xxvi
 microenvironment, 132
 plate, xxii
 position, 131
 sap, x, 8
 shapes, 131, 142, 143
 types, vascular plants, 17
 wall, viii, x, xi
  primary, xxi
  resistance to turgor pressure, xxv
 wet, 6–8
Cell division, stages, xxvii
 blue-green algae, 209, 210
 bacteria, 198–99
 *Chlamydomonas*, 138
 desmids, 145
 diatoms, 223–24
 yeast, 172
Cellulose, viii, xxi, 177
Central body, blue-green algal cells, 209
Central vacuole, ix, x, xii
 in some green algae, 142, 143, 144
Centric diatom, 225–27
Centroplasm, blue-green algae, 208
Centrosome, xxviii
*Ceratium*, 232, 233
*Chara*, 153–55
Charophyte, 144, 153–55
Checkerboard diagram, xxxx
Chelate, 205
Chemical elements, plant growth, 42
Chemical energy, xxiii, 201
Chemosynthetic bacteria, 201
Chiasmata, xxxv, xxxvi, xxxxiii
*Chlamydomonas*, 136–40

# INDEX

asexual reproduction, 138
cell structure, 136–38
effects of environmental changes, 139
sexual reproduction, 139–40
syngamy, 139
*Chlorella*, 144, 145
Chlorenchyma, 4, 17
Chlorophyll
  *a*, xii, xiii, xiv
    blue-green algae, 209
    green algae, 136
    molecular structure, xiv
  *b*, xii, xiii, xiv
    euglenoids, 136
    green algae, 136
  bacterial, 198
  *d*, 214
  fluorescence, xviii
  photosynthesis, xvi, xviii, xix
Chloroplast, xii, xiii, 137
  membrane, xi
  shapes, 137, 143–44
  types, xii, 143–44
*Chondrus crispus*, 215
Chromatid, xxviii, xxxi, xxxv
  chromosomes, xxxvi, xxxvii
  duplication, xxxi, xxxiv
Chromatin, ix, xi, xxvii, xxviii
Chromatin body, 198
Chromatography, xiii, xiv
Chromatophore, xii, 201
Chromonema, chromonemata, xxviii
Chromoplast, xii, xiii
Chromosome number, xxxiii, 140
Chromosomes, viii, xxvii, xxviii, xxix
  and chromatids, xxxiv
*Chroococcus*, 211, 212
*Chrysanthemum*, 132
*Chrysolaminarin*, 227
Chytrid, 179
Citric acid, 177
*Cladophora*, 152
Class, 236
Classification systems, 236–44
*Clematis*, tendrils, 52
Climbing plant, 52–53
Clone, 141
*Closterium*, 145
*Clostridium*, 203
Club fungi, 239, 242
  mycelium growth, 190, 191
  reproduction, 189
  sporophores, 190, 191
Club mosses, 64, 97, 240
Cluster cup, *Puccinia graminis*, 194, 195
Cool Age swamp forests, 69
*Coccochloris*, 211
Coccolithophorids, 231
*Cocconeis*, 228, 229
Coccus, 199
Cocklebur, 127
*Coelosphaerium*, 211
Coleoptile, 122, 130
Coleorhiza, 122
*Coleus*, xiii, xvi
Collenchyma, 17–18

Columella
  *Anthoceros* sporophyte, 90
  bread mold sporangium, 174, 180
  moss capsule, 79
Commensalism, 205
Common barberry, host of *Puccinia graminis*, 193
Common vs. scientific names, 236, 237
Companion cell, 19–20
Competition, xxxxiii
Composite family, flowers, 167
Compound pistil, 115, 116, 124
Conceptacle, *Fucus*, 220–21
Cone, club mosses, 65, 98–100. See also *Equisetum*; Pine
Conidia
  *Alternaria, Penicillium, Venturia, Aspergillus,* and *Fusarium*, 182
  brown rot of stone fruits, 188
  nature and formation, 181–82
  plant diseases, 182
Conidiophore, 181–82
Conifer, 104, 241
Conjugation tube, *Spirogyra*, 150
*Conocephalum*, 189
Contractile vacuole, 137, 138
Copper, 42
Coral fungi, 191
Coralline red algae, 215
Cork
  stems, 27, 30–34
  roots, 40, 41
Corm, 54
Corn
  cell of leaf, xii
  coleoptile, 130
  embryo, 121–22
  grain or caryopsis, 122
  phloem and xylem, 19, 25, 26
  smut disease, 191, 192
  stem structure, 26
  vascular tissue differentiation, 25, 26
Corolla, 113
Cortex
  root, 35–41
  stem, 15, 23
Corticating cell, *Chara*, 153–55
*Cosmarium*, 145
Costae
  diatom valves, 226
  moss leaf, 81
Cotyledon, 46
  bean embryos and seedlings, 129–31
  origin in embryo, 120–21
  pine embryo, 112
  storage organs, 121
Cristae, x
Crystals, x
*Cuscuta*, 170
Cuticle, 3
  early land plants, 60, 62
  stem, 30
Cuticular transpiration, 9
Cutin, 3, xxi
Cutinized epidermis

*Anthoceros* sporophyte, 90
moss capsule, 81
primitive land plants, 61
stem, 27, 28, 30
Cyanophycean starch, blue-green algae, 210
Cycads
  classification, 241
  megasporophylls, 163
  microsporophylls, 163
  motile sperm, 161
Cyclosis, ix, 154
*Cyclotella*, 226, 228, 229
*Cymatopleura*, 230
*Cymbella*, 228, 229
Cystocarp, *Nemalion*, 217
Cytokinesis, xxxi
Cytokinin, 133
*Cytophaga*, 199
Cytoplasm, viii, ix, x
  division, xxxi
  enclosures, x
  green algal cell, 137
  inheritance, xxxxiii
  membranes, viii, xxvi–xxvii
Cytoplasmic streaming, ix, 154
Cytosine, xxxii

## [D]

Daisy, 167
Dandelion, 127, 167
Darwin, Charles, xxxxiii
Daughter chromosome, xxxi, xxxvii
Decay and decomposition, 177–78, 202–5
Deciduous woody plant, 52
Decussate leaf arrangement, 47, 48
Dehiscent fruit, 122
Denitrification, 203
Dentate leaf margin, 49
Deoxyribose nucleic acid (DNA), xxxii
  in blue-green algae, 209
Desmids, 144, 145
Devonian swamps, plants, 62
Diaminopimelic acid, 198
Diatom, 222–30, 237
  auxospore formation, 224, 225
  cells, 222–24
  movements, 227
  pigments, 222
  populations, 224
  recognition features, 225–29
  silica in walls, 223
  slide preparation, 225
  storage of oil in cells, 223
  symmetry, 226, 227
  valves and girdle bands, 222–24
  zygotes, 225
*Diatoma*, 228, 229
Diatomaceous earth, 223
Dicot flower (vs. monocot), 166, 167
Dicot stem, herbaceous, 26–29
Dicotyledon, 121

Dictyosome, xi, 138
*Dictyostelium*, 173
Differentiation, 12, 25
Diffusion
 gases, xxiv
 gradients, xxvii, 132
 nature, xxiii, xxiv
 solutes and solvents, xxiv
 sperm attractants, 77, 92
Digestion
 fats, xxii
 proteins, xxii
 during seed germination, 129
 starch and cellulose, xxi
Digestive enzyme, 177, 200
Dihybrid cross, xxxx–xxxxii
Dikaryotic condition, 189
 establishment, club fungi, 190
 *Puccinia graminis*, 193
*Dinobryon*, 231, 232
Dinoflagellate, 232, 233, 238
Dioecious, 142
*Diplococcus*, 199
Diploid (2n), xxxiii
 condition, 156
 plant body, *Fucus*, 220
 protonema, 158
Disease forecasting service, 183
Division, 236
DNA. See Deoxyribose nucleic acid
Dodder, 170
Dominance vs. recessiveness, xxxviii, xxxx
Downy mildew, cucurbits, 183
Drupe, 124
Dry fruit, 122
Dung fungus (*Pilobolus*), 181
Dwarf male, *Oedogonium* species, 151

## [E]

Early land plants, 158–60
Early wood, annual ring, 31–34
Earth stars, 191
*Ectocarpus*, 218–19
Egg
 embryo sac, flowering plants, 117–20
 female gamete, xxxx
 female gametophyte, pine, 109–10
 *Laminaria*, 219, 220
 moss, 75–77
 *Oedogonium*, 151
 *Volvox*, 147
Elater
 *Equisetum* spores, 96, 98
 *Marchantia* capsule, 85–89
Electron
 acceptors, photosynthesis, xviii, xix
 emissions, chlorophyll, xviii
 microscope, viii, xi
Elm twig and bud, 16
*Elodea* (= *Anacharis*), xxv
Embryo
 axis, 121
 corn, 121, 122
 dicot, 120, 121
 fern, 92, 93
 flowering plants, 120–22
 *Marchantia*
 mitosis, 77
 monocots, 121, 122
 moss sporophyte, 76–78
 pine, 112
Embryo sac, xxxx
 free nuclear stage, 116–17
 in ovules of flowering plants, 116–20
Embryo stage, 160
Emergent vascular plants, 62
Endodermis
 primitive stems, 23–24
 roots, 38–40
Endoparasitic slime mold, 173
Endoplasmic reticulum, xi, 138
Endosperm, xxxiii
 cellular stage, 120
 corn grain, 122
 free nuclear stage, 120
 innovation, 162
 storage organ, 120–22
Endospore, bacteria, 201
Energy
 heat, xv
 light, xvi
Enzymes, x
*Ephedra*, 241
Epicotyl
 arch, pea seedling, 129, 130
 corn embryo, 122
 corn seedling, 130
 dicot embryo, 120, 121
 pine embryo, 112
Epidermal hairs, 18
Epidermis. See also Stem; Leaf
 moss capsule, 79
 origin from protoderm, 15
 root, 35–38
Epigyny, 165
*Epithemia*, 228, 229
Epivalve, 222, 223
Equatorial plate, xxvii, xxxv, xxix
*Equisetum*
 classification, 240
 fossil relatives, 69, 70
 gametophyte and gametangia, 96, 97
 general nature and structure, 67, 68
 life cycle, 96–98
 sporangia, spores, elaters, 96–98
 sporangiophores, 96, 98
 strobilus (or cone), 97–99
*Erysiphe*, 174, 175
*Escherichia coli*, 205–6
Essential elements, plant growth, 42–43
Ethylene, 133
Eucaryotic cell, viii
*Euglena*, 232, 233
*Eunotia*, 228, 229
Eusporangiate ferns, 241
Evaporation, wet cell walls, 7
Evolutionary relationship chart, 244
Evolution, xxxiii–xxxxiv
Exarch xylem, 24
Extracellular digestion, 172, 177
Eyes, potato tuber, 54
Eyespot
 *Chlamydomonas*, 137
 *Euglena*, 233

## [F]

$F_1$, $F_2$, $F_3$ generation, xxxx
Facultative aerobe, anaerobe, 201
Fairy ring, 191
False branching, blue-green algae, 212, 214
Family, 236
Fats, xxii, xv
Fatty acid, xxii
Fatty substance, membranes, xxvii
Female cone, 104, 105
Female gametophyte
 evolutionary loss of independence, 161
 food storage function, pine, 112
 free nuclear stage, pine, 108–9
 ovules of flowering plants, 116–20
 pine, 108–12
 *Selaginella*, 100–101
Fermentation, xxiii, 204
Fern
 ancient, 240–41
 antheridium, 99
 archegonium, 92, 93
 embryo, 92, 93
 evolution, leaf, 68
 gametophyte, 91–93
 leaf cells, xii
 life cycle, 91–98
 prothallus, 91–93
 sorus (sori), 94, 95
 sporangium, 94, 95
 spore discharge, 95
 spore germination, 91
 spore transport, 68
 sporophyll, 94, 95
 sporophyte, 70–72
 vegetative structure, 68–71
Ferns and fern allies, 236
Fertilization, xxxiii
 double, 119–20
 ferns, 92
 *Fucus*, 222
 *Marchantia*, 83
 *Oedogonium*, 151
 ovules of flowering plants, 119, 120
 pine, 111–12
 *Selaginella*, 101
Fertilization tube, *Saprolegnia*, 184
Fibers
 flax, 17
 phloem, 19
 xylem, 20
Fiddlehead (young fern leaf), 71
 green algae, 142–43

# INDEX

Filament, stamen, 113
Flaccid, xxvi
Flagellum (flagella), 136–37, 199
Fleming, Sir Alexander, 178
Floral bud, 51
Floral induction substance, 134
Floridean starch, 215
Flower
  color, example of inheritance, xxxviii
  definition, 112
  parts, 113–15
Flowering plants
  evolutionary trends, 164–68
  life cycle, 112–22
Follicle, 123
Foot
  fern embryo, 92, 93
  *Marchantia* embryo, 83, 88
  moss embryo, 76–78
Fossil record, primitive land plants, 62
*Fragilaria*, 228, 229
Free nuclear stage
  embryo sac, 116–17
  endosperm, 120
  female gametophyte of pine, 108–9
Frond, fern, 68
Fructose, xx
Fruit
  ripening, xiii
  true, 122
  types, 122–26
Fruiting body, slime bacteria, 199
Fruiting tips, *Fucus*, 220, 221
Frustule, diatom, 223
*Fucus*, 220–22
Fundamental tissue, 15
Fungal component, lichens, 175, 176
Fungal metabolism, 177
Fungi
  asexual reproduction, 178
  definition, 170–71
  industrial uses, 177–78
  possible origins, 171
  presence in *Psilotum* gametophyte, 63
  sexual reproduction, 183
  spread by spores, 179
  true, 238
  zoospores in some forms, 179
*Fusarium*, 182

## [G]

Gametangium (gametangia). See Accial initials, Antheridia, Archegonia, Ascogonium, Oogonia, etc.; Life cycle; Sexual reproduction
Gamete, xxxiii. See also Life cycle
  discussion, 156
  gene makeup, xxx
  sexual differences, xxxx, 140–42

Gametocytic meiosis, 157
Gametophyte, xxii. See also Horsetail; Laminaria; Life cycle
  general discussion, 156
Garden pea, genetic studies, xxxxii
Gas exchange, 2, 4–10
Gas production, bacteria, 200
Gas vacuole, blue-green algal cells, 210
Gelatin-digesting bacteria, 200
Gel (state of cytoplasm), viii
Gemmae and gemmae cups, *Marchantia*, 83, 85
Generative cell, 114, 119
Genes
  chromosomes, xxxi
  DNA, xxxii
  flower color, xxxiv
  nature, xxxii
  recombinations, xxxx
  role, xxxii
  segregation and recombination, meiosis and fertilization, xxxvii
Genetic code, xxxii
Genetic symbols, xxxviii
Genotype, xxxx
Genus, plant classification, 236
Geotropism, 133
Geranium family, 165–66
Germination
  moss spores, 74
  seeds, 128, 129
Gibberellin, 133
Gill, mushroom sporophore, 190
*Ginkgo*, 241
Girdle band, diatom cells, 222–23
Girdled stem, 20
*Gleotrichia*, 212, 214
Gliding movement
  bacteria, 199
  blue-green algae, 211
Globule, charophytes, 155
*Gloeocapsa*, 211
Glucose, xx, xxii, xv
Glycerine, xxii
Glycine, xxii
Glycolysis, xxiii
Glycoside, xv
Glycyl glycine, xxii
*Gnetum*, 241
Golden-brown algae, 231–32
  classification, 237
  pigments, 231
Golgi membranes, xi, 138
*Gomphonema*, 228, 229
*Gomphosphaeria*, 211
Grain, corn, 122, 123
Gram staining technique, 200
Grana, xi, xii
Granule, cytoplasm, x
Grape tendril, 52
Grass family, 167
Grass-green algae, 136
Green algae
  classification, 237
  nature, xii, 136
Green cell, xv

Green leaf color, xviii
Green sulfur bacteria, 201
Ground meristem, 15
  cell divisions, 16
  roots, 35, 36, 37
Ground parenchyma, 26
Ground tissue, 15
Growth and differentiation, 130–35
Growth regulating substance, 132–35
Guanine, xxxii
Guard cell, 3–5, 8–9
Gullet, *Euglena*, 232
Gymnosperm. See also Life cycle, pine
  classification, 241
  vs. angiosperm, 72, 104
*Gyrosigma*, 230

## [H]

Half-cell (semicell), desmids, 145
*Hantzschia*, 228, 229
Haploid condition, 150
  nuclei (*n*), xxxiii
  sporophytes, 158
Haustoria
  dodder, 170
  fungi, 173–75, 178
Hazelnut, 123
Heartwood, 32, 35
Heavy oxygen, xviii
Helix, double-stranded, xxxii
Helotism, 205
Hematochrome, xiv, 233
Hen and chicken plant, 55
Hereditary changes, xxxxiii
Heterocyst, 212–14
Heterogamy, 141
Heteromorphic alternation of generations, *Laminaria*, 219
Hetcrospory
  *Selaginella*, 66, 97
  sexuality of gametophytes, 101–2
  water ferns, 96
Heterothallism
  algae, 140–42
  black bread molds, i.e., *Phycomyces*, 184, 188
  *Marchantia*, 85–86
  *Puccinia graminis*, 194
Heterotrophic bacteria, 201–5
Heterozygous condition, xxxviii, xxxx
Hickory nut, 123
Holdfast
  *Laminaria*, 219
  tendrils, 52
Homologous chromosomes, xxxiv, xxxvii
Homosporous ferns (vs. heterosporous), 96
Homospory, *Lycopodium*, 102
Homothallism, 140–42
Homozygous condition
  sexual life cycles, xxxviii

Homozygous condition *(continued)*
  significance, xxxvii, xxxx
Hormone, plant, 132–35
Hornwort, 74, 89–91, 239
Horsetail, 64, 67–70, 96–98. *See also Equisetum*
  classification, 240
Humic acid, 42, 43, 205
Hunger, xv
Hyaloplasm, viii
Hybrid, xxxviii
Hydrocarbon gas, 2
*Hydrodictyon*, 148
Hydrogen acceptor, xviii
Hydrogen sulfide
  bacterial photosynthesis, 201
  production by bacteria, 204
Hydrophyte, 58, 59
Hydroxyl (¯OH) ions, xix
Hypha (hyphae), 173
Hypocotyl, 46
  bean seedling, 129
  dicot embryo, 120, 121
  pine embryo, 112
Hypogyny, 165
Hypovalve, diatom cell, 222–23
Hyrax, 225

[I]

Imbibition, xxvii, 127, 128
Immature embryo, seeds, 128
*Impatiens*, 126
Imperfect fungi, 239
Impermeable seed coat, 128
Indehiscent fruits, 122
Indeterminate flowering, 134
Indian pipe *(Monotropa)*, 170
Indole acetic acid, 132
Indusium, fern sorus, 94
Inflorescence, 113, 167
Inheritance, xxxvii–xxxxiii
Inorganic nutrients, xv
Insect pollinator, 168
Integument of ovule
  flowering plants, 116
  pine ovule, 108–9, 110, 111
Intercellular space, ix
Interfascicular cambium, 30
International Code of Botanical Nomenclature, 236
Internodal cell, charophyte, 153–55
Internode, stem, 12, 13, 46
Interphase
  meiosis I, xxxvi
  mitosis, xxxi, xxxii
Iodine, xvi, 42
Ion exchange, 37
*Irisene*, xiii
"Irish Moss," 215
Iron
  availability, 204, 205
  bacteria, 202
  compounds, 202
  micronutrient element, 42
*Isoetes*
  classification, 240

heterosporous nature, 67
structure of quillwort, 66–67
Isogamete, green algae, 140
Isogamy, 140–42
  *Chlamydomonas*, 139
  *Ectocarpus*, 218
  *Phycomyces*, 184–85
  *Spirogyra*, 150
  *Ulothrix*, 149
Isomorphic alternation of generations, *Ectocarpus*, 218
Isotope, oxygen, xviii
Isthmus, desmid cell, 145

[J]

Jerusalem artichoke, 55

[K]

Karyogamy
  ascomycetes, 185–86
  basidiomycetes, 139–40, 189, 194
Karyon, viii
Keel, diatom valve, 227
Kelp, 217–18
Kinetochore, xxxv–xxxvii

[L]

*Laminaria*, 219–20
Land plants, 60, 74
Larch, 164
Late blight disease, 183
Lateral bud, 50, 51
Lateral meristem. *See* Cambium; Cork
Lateral root, 46
Late wood, 31–34
Latex, tubes, 18
Latin, plant classification, 236
Leaf
  arrangement, 14, 47–48
  axils, 14
  bud, 51
  coloration, xiii
  compound vs. simple, 47, 48
  evolution, 69
  gap, 66, 69–71
  internal atmosphere, 4–8
  margins, 49
  moss, 75, 80, 81
  palmate, 47–49
  peltate, 47
  pinnate, 47–49
  primordia, 14, 15
  scar, 51
  skeletons, 5
  *Sphagnum*, 82
  structure and function, 3–10
  succulent, 55
  trace, 68, 71
  types, 46
  venation, 49
  vascular tissues, 24

Leaflet, 47, 48
Leafy liverworts, 86–89, 239
Legume, 123
Legume family, 165, 167
Lemma, grasses, 167
Lenticel, 51–52
*Lepidodendron* (fossil club moss), 66–67
Leptosporangiate fern, 241
Leucoplast, xii, xiii
Leucosin, 227
Lichen, 175–76
  algal component, 175–76
  ascocarps, fungal component, 188
  classification, 239
Life cycle
  basic discussion, xxxiii
  brown algae, 218–22
  bryophytes, 74
  club mosses, 97–102
  diagrams
    pregametic meiosis, 157
    presporic meiosis, 157
    zygotic meiosis, 156
  ferns, 91–98
  flowering plants, 112–22
  fungi, 186
  green algae, 138–55
  pine, 104–12
  red algae, 215–17
  synopsis of nuclear events, xxxiii
Light
  absorption by chlorophyll, xviii
  flower induction, 134–35
  growth responses, 132–33
  guard cell response, 9
  photosynthesis, xviii
  seed germination, 128
  spectrum, xviii
Lignin, xvi, 17, 21
Ligule, *Selaginella*, 65
Lilac leaf, powdery mildew, 178
Lily flower, 166
Limiting factors, theory, xix, xx
Linear tetrad, megaspores, 108–9, 117
Linen fiber, 17
*Linnaeus*, 236
Lipopolysaccharide, 198
Lipoprotein, 198
Liverwort, 74, 82–89, 239
Locules, diatom septa, 227
Lodicule, grass flower, 167
Long day or long night plant, 134
Lorica, *Dinobryon* cell, 231, 232
Lower green land plant, 74
Luciferin-luciferase enzyme system, 232
Lycopene, xiii
*Lycopodium* (homosporous club moss), 64–65, 102, 240
*Lyngbya*, 211

[M]

Macerated tissues, 21
Macrophyll, 68–71

# INDEX

Macrophyllous plants, 240, 241
Magnesium, xiv, 42
*Magnolia*, 165
Male gametes, xxxx. See also Sperm.
  motile types. See *Chlamydomonas*; Life cycle; *Oedogonium*; *Ulothrix*
  pollen tubes, 111, 119, 120
  special types in fungi. See Ascomyces; *Puccinia graminis*; *Saprolegnia*
Male gametophytes
  flowering plants, xxxx, 118-20
  *Marchantia*, 85-86
  pine, 111
  *Selaginella*, 100-101
Male pine cones, 104-6
*Mallomonas*, 231-32
Manganese, 42
Maple, 16, 20
*Marchantia*, 82-89
  air chambers and pores in thallus, 83-84
  antheridia and archegonia, 83, 86-87
  classification, 239
  cutinized epidermis, 82, 84
  gametophyte body, 82-86
  gemmae and gemmae cups, 83, 85
  genetic control of sexuality, xxxviii, 85
  male and female plants, 83, 85-89
  sperm discharge, 83
Marigold, 167
Marl formation, 154
*Marsilea*, 96
Mass flow hypotesis, 43
Matzke, E. B., vii
Medullary ray, 27
Megasporangiate cone of pine, 104, 105
Megasporangium
  ovule of flowering plants, 116, 117, 119
  ovule of pine, 108-10
  *Selaginella*, 99, 100
Megaspores, xxxx
  heterosporous ferns, 96
  linear tetrad, 108-9, 117
  mother cells, xxxix, xxxx
  ovule of flowering plants, 116-17
  ovule of pine, 108-10
  *Selaginella*, 99-101
Megasporocyte. See Megaspores, mother cells
Megasporophylls
  cycads, 163
  primitive flowering plants, 115, 164
  *Selaginella*, 99, 100
Meiosis, xxxiii
  algal zygotes, 139-40
  basidia, 189
  chance distribution of genes, xxxx
  diatoms, 224-25
  *Ectocarpus*, 219
  flowering plants, 114, 116
  *Fucus*, 222
  life cycles, 140
  moss, 80
  phases, xxxix-xxxxiii
  pine, 107, 108
  pregametic, 157
  presporic, 157
  *Puccinia graminis*, 194
  red algae, 215, 217
  sac fungi, (young asci of), 185
  segregation of genes, xxxviii, 85
  *Spirogyra*, 150
  summary, xxxvii
  zygotic, 156
Meiospores, xxxiii, 74, 79, 140, 157
Meiosporophyte, 157
*Melosira*, 226, 228, 229
Membrane
  double, x
  differential permeability, xxiv-xxv
  type of plant body, 144, 153
Mendel, Gregor, xxxxii
*Meridion*, 228, 229
*Merismopedia*, 211
Meristem, xxvii, 12, 13. See also Cambium; Cork; Stem tip
Meristematic cell, 75, 131
Meristematic zone, *Anthoceros* sporophyte, 90
Mesophyll, 4, 15
Mesosome, bacterial cell, 198
Messenger RNA, xxxii
Metaphase
  meiosis I (metaphase I), xxxv
  meiosis II (metaphase II), xxxvii
  mitosis, xxviii, xxxi
Metaphloem
  corn leaf, 25
  root, 38
Metaxylem, 22-24
  root, 38, 39
  vessels of corn, 25, 26
Methane, 2
Methane-producing bacteria, 212
Micelles, xxi
*Micrasterias*, 145
Microbodies, xi
*Microcystis*, 211, 212
Microcysts, slime bacteria, 199
Microfibril, xxi
Micronutrient element, 42, 204
Microorganisms in soil, 42, 43
Microphyll, club mosses, 66
Micropyle
  flowering plant ovule, 116-20
  pine ovule, 108-12
Microsporangia
  flowering plants, 113-14
  pine, 106-8
  *Selaginella*, 97-100
Microsporangiate cone, pine, 104-6
Microspores, xxxx
  flowering plants, 114
  heterosporous ferns, 96
  mother cells, xxxix, xxxx
  pine, 106-7
  *Selaginella*, 97-100
Microsporocytes. See Microspores, mother cells
Microsporophylls
  cycads, 164
  flowering plants. See Stamens
  pine, 106-8
  *Selaginella*, 99-100
Microtubule, xi
Middle lamella, viii, ix, xxii
Midrib, 49
Milkweed follicle, 123
Mitochondria, viii, ix, x, xi, 138
Mitosis, xxviii, xxix, xxx
Mitospore, 138
Mixed bud, 51
*Mnium*, 80
Modified stem, 54-55
Molybdenum, 42
Monera, 243
*Monilinia fructicola*, 187
Monocot embryo, 121-22
Monocot vs. dicot flower, 166, 167
Monocotyledon (monocot), 121
Monohybrid cross, xxxviii
Monokaryotic condition, 189
Monokaryotic mycelium, *Puccinia graminis*, 193-95
Monoploid (n), xxxiii
Monospore, red algae, 215
*Monotropa*, 170
*Morchella esculenta*, 187-88
Morel, 187-88
"Moss-back turtle," 152
Mosses
  adaptations, 80-81
  antheridia and archegonia, 75-77
  capillary water movement, 80
  capsule, 79
  cell of leaf, xii
  classification, 240
  dormancy of spores, 74
  embryo, 76-78
  life cycle, 74
  meiosis and spore formation, 80
  spore germination, 76, 82
  sporophyte, 77, 81
Motile colony, 144-47
Motile unicell, 144, 136
Motility, bacteria, 199
*Mougeotia*, 143, 144
Movement
  in blue-green algae, 211
  of diatoms, 227
Mucopeptide, bacterial cell wall, 198
Multiple fruit, 122
Mushroom
  growth, 190-91
  life cycle, 190
  sporophores, 190-91
  toxins, 191
Mustard family, 165, 166
Mutations, xxxiii, xxxxiv

Mutualism, 205
Mycelial bacteria, 200
Mycelial pseudoparenchyma, 175
Mycelial strand, 175
Mycelium (mycelia), 173
Mycorrhizal relationship, 178
Myxobacteria, 199

## [N]

NADP, NADPH, xix
Naked bud, 50
Naked sorus, 95
Nannandrous, species of *Oedogonium*, 151
Natural selection, xxxxiii–xxxxiv, 60
*Navicula*, 223, 224, 226, 228, 229
Neck, archegonium, 75–77
Neck canal, moss, 75–77
Nectar, 18
*Neidium*, 228, 229
*Nemalion*, 215–17
*Neocosmospora vasinfectans*, 185, 187, 188
*Nepenthes*, 59
Netted venation, 49
*Neurospora*, 185–87
Nicotinamide-adenine dinucleotide phosphate (NADP), xix
*Nitella*, 153, 154
Nitrification, 202
Nitrogen, xv
 bases, xxxii
 cycle, 203
 essential element, 42
 fixation
  blue-green algae, 213, 214
  free-living bacteria, 203
  lightning, 203
  manufacturing processes, 203
  root nodule bacteria, 203–4
*Nitzschia*, 228, 229
Nodes
 charophytes, 153–55
 internodes, 46
 stems, 12, 13, 26
Nongreen algae, 208
Nonmotile colony, 144, 148
Nonmotile unicell, 144, 145
Nonseptate basidia, 189–90
Nonseptate hyphae, 173
*Nostoc*, 212, 214
Nucellar cap, (pine seed), 112
Nucellus. See Megasporangium, flowering plants, pine
Nuclear membrane, viii, ix, xi, xxvii, xxxi, xxxvi
Nuclear sap, xi, xxvii
Nucleic acid, xv
Nucleolus, ix, xi, xxvii, xxviii
Nucleoprotein, xxii
Nucleotide, xxxii
Nucleus, xviii, xxvii, 137. *See also* Meiosis; Mitosis
Nucule, charophytes, 155
Nut, 123.
Nutrient availability, 131–32

Nutrient sink, 43
Nutrition, fungi, 176–78

## [O]

Obligate aerobes and anaerobes, 201
Obligate parasites, 205
*Oedogonium*
 cell and chloroplasts, 143
 filamentous green algae, 150–52
 sexual reproduction (oogamy), 151
 zoospores and asexual reproduction, 150–51
Oil
 ducts, 18
 storage in diatoms, 223
*Oleander*, 59
Onion root tip, xxviii
*Oocystis*, 143, 144
Oogamy, 140–42
 discussion, 156
 *Oedogonium*, 151
 *Saprolegnia*, 183–84
 *Volvox*, 147
Oogonia
 charophytes, 155
 *Fucus*, 221, 222
 green algae, 141,
 *Laminaria*, 219–20
 *Oedogonium*, 151
 *Saprolegnia*, 183–84
 *Vaucheria*, 230–31
Oospores
 algae, 140
 charophytes, 155
 *Oedogonium*, 151
 *Saprolegnia*, 184
 *Vaucheria*, 231
 *Volvox*, 147
Operculum, 79
*Ophiocytium*, 231
Order, classification system, 236
Organic acid, x, xxiii
Organic autotroph, 202
Organic nutrient, xv, 43
Organic peroxide, xix
*Ornithogalum*, 124, 166
*Oscillatoria*, 211
Osmosis, xxiv
Osmotic pressure, xxv
Osmotic water uptake, germinating seeds, 128
Ostiole, *Fucus* conceptacle, 220
Ovary, 115
Ovule-bearing scales
 gymnosperms, 163–64
 pine, 108–9
Ovule, 115, 116–18
 cycads, 163
 development, flowering plants, 116–17
 origin, modification of megasporangia, 161
 pine, 108–12
*Oxalis*, 166

Oxycarotene, xiv, xv
Oxygen
 burning and respiration, xv, xxiii
 isotope form ("heavy oxygen"), product of photosynthesis, xvi, xvii
 normal concentration, atmosphere, xvi
 requirements by bacteria, 200
 in solution, 3
Ozone "shield," 60

## [P]

Palea, 167
Palisade parenchyma, 4
Palmate venation, 47, 48, 49
*Pandorina*, 146
Papillae, conjugation, 150
Parallel veins, corn leaves, 5, 6, 49
Paramylum, 232
Paraphyses
 *Fucus*, 221, 222
 moss, 76
Parasite vs. saprophytes, 178
Parasitic bacteria, 205
 flowering plants, 170
 fungi, 173
Parenchyma
 cell to cell conduction, 17
 factor in plant growth, 17
 maintenance of shape, 17
 nature of cells, 17
 phloem, 19
 photosynthesis, 17
 xylem, 20
Partial dominance, xxxviii, xxxx
Passion flower tendril, 52, 53
Pea, 129–30
Peach
 brown spot disease, 187–89
 fruit (drupe), 124
 leaf curl disease, 187
Peat moss *(Sphagnum)*, 81, 82, 240
Pectin, xxii
*Pediastrum*, 144, 147–48
Pedicel, 113
Peduncle, 113
*Pellionia*, xx, xxi
Peltate leaf, 47
Penetration tube, *Saprolegnia*, 183, 184
Penicillin, 178
*Penicillium*
 conidia and conidiophores, 182
 mycelium and spores, 181
 source of penicillin, 178
Pennate diatom, 222, 225–27
Perennial, 53
Perianth, 113
Pericarp, 123
Pericycle, 23, 24
 functions, 40, 41
 location and structure, 38, 39, 40

# INDEX

Peripheral fiber, xxxi
Periplasm, blue-green algal cells, 208
Peristome teeth, moss capsule, 79, 80
Perithecium, ascocarp, 187
Permeability, membranes, xxiv, xxv
Petal, 113
Petiole, 46
*Petunia,* 167
*Phacus,* 233, 234
Phenotype, xxxx, xxxxii
Phloem
 cell types, 19 20
 conduction, 19
 differentiation from procambium, 24–25
 leaf veins, 5
 movement of solution, 43–44
 rays, 27, 29
 roots, 38, 39
 secondary, 28, 29–34
 stem, 19
Phosphate group, xxxii
Phosphate recycling, bacteria, 204
Phosphorescence, marine dinoflagellates, 232
Phosphorus, xiii, 42
Phosphorylation, xx
Photon, xviii
Photoperiod, 134
Photophosphorylation, xix
Photosynthesis, xii, xvi–xix, 2
 evolutionary advent, 3
 importance of oxygen produced, 3
 mechanism, xviii
 oxygenation of water by algae, 60
 products, uses, xx–xxiv
 rates and limiting factors, xx
 summary reaction, xviii
Photosynthetic bacteria, 201
Photosynthetic lamellae, xii
Phototropism, 133
Phycobilin pigment, blue-green algae, 209
Phycocyanin, 209, 214
Phycoerythryn, 209, 214
*Phycomyces,* 185, 186
Phylum, 236
*Physarum polycephalum,* 172
Phytochrome, 134, 135
Phytol, xiv
Pigment
 algae. *See* various groups of algae
 green plants, xii–xv
 higher plant chloroplasts, xii–xiv
 plastids, xii–xiii
*Pilobolus,* 181
Pine
 cones, 104–8
 leaf cell, xii
 life cycles, 104–12
 seed, 112
Pineapple, 125
Pinnate leaf, 47, 48, 49

*Pinnularia,* 228, 229
Pitcher plant, 59
Pith, 15, 26–34
Pithlike tissue, large roots, 39
Pits, 20–23
Pitted thickening, 22
Plane of cell division, meristematic cell, 15
Planktonic algae, xvi
Plant
 body type, 46
 fungi, 171
 green algae, 144
 selected blue-green algae, 211–12
 vascular plants, 46
 breeding techniques, xxxxii
 cell, vii
 classification, 71–72, 236
 coloration, xiii–xv
 diseases, nature and control, 187
 foods, xv
 height, xxxx
 hormones, 32–35
Plant Kingdom, 236, 242–43
Plasma membrane, viii–xi
Plasmodesm, plasmodesmata, x, xi
Plasmodium, 172
Plasmogamy
 ascomycetes, 185
 part of syngamy, 139, 140
 *Puccinia graminis,* 195
Plasmolysis, xxv
Plastids
 nature, functions, and pigments, viii–xii
 starch formation, xx, xxi
Plum pocket disease, 187
Plumule, embryo, 121
Plurilocular gametangia, *Ectocarpus,* 218
Plurilocular zoosporangia, *Ectocarpus,* 218
Plus and minus mating types (= plus and minus strains)
 algae, 141, 142
 fungi. *See Phycomyces; Puccinia graminis; Rhizopus*
Poison ivy, 52
Polar nuclei, embryo sac, 117–20
Poles, dividing cells, xxviii
Pollen cone, pine, 104–6
Pollen discharge, pine, 108, 118
Pollen grains, xxxx
 flowering plants, 115
 pine, 107–8
Pollen tube
 eliminating need for swimming sperm, 161
 flowering plants, xxxx, 118–20
 pine, 109–12
Pollination
 agents, 118
 flowering plants, 118–19
 mechanisms, 161
Polysaccharide, xx
*Polytrichum,* 78
Pome, 124

*Porella,* 89
Pores
 diffusion, 8
 pollen grain wall, 114, 118
Potassium, 42
Powdery mildew, 178
Pregametic meiosis, 157
Presporic meiosis, 157
Prickles, 50
Primary cell wall, viii–ix, xxxi
Primary endosperm cell, xxxiii, 120
Primary meristem, 15
Primary phloem
 roots, 38, 39
 stem, 19
Primary root
 fern embryo, 92, 93
 seedlings, 129
Primary tissue
 displacement during growth, 31–34
 roots, 35–41
 stems, 28
Primary xylem
 roots, 38, 39
 stems, 24, 27
Primitive emergent plants, 60, 61
Primitive flower, 164, 165
Primitive vascular plant, 23, 62
Procambium, 15, 16, 23
 cell divisions, 16
 corn leaf, 24, 25
 roots, 35–37
Procaryota, 243
Procaryotes, 198
Progametes (progametangia), black bread molds, 184–86
Promeristem, 13, 35, 36
Prophase
 meiosis I (prophase I), xxxiv–xxxv
 meiosis II (prophase II), xxxvi
 mitosis, xxviii
Prop roots, 56, 57
Proteins, xv, xxii, xxvi
Prothallial cells
 male gametophytes, *Selaginella,* 100
 pine pollen, 107
Prothallus, fern, 91–93
Protoderm, 15
 cell division, 16
 roots, 35–37
Protonema, moss, 74
Protophloem
 differentiation, 25, 26
 roots, 38
Protoplasm, viii
Protoplast, viii
Protostele, 23, 24,
Protoxylem
 dicot stem, 27
 differentiation, 25, 26
 root, 38, 39
Pseudoplasmodium, 173
Pseudoraphe, diatom valves, 227
Pseudovacuole, blue-green algal cells, 210

Psilophyte
  classification, 240
  fossil and living, 62–64
  gametophytes of fossil forms unknown, 63
*Psilotum*
  classification, 240
  gametophyte, 63
  sporangia and stem structure, 63, 64
Pteridophyta, 236
*Puccinia graminis*, 191–97
  black stem rust of cereal grains, 191–97
  breeding for resistance, 196
  cluster cups, 194, 195
  life cycle, 192, 193
  races, 196
  spore types, 194
  varieties, 196
Puffball, 191
Punctae, diatom valves, 226
Purple sulfur and nonsulfur bacteria, 201
Push-pull system, 43–44
Pustule, 192
Putrefaction, anaerobic bacteria, 204
Pyrenoid, xii, 137, 138
Pyrrol ring, xiv
Pyruvic acid, xxiii

## [Q]

Quillwort, 64, 66–68, 240

## [R]

Radicle, 120, 121
  corn embryo, 122
  emergence from seed, 129
Raphe, diatom valves, 227
Raspberry, 124
Ray initial, 29, 31
Receptacle, flower, 113
Receptive hyphae, *Puccinia graminis*, 193–95
Recessive lethal gene, xxxxiii
Recessiveness, genetic, xxxx
Red algae, 214–17
  classification, 238
  food storage products, 215
  monospores, 215
  "nature printed" illustrations, 216
  pigments, 214
  selected life cycle, 215–17
  types, 215
  uses, 215
Red rust stage, black stem rust, 192
Red tide, 232
Relative humidity, transpiration, 7–9
Resin, resin duct, resin canal, 18
Respiration, xxii, 2

aerobic, xxiii
anaerobic, xxiii
and burning, xxiii
and energy release, xxiii
fungal by-products, 177
in green cells, 7
mechanisms, x
and photosynthesis, xvii
summary of reactions, xxiii
Reticulate chloroplast, 143, 144
Reticulate thickenings, 22
*Rhizoclonium*, 152
Rhizoidal holdfast, *Laminaria*, 219
Rhizoidal mycelium, chytrids, 174
Rhizoids
  black bread mold, 174
  early land plants, 61
  fern gametophyte, 91–93
  *Marchantia*, 82, 82
  mosses, 74
Rhizome, 54
  derived from fern embryo, 93
  ferns, 68, 69
Rhizophores, 66
*Rhizophydium*, 174
*Rhizopus*, 184
*Rhizopus stolonifer*, 184
*Rhopalodia*, 228, 229
Ribonucleic acid, xxxii
Ribose, xxxii
Ribosome, xi, xxxii
  algal cells, 138
  blue-green algae, 210
*Ricciocarpus*, 83
Rickets, 134
*Rivularia*, 214
RNA, xi, xxxii
Rockweed, 217, 222
Root, 34–41
  absorption by, 36, 37
  adventitious, 50
  cap, 35
  fibrous, 55
  functions, 41, 42
  grafts, 20
  hairs, 46
  meristems, 35, 36
  origin and nature, 35–36, 66
  prop, 56
  relation to soil, 36
  seminal, 56
  soil conditions needed, 36, 37, 41
  storage, 56
  tip, 35
  tissue systems, 35–41
  types, 55
Rose family, 165
Rubber, latex, 18
Rubber plant, 58
Rumen of cow, 205
Runner, 54
Rust fungi, 196
Rusts and smuts, 191

## [S]

Sac fungi (ascomycetes)

classification, 239
sexual cycles, 185–89
*Salvinia*, 96
Sap flow, 20
*Saprolegnia*, 179, 180
  asexual reproduction, 180
  sexual reproduction, 183, 184
  sporangiospores, 179, 180
  zygotes and oospores, 184
Saprophytes
  vs. parasites, 178
  recycling organic matter, 205
Sapwood, 32, 35
Sarcinae, 199
*Sarcoscypha*, 187
Scalariform thickening, 22
Scarification of seed, 128
Scarlet cup fungus, 187
*Scenedesmus*, 144, 147, 148
Scientific name, binomial, 236
Sclerenchyma, 23, 24
  bundle cap, 17
  fibers, 17, 26
  function in mechanical support, 17
  larger veins, 17
  nature, 17
*Sclerotinia fructicola*, 187
Scouring rush, 67
Scurvy, 134
Scutellar node, 56, 57
Scutellum, 122
Sea "lettuce," 153
Secondary growth
  herbaceous dicot stems, 30
  roots, 40–42
  woody stems, 31, 34
Secondary phloem, 28
Secondary xylem, 28, 31–34
Secretion, 18
Seed-bearing scale, pine, 104, 105
Seed coat
  origin from integument, 108
  pine seed, 112
  relative impermeability, 128
Seed cone, pine, 104, 105
Seed dispersal
  birds and animals, 125
  hooks and spines, 126
  parachutes, etc., in the wind, 126
  shattering fruits, 126
Seed dormancy, 126, 128
Seed fern, 72
Seed germination, 126–30
Seed habit, 160, 162
Seed plant, 72
*Selaginella*, 64–66
  classification, 240
  cone of, 65, 97
  heterospory, 99
  life cycle, 97–102
  ligule of leaf, 65
  microsporophylls, megasporophylls, etc., 99–100
  nature of leaf, 65
  sporophylls, 97
Selenium, 42
Self-pollination, xxxx
Seminal root, 56
Sepal, 113

# INDEX

Septa, 227
Septate basidium, 189–90
Septate hyphae, 173
Serrate leaf margin, 49
Sessile leaf, 47
Seta (stalk), moss sporophyte, 78
Sewage fungus, 204
Sex cell (gamete), xxxiii
Sexual reproduction. See Life cycle
Sheath, blue-green algae, 208, 209
Shellfish, 232
Shoot system, 46
Short day and short night plants, 134
Sieve plate and tube, 19
*Sigillaria*, 67
Siliceous scale, chrysophytes, 232
Silicoflagellate, 231
Simple fruit, 122
Simple leaf, 47, 48
Simple pistil, 115, 164
Slime bacteria, 199
Slime molds
  cellular, 173, 242
  classification, 238
  general nature, 172, 173
  plasmodium, 172
  sporangia, 172, 173
  true, 172
Smooth leaf margin, 49
Smut fungi, 191
Snapdragons, 4, 115
Soil
  effect on root growth, 42
  interdependence of organisms, 176, 177
  management, 42
  nutrients, 41–42, 47
  weathering of particles, 43
*Solanum* tendrils, 52
Solomon's seal, rhizome, 54
Sol state of cytoplasm, viii
Sorus (sori)
  ferns, 94, 95
  *Laminaria*, 219
Species, unit of classification, 236
Spectrum, white light, xviii
Sperm, 140–42
  *Fucus*, 220, 221, 222
  *Laminaria*, 219, 220
  *Marchantia*, 83
  moss, 75–77
  pine, 111
  *Selaginella*, 100
  *Vaucheria*, 230
  *Volvox*, 147
Spermagonium, *Puccinia graminis*, 193–95
Spermatia
  *Puccinia graminis*, 193–95
  red algae, 217
  transfer to receptive hyphae, 194–95
Spermatophyta, 236
*Sphaerocarpos*, 85, 86
Sphaerosome, xi
*Sphaerotilus*, 204
*Sphagnum*, 81, 82
  classification, 240
  green cells in leaf, 82

sporophyte, 82
  water storage cells in leaf, 82
*Sphenophyllum*, 68
Spikes, grasses, 167
Spindle fibers, xxviii, xxxi, xxxv–xxxvii
Spine, 50
Spiral chloroplast, 143
Spiral thickening, 22
*Spirillum*, 199
Spirochete, 199
*Spirogyra*, 143, 154
  demonstration of plasmolysis, xxv–xxvi
  plus and minus strains, 150
  sexual reproduction (conjugation), 150
Splash platform, *Marchantia*, 83
Spongy parenchyma, 4
Sporangia
  black bread mold, 174
  early land plants, 6, 7
  *Equisetum*, 96, 98
  ferns, 94, 95
  *Selaginella*, 97
  slime molds, 172, 173
  sporangiophores, *Phytophthora infestans*, 182, 183
  types, 158
Sporangiophores
  bread mold, 174
  *Equisetum*, 96, 98
  *Saprolegnia*, 179, 180
Spore discharge
  ferns, 95
  *Marchantia* capsules, 85, 87, 88
  moss capsules, 79, 80
  *Selaginella* sporangia, 100
Spore mother cells, xxxiii, xxxiv, 7, 9. See also Megaspores, mother cells; Microspores, mother cells
Spores (meiospores), xxxiii
  black bread mold, 180
  *Equisetum*, 96–98
  *Lycopodium*, 65, 102
  mosses, 76
  nutrition, and heterospory, 99–101
  types, 158
  wind dispersal, 61, 62, 158
Sporocyte, xxxiii
Sporophore, 190, 191
Sporophyll,
  evolution, 162–64
  ferns, 94, 95
  segregation, vegetative leaves, 162–64
  *Selaginella*, 97
Sporophyte, xxxiii, 157
  ferns, 70–72
  leafy liverworts, 87–89
  *Marchantia*, 83, 87, 88
  moss, 77–78
  *Sphagnum*, 82
  tetraploid, experimental generation, 158
Spring wood, 31–34
Stalked bacteria, 199
Stamen, 112–14, 118, 164
*Staphylococcus*, 199

Starch, xii
  digestion, xxi, 200
  formation, xiii, xx, xxi
  grains, in plastids, xx, xxi
  iodine test, xvi
  in stem sections of *Pellionia*, xxi
Statospore, golden-brown algae, 231, 232
*Staurastrum*, 145
*Stauroneis*, 228, 229
Stauros, 229
Stele, 23, 24
  centripetal differentiation, 23, 24
  root, 35, 41
Stellate chloroplast, 143, 144
Stem modification, 54, 55
*Stemonitis*, sporangia, 173
Stem tip, 13, 46
*Stephanodiscus*, 226
Sterigma (sterigmata), 189, 190, 194
Sterile basal leaf, 164
Sterile jacket cell, 76
*Stigeoclonium*, 152
Stigma
  algae, 137
  pistil, 113, 114
Stigmarian appendage, 66
*Stigonema*, 212, 214
Stipe, *Laminaria*, 219
Stipule, 46
Stipule scar, 46, 51
Stolon, 54, 173, 174
Stomate, 3, 4
  evolution, 61
  grasses, 4
  guard cells, 8–9
  leaf epidermis, 3, 4, 9
  moss capsules, 79, 81
  rates of diffusion, 8, 9
  role, transpiration, 8
  stem epidermis, 28
  theories of opening and closing, 8, 9
Stone cell, 17
Stonewort, 154, 237
Storage root, 56
Strawberry, 124, 125
*Streptobacillus*, 199
*Streptococcus*, 199
Streptomycin, 200
Striae, diatom valves, 226
Strobilus
  club mosses, 65
  *Equisetum*, 96–98
  *Selaginella*, 97–102
Stroma, xii
Structured granule, blue-green algae, 210
Style, 115
Suberin, xxi, 30
Succulent leaves and stems, 55
Sucrose, xx
Sulfur
  deposition by bacteria, 201
  essential element, 42
  reduction by bacteria, 204
Summer wood, 31–34
Sunflower "seeds" (achenes), 123
Sunlight, energy, xv

*Surirella*, 228, 229
Survival, xxxxiii–xxxxiv
Suspensor, embryo, 120, 121
Sweet pea flower, 115, 165, 167
Symbiotic relationship, 205
*Synedra*, 228, 229
Synergid, embryo sac, 117–20
Syngamy, 139, 215, 217
Synthetic plant growth hormone, 123
*Synura*, 231, 232

## [T]

*Tabellaria*, 227, 230
Tapetum, 114
*Taphrina*, 187
Tap root, 55
Tassel, 13
Teleological conclusions, 60
Teliospore, *Puccinia graminis*, 193, 194
Telophase
  meiosis I (telophase I), xxxvi
  meiosis II (telophase II), xxxvii
  mitosis, xxxi
Tendril, 52, 53
Terminal bud, 50, 51
Tetrads of spores, 79. See also Life cycle
Thallophyta, 236
Thallose liverwort, 239
Thorn, 50
Thylakoid, xii, 209
Thymine, xxxii
Tippo, Oswald, 242–43
*Tmesipteris*, 63, 240
Tobacco, blue mold, 183
*Tolypothrix*, 212, 214
Tomato
  flower, 166
  fruit (berry), 124
  leaf, xiii
Torus, 23
Toxic algae, 214
Toxic substance, fungi, 178
Tracheid, 20, 21
*Trachelomonas*, 233, 234
Transfer RNA, xxxii
Transpiration, 7–10
  benefits, 10
  cuticular, 8
  effect of temperature, relative humidity, wind, leaf structure, 9–10
  environmental significance, 10
  pull, magnitude, 8
  relation to stomates, 8–9
*Trebouxia*, 145
*Tribonema*, 227, 231
Trichogyne
  red algae, 215, 217
  sac fungi, 185–86
Triploid *(3n)*, xxxiii
*Tropism*, 132–33
True branching, blue-green algae, 212, 214
Tube nucleus, 114, 118, 119
Tuber, 54

Tubular filament body type, 144, 230–31
Turgid, xxv
Turgor pressure, x, xxv
Twig, 16, 51
Twining stem, 52, 53
2,4-D, 132
Tyloses, 32, 35

## [U]

*Ulothrix*
  cell and chloroplast, 143, 144
  filaments, asexual and sexual reproduction, 148–50
Ultraviolet radiation, 60
*Ulva*, 153
Unbound water, xxvi
Unbranched filament, 144, 149–52
Unilocular zoosporangium
  *Ectocarpus*, 218
  *Laminaria*, 219
Unit factor, xxxxii
Uracil, xxxii
Uredospore, 182
  germination, 192, 193
  *Puccinia graminis*, 192, 193, 196
  spread, 196
*Ustilago maydis*, 191

## [V]

Vacuolar membrane, ix–xi
Vacuole, ix–xi
Valve mantle, 223, 224
Variegated leaves, xiii
Vascular bundle
  comparison to vein of corn leaf, 25
  corn stem, 26
  dicot stem, 26, 27
  scars, 51
Vascular cambium
  roots, 40–41
  stems, 27
Vascular ray, 27
Vascular tissue system, 15
*Vaucheria*, 227, 230, 231
Veins
  comparison, vascular bundles, 26
  endings, 5
  large and small, 5, 6
  sheath cells, 6, 7
Venter, archegonium, 76
Ventilated chlorenchyma, 5, 62
  *Anthoceros* sporophyte, 90
  cortex of stems, 27, 28
  moss capsule, 79, 81
*Venturia*, 182
Vessel and vessel segment, 20
*Vibrio*, 199
Vine, 53, 54
Vitamins, 133, 134
*Volvox*, 146, 147

## [W]

Wallace, Alfred, xxxxiii
Water
  capillary movements, 62, 80, 81
  cohesion of molecules, 8
  deficits, in leaf cells, 8
  diffusion, 8, 9
  movements, in xylem, 7–8
  in photosynthesis, xviii, xix
  storage cells, *Sphagnum*, 81, 82
  temperature, as related to $CO_2$ absorption, 2
Water fern, 96
Water lily, 59
Water net, 148
Water vapor, xv, 3, 7, 8
Weathering, carbonates, 2
Weed killer, 132
*Welwitschia*, 241
Wet cell wall, significance in leaf functions, 6–8
Whorled leaf arrangement, 48, 49
Wood, 28, 31–34
Woody perennials, 53
Woody twig, 16

## [X]

Xanthophyll (oxycarotene), xii, xiii
  green algae, 136
  molecular structure, xiv
Xerophyte and xerophytic modification, 56–58
Xylem
  arrangement in primitive stems, 23–24
  possible evolutionary origins, 62
  rays, 21, 27, 29, 34
  roots, 38, 39
  secondary, 28, 31–34
  stems, 24, 27
  structure and cell types, 20–23
  veins, 5

## [Y]

Yeast, 171, 172
Yellow-green algae, 227, 230–31
  cell structure, 227
  classification, 237
  pigments, 227
  plant body types, 227, 230, 231
Young leaves, 14
*Yucca*, 168

## [Z]

Zinc, 42
Zone of elongation, roots, 35, 36
Zoochlorellae, 145
Zoosporangia
  brown algae, 218, 219
  *Saprolegnia*, **179**–80

Zoospores
  brown algae, 218, 219
  *Chlamydomonas*, 138
  chytrids, 179
  *Oedogonium*, 150, 151
  *Saprolegnia*, 179, 180
  *Vaucheria*, 231
Zygnema, 143, 144

Zygospores
  *Chlamydomonas*, 140
  *Phycomyces*, 185, 186
  *Spirogyra*, 150
Zygote germination, *Spirogyra*, 150
Zygotes, xxxiii, 76–77
  fern, 92
  flowering plants, 119, 120
  *Fucus*, 222
  pine, 111, 112
  red algae, 215, 217
  tetraploid, experimental generation, 158
  zygospores, 140
Zygotic meiosis, 156–57

## Date Due

| | | |
|---|---|---|
| JA 5 88 | | |
| ~~NO 17~~ | | |
| NO 1 89 | | |
| ~~Fac~~ | | |
| | | |
| OC 22 '92 | | |
| | | |
| | | |
| | | |
| | | |
| | | |

```
581
D                                37747

Dodd, John Durrance

Course book in general botany
```